Hybrid Intelligent Techniques for Pattern Analysis and Understanding

Hybrid Intelligent Techniques for Pattern Analysis and Understanding

Edited by
Siddhartha Bhattacharyya, Anirban Mukherjee,
Indrajit Pan, Paramartha Dutta,
and Arup Kumar Bhaumik

CRC Press is an imprint of the
Taylor & Francis Group, an **informa** business
A CHAPMAN & HALL BOOK

MATLAB® is a trademark of The MathWorks, Inc. and is used with permission. The MathWorks does not warrant the accuracy of the text or exercises in this book. This book's use or discussion of MATLAB® software or related products does not constitute endorsement or sponsorship by The MathWorks of a particular pedagogical approach or particular use of the MATLAB® software.

CRC Press
Taylor & Francis Group
6000 Broken Sound Parkway NW, Suite 300
Boca Raton, FL 33487-2742

CRC Press is an imprint of Taylor & Francis Group, an Informa business

Printed on acid-free paper

International Standard Book Number-13: 978-1-4987-6935-8 (Hardback)

Library of Congress Cataloging-in-Publication Data
LoC Data here

Visit the Taylor & Francis Web site at
http://www.taylorandfrancis.com

and the CRC Press Web site at
http://www.crcpress.com

Contents

Foreword

Many real-life problems suffer from uncertainty, imprecision, and vagueness; and because of this, conventional computing paradigms often fall short of a solution. Even the latest intelligent computing paradigms are frequently insufficiently robust. Hybrid intelligent computing is an alternative paradigm which addresses these issues upfront.

An intelligent machine is called so because of the various methodologies offered by the soft computing paradigm, including fuzzy and rough set theory, artificial neurocomputing, evolutionary computing, and approximate reasoning. At times, none of these techniques meet the demands of the real situation, but an effective symbiosis of more than one can offer a formidable and/or robust solution. This is because all of these soft computing techniques suffer from limitations. Thus the impetus is for hybrid methodologies, the spirit of which is to exploit the strengths of one technique to supplement the limitations of another. Of late, there is enormous growth of research exploration of injecting elements of intelligence using efficient hybrid techniques. All these initiatives indicate that the individual soft computing techniques do not conflict but rather complement one another. Hybrid intelligent systems stem from the synergistic integration of the different soft computing tools and techniques. The fusion of these techniques toward achieving enhanced performance and more robust solutions can be achieved through proper hybridization. In fact, recent reports reveal the inherent strength of such hybridization of computational methods.

Pattern analysis and understanding have been daunting tasks in the computer vision research community given the vast amount of uncertainty involved. Proper analysis of patterns plays a key role in many real-life applications. Traditional applications include pattern recognition, image processing, image mining, object recognition, video surveillance, and intelligent transportation systems. As an example, wrinkles due to aging can often be used for the estimation of age progression in face images. This can be an enormous help in tracing unknown or missing persons. Images exhibit varied uncertainty and ambiguity of information and hence understanding an image scene is far from being a general procedure. The situation becomes even more difficult when the images become corrupted with noise artifacts.

Chapters 1, 2, and 3 all deal with handwriting recognition, in Bangla characters, digits, and signatures, respectively. Handwriting recognition is a very important application for identity verification and for any kind of machine input such as with PDAs, postal addresses, or data entry. As integration of the Web and computing in general becomes more and more common, these problems of interface will only become more important. The authors use a number of different hybrid methodologies, and, in general, show that hybrid approaches can enhance distinction accuracy and effectiveness when compared to current state-of-the-art approaches.

Chapter 4 presents a new strategy for the detection of simple parametric objects in images, called estimation of distribution algorithms (EDAs). The authors show that EDA compares favorably with two variants of the Hough transform method in terms of computational time necessary.

Chapters 5 and 6 both deal with medical applications of machine learning. In Chapter 5, the different phases involved in detection of the malarial parasite in blood smears are dealt with using appropriate methods tailored to the challenges of each stage. In Chapter 6, the authors specifically look at image improvement in MRIs, mammograms, and ultrasound, which is necessary due to different types of noise that can occur, using curvelet transform thresholding combined with a Weiner filter.

In Chapter 7, the authors consider the important problem of face recognition and propose an improved optimal feature selection procedure for better recognition accuracy and demonstrate its utility in comparison with existing techniques.

Chapter 8 deals with motion recognition for use in gaming interface applications. Combining Markov models to find feature points with an artificial bee colony algorithm for optimization improves performance compared to existing techniques.

Chapter 9 deals with smart visual surveillance using a contextual information-based anomalous activity detection and localization algorithm. Unusual event detection is based on a probabilistic estimation of regions being normal. Adaptive inference mechanisms are used to determine the decision-making thresholds for anomalous event detection.

In Chapter 10, the authors introduce a new hybrid method combining particle swarm optimization with cellular automata. The method uses fuzzy membership values to identify clusters, then 2D interconnected cell processing to prioritize exclusion of outliers. The authors apply the method to pixel classification and show improvement over several current methodologies.

Chapter 11 deals with segmentation methods. In this chapter, the authors provide a comprehensive review of the existing segmentation techniques in the analysis of brain MR images including hybrid procedures.

Chapter 12 uses a computer aided classification system for renal ultrasound images, for the detection of renal disease. Three methods are compared using both the signal and transform domains, and an optimal design is constructed for clinical use.

The book concludes with two chapters dealing with big data. In Chapter 13, the authors propose a new data clustering method for large datasets applying a self-organizing map algorithm to cluster data by using the Laplacian spectra of a graph. The method is compared with current data clustering models on standard datasets. In Chapter 14, the authors propose a hybrid quantum algorithm for sequence clustering based on quantum K-means clustering and a hidden Markov model. Evidence for an exponential speedup in run-time complexity compared to classical approaches is provided.

The editors have done a commendable job in bringing this volume as a rare collection of the applications of hybrid intelligence to pattern recognition and analysis. I hope that the present volume will come in good stead for young researchers in this domain.

Elizabeth Behrman
Kansan, USA
April, 2017

Preface

With the extensive research in the intelligent computing paradigm applied to various problems over a few decades, the shortcomings of every component of this paradigm have gradually become prominent. It is an undeniable fact that each individual ingredient of intelligent computing, be it fuzzy mathematics or evolutionary computation, if not an artificial neural network, has its own points in its favor. However, with the passage of time, human civilization starts appreciating the drawbacks of individual components, thereby necessitating the need of up-gradation into the next-generation computational platforms. Hybrid intelligent computing appears as the natural extrapolation of the field of computational intelligence. Through the process of hybridization, the limitation of one ingredient is supplemented by the strength of the other. Because of this collaborative performance, a surge of research initiatives is observed nowadays. Researchers have put in all their efforts to explore and exploit the power of hybrid intelligence as an alternative paradigm altogether. In other words, there has been a paradigm shift in the effort of research in the relevant field. In the present scope, the editors are trying to bring some recent findings of researchers engaged in hybrid intelligence research by identifying pattern recognition and related issues as the application domain. The spectrum is kept balanced. It is not too broad, with an aim to encourage contribution focused in and around pattern recognition. On the other hand, it has been meticulously maintained that the spectrum must not be too limited, so that a researcher engaged in research in a relevant field may enjoy every scope to ventilate his/her ideas in the form of tangible contribution. It is indeed encouraging for the editors to come across fourteen chapters, which have been thoroughly peer reviewed and accommodated in the present treatise.

In Chapter 1, the authors verify a writer from his/her handwritten text with adequate empirical justification by using an efficient combination of the dual tree complex wavelet transform (DTCWT) and gray level co-occurrence matrix (GLCM) to enhance the relevant texture feature.

A similar problem of handwritten digit recognition is the domain of investigation of the authors of Chapter 2. They offer an effective hybridization to meet this goal. Elaboration in the form of discussion on other techniques for the said problem is an added point of interest of this chapter.

A very close application of handwritten signature recognition comes under the purview of Chapter 3, where the authors offer an offline classification algorithm based on bio-inspired ingredients and justify its efficacy on available benchmark datasets.

Effectiveness of estimation of distribution algorithms (EDAs) for detection of geometric objects such as line, circle, parabola, etc. as well as retinal fundus images, is substantiated with ample experimentations by the authors of Chapter 4. They have worked on two variants for this purpose, justifying impressive performance primarily with respect to computational time.

A comprehensive machine learning approach is offered by the authors of Chapter 5. They have shown the significance of their approach by dint of their thorough empirical experiments in the context of detection of the malaria parasite in blood smear.

Noise classification is very sensitive and crucial in medical image processing mainly because of the associated medical implication. The result of a wrong classification at times may be catastrophic in certain contexts. The authors address this important issue in Chapter 6 with respect to different medical imaging modalities and share some of their impressive findings in this regard.

Chapter 7 handles how faces recognition may offer an avenue in respect to biometric authentication. The authors elaborate the problem pertaining to face recognition for this purpose, duly addressing the sensitivity associated therein.

The hand motion recognition system (HMRS), an indispensable component of gesture recognition comprising multiple steps to achieve its goal, needs effective optimization of the Markov state sequence obtained in some intermediate step in order to ensure subsequent operative performance of the system. In Chapter 8, authors offer the artificial bee colony approach to achieve this goal.

A very timely and important challenge of developing a smart visual surveillance methodology aiming at effective crime management is the contribution of the authors in Chapter 9.

The authors in Chapter 10 exhibit the effectiveness of hybridization in watershed image analysis. They justify it by means of extensive experimentations which also indicate the performance supremacy of such a hybrid technique over prior arts.

Medical imaging plays a crucial role in diagnosis and treatment of diseases. Magnetic resonance imaging (MRI) nowadays, is an indispensable tool used for medical purposes. The authors take up the sensitive issue of segmentation of such images in Chapter 11, which also supplements relevant information as an effective compilation.

An important task of classification of renal ultrasound images is taken up by the authors in Chapter 12. They use Haralick's texture descriptors to achieve this purpose, whose effectiveness is further justified by the extensive experimentation.

Effective application of the Self-organizing map (SOM) using Laplacian spectra of a graph to achieve clustering of large non-categorical datasets is what is contributed by the authors in Chapter 13.

In the final chapter, the authors deal with the problem of clustering in the big data paradigm by using quantum K-means clustering applied on big data sequences represented in terms of quantum states with the help of quantum RAM (QRAM).

The editors of this volume leave no stone unturned to make a meaningful compilation of recent research trends. The treatise contains fourteen chapters encompassing various application domains. The authors of different contributing chapters share some of their latest findings in the present scope. It is needless to mention that such an effort on the part of the editors to come out with an edited volume is possible only because of the efforts rendered by the authors. The editors also want to avail this opportunity to express their thanks to CRC Press/Taylor & Francis Group as an international publishing house of eminence to provide the scope to make the efforts of the present editors to bring out such a compilation. The editors would also like to take this opportunity to render their heartfelt thanks to Ms. Aastha Sharma for her continuous guidance during the project tenure. But for their active cooperation, this would not have met such a success. Last but not the least, the editors earnestly feel like expressing their gratitude to those whose names may not figure here but who have helped in converting the present initiative into a fruitful one.

Siddhartha Bhattacharyya
Kolkata, India

Anirban Mukherjee
Kolkata, India

Indrajit Pan
Kolkata, India

Paramartha Dutta
Santiniketan, India

Arup Kumar Bhaumik
Kolkata, India

April 2017

MATLAB® is a registered trademark of The MathWorks, Inc. For product information, please contact:

The MathWorks, Inc.
3 Apple Hill Drive
Natick, MA 01760-2098
USA Tel: 508-647-7000
Fax: 508-647-7001
Email: info@mathworks.com
Web: www.mathworks.com

Contributors

Swati Adhikari
Department of Computer Science
University of Burdwan
Burdwan, India

Samir Kumar Bandyopadhyay
Department of Computer Science & Engineering
University of Calcutta
Kolkata, India

Nabanita Basu
Department of Computer Science & Engineering
University of Calcutta
Kolkata, India

Arit Kumar Bishwas
Department of Information Technology
Amity University
Noida, India

Ranjit Biswas
Department of Computer Science and Engineering
Assam University
Silchar, India

Indu Chhabra
Department of Computer Science and Applications
Panjab University
Chandigarh, India

Youcef Chibani
Department of Electronic and Computer Sciences
University of Sciences and Technology Houari Boumediene
Algiers, Algeria

Ivan Cruz-Aceves
Departamento de Ingeniería Electrónica Centro de Investigación en Matemáticas
Guanajuato, Mexico

Rajib Das
School of Water Resources Engineering
Jadavpur University
Kolkata, India

Subhasish Das
School of Water Resources Engineering
Jadavpur University
Kolkata, India

Jesus Guerrero-Turrubiates
División de Ingenierías Campus Irapuato-Salamanca
Universidad de Guanajuato
Guanajuato, Mexico

Chayan Halder
Department of Computer Science
West Bengal State University
Barasat, India

D. Jude Hemanth
Department of Electronics and Communication Engineering
Karunya University
Coimbatore, India

Hemant Jain
Division of Software Systems
VIT University
Vellore, India

Kalyan Mahata
Information Technology Department
Government College of Engineering and Leather Technology
Kolkata, India

J. K. Mandal
Department of Computer Science & Engineering
University of Kalyani
Kalyani, India

Ashish Mani
Department of Electrical & Electronic Engineering
Amity University
Noida, India

Raghav Menon
Department of Electronics and Communication Engineering
Kumaraguru College of Technology
Coimbatore, India

Sanjay Nag
Department of Computer Science & Engineering
University of Calcutta
Kolkata, India

Hassiba Nemmour
Department of Electronic and Computer Sciences
University of Sciences and Technology Houari Boumediene
Algiers, Algeria

Vasile Palade
Faculty of Engineering and Computing
Coventry University
Coventry, UK

Jaya Paul
Information Technology Department
Government College of Engineering and Leather Technology
Kolkata, India

Tuhin Utsab Paul
Department of Computer Science & Engineering
Amity University
Kolkata, India

Jeny Rajan
Department of Computer Science and Engineering
National Institute of Technology Karnataka
Mangalore, India

Yadav Nitesh Ramprasad
Department of CSE
Karunya University
Coimbatore, India

Narendra Rao T. J.
Department of Computer Science and Engineering
National Institute of Technology Karnataka
Mangalore, India

Kaushik Roy
Department of Computer Science
West Bengal State University
Barasat, India

Parthajit Roy
Department of Computer Science
University of Burdwan
Burdwan, India

Sudipta Roy
Department of Computer Science and Engineering
Assam University
Silchar, India

K. Martin Sagayam
Department of ECE
Karunya University
Coimbatore, India

Anasua Sarkar
Computer Science and Engineering Department
Jadavpur University
Kolkata, India

Yasmine Serdouk
Department of Electronic and Computer Sciences
University of Sciences and Technology Houari Boumediene
Algiers, Algeria

Ryan Serrao
Division of Software Systems
VIT University
Vellore, India

Komal Sharma
Department of Electrical and Instrumentation Engineering
Thapar University
Patiala, India

Juan Manuel Sierra-Hernandez
División de Ingenierías Campus Irapuato-Salamanca
Guanajuato, Mexico

Geetika Singh
Department of Computer Science and Applications
Panjab University
Chandigarh, India

B. K. Tripathy
Division of Software Systems
VIT University
Vellore, India

Jitendra Virmani
Council of Scientific Industrial Research
Central Scientific Instruments Organization
Chandigarh, India

Chapter 1

Offline Writer Verification Based on Bangla Handwritten Characters Using Enhanced Textual Feature

Jaya Paul, Anasua Sarkar, Chayan Halder, and Kaushik Roy

1.1 Introduction

Writer verification has been magnetizing the attraction of the researchers as a most important role play for writer authentication. The last few years, many algorithms and systems with high authentication rates have been introduced. Various approaches for handwritten character authentication problem have been developed [1–4]. Some approaches are worked in Bangla characters [5]. In one of the relevant early works, the likelihood ratio is used in forensic document identification [6]. The statistical model [7,8] is used for writer verification and also includes a mathematical formula for calculating the strength of evidence of same/different writer. The approaches are classified into two categories, namely online systems and offline systems. The handwritten shape and stock information are available in online verification system [9], but this information is unavailable for the offline method. The offline system is more complex than the online system. There has not been much work on the verification of handwritten Bangla characters [9,10]. Therefore, there is a pressing need for such a system. The present paper deals with writer verification based on handwritten Bangla characters.

1.2 Literature Survey

In the literature, different works can be found on writer verification. Text-independent and text-dependent data sets are available in the writer verification system. In the text-dependent method, the text contents are same for known and unknown writers. The text-dependent system for writer verification offers higher discriminative power using small amount of data, as compared to text-independent methods [11]. The text-independent system finds similarity between writing patterns to verify the writers. Forgery possibility is reduced for text-independent and needs a large volume of data. But the text-dependent method increases the forgery case due to copy of same text. But it is very useful in signature verification where most of the time the genuine writer uses the system and not suitable for frequent rejection. In the text-independent method forgery is reduced, but it needs more data samples. In this case it is possible that the genuine writer can also be rejected. The writer verification uses in different purposes, like historical document analysis [12], graphology, mobile devices [13], DNA and fingerprints are also used in terms of verification [14].

1.2.1 Offline and pseudo-dynamic approaches

Earlier works [15] and [16] have proposed many methods for writer verification. Handwriting shape information has been widely used in offline writer verification mode. Primary methods that are used for character shape information of handwritting are referred to as offline approaches [15,16]. Gray level co-occurrence matrix (GLCM)[17] and dual-tree complex wavelet transforms (DT-CWT) features are frequently used in Bangla character and writer verification [18–20]. In [21], the researchers propose a method that estimates dynamic information from static handwriting. All these methods are calling pseudo-dynamic approaches and are resistant to forged handwriting. The methods are categorized into used motor control theory [22] and temporal order of stroke production. Ferrer et al. [23] propose methods that explain intensity variation and stroke information. Earlier forensic science studies [24,25] report the usefulness estimate of dynamic information from static handwriting.

To enhance the performance of the writer verification system, we have proposed a new approach to extract texture information from the handwriting sample [18,20]. The survey [19,20] show texture descriptors, texture based features and dissimilarity. Local binary patterns and phase quantization have been used in [19] writer identification. In [20] the scientists achieve the performance of 96.1%. Some examples of this scheme can be found in Marti et al. [26], Srihari et al. [27], Bulacu et al. [28], Bensefia et al. [29], Siddiqi and Vincent [30].

1.2.2 Approaches of writer-dependent and writer-independent methods

Writer-dependent builds on one specific model per writer. The drawback of this method is that a new model needs to be built for a new writer [15,16]. Another important issue in this method is that usually a substantial amount of data is need to construct a well-founded model. However, this type of approach used for signature verification.

The writer-independent method is based on the FDE (forensic document examination) approach, which considers the intraclass (writer) and interclass (writer) similarities of handwritten characters. This model classifies the handwriting into two authenticity classes. The two classes are genuine and forgery. This approach is used in various work [31]. The probability distributions (PDF) of interclass (writer) and intraclass (writer) are collected from training samples in the writer-independent approach. Comparing these probability density function values, the original author of that document is decided by dissimilarity. Based on the concept of dissimilarity, the writer-independent method used a distance metric [32]. Muramatsu and Matsumoto [33] propose a model which used user specific dissimilarity vectors that combine original dissimilarity and used mean vector for each writer. It is considered as a distance of different writers and individual characteristics in handwriting.

Most of the identification and verification works on western script (like Roman script). Srihari et al. [34], Bulacu et al. [28], Bensefia et al. [29], Siddiqi and Vincent [30], etc. Doermann and Jain propose an identification system which used the multi-script writers on the English, Arabic and Greek languages [35].

1.2.3 Writer identification (many to one)

Writer identification systems must recognize a writer among N number of writers in a large dataset. The process of identifying a person involves performing comparisons among multiple handwritings. These systems aim to confirm identity and build on writer information. A variety of features have been proposed, like structural or statistical and local or global. All these features are served to distinguish the individual writer from other writers. Said et al. [3] used Gabor filters and co-occurrence matrices to identify the writer and consider each writer's handwriting as a different texture. Macro and micro features are extracted [34] from the document and establish the individuality of handwriting. The research pursued in [36] proposes an identification system [26] which extracts structural features from each line of handwritten documents.

1.2.4 Writer verification (one-to-one matching)

A writer verification system decides whether a handwritten document belongs to the same writer or different writers. It is the validity checking procedure for the claimed identity. Naturally, the writer verification issue is a true/false problem. Signature verification is one of the ways which is used with many different methods of classification as reported in the survey, namely distance measures [1,37], dissimilarity [38], hidden Markov models [39], Bayesian classifiers [40], and grapheme clustering [29].

The manual process of document examination is a tedious job and is also very time-consuming. Srihari et al. [41] introduce a handwriting examination system, which is known as CEDAR-FOX. This system can be used in writer verification as well as in signature verification.

1.2.5 Biometric property

Nowadays the biometric property is used in different areas in our daily life. This property is a human identification system by determining the authenticity of particular behavioral features [42]. A small portion of the biometric properties of a human are used in different applications. A general biometric system is shown in Figure 1.1.

A number of biometric properties have been used in several areas [43]. For the requirements of the system, a specific biometric property is used. Some applications of biometric properties are found in the face recognition system [44], with two phases—face detection and tracking. Some other methods are—facial matching, fingerprint recognition [45], hand geometric recognition [46], personal recognition [47], retina recognition [48], voice recognition [49], sclera recognition [50], and keystroke dynamics [51]. In Figure 1.2, we show different examples of the biometric characteristics.

1.2.6 Handwriting as biometric

Handwriting biometric is the authentication process of identifying the author of a given handwritten text, offline or online. Automated handwritten documents and signatures (online, offline) are considered the most socially and legally accepted property for authentication. Writer verification and writer identification are methods for writer authentication based on the system. Signature identification [52] and verification are also authentication methods based on the system. Signature verification [53] and identification can be either dynamic or static. In the offline mode, digital images of handwritten documents are available. In the online mode, documents are acquired by means of pen-sensitive computer displays or graphic tablets. A handwritten document is a biometric property generated by a composite method arising in the writer's brain [54]. Handwriting-based authentication is commonly used in forensic document examination [25]. The main aim of the forensic document examination is to provide enough evidence about questionable documents, using various scientific methods.

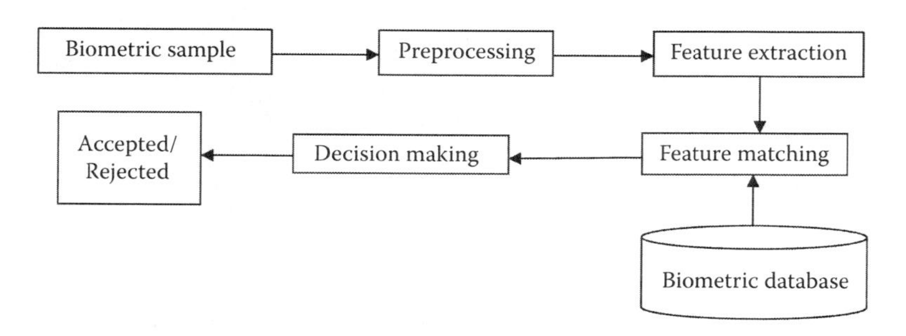

FIGURE 1.1: General biometric model.

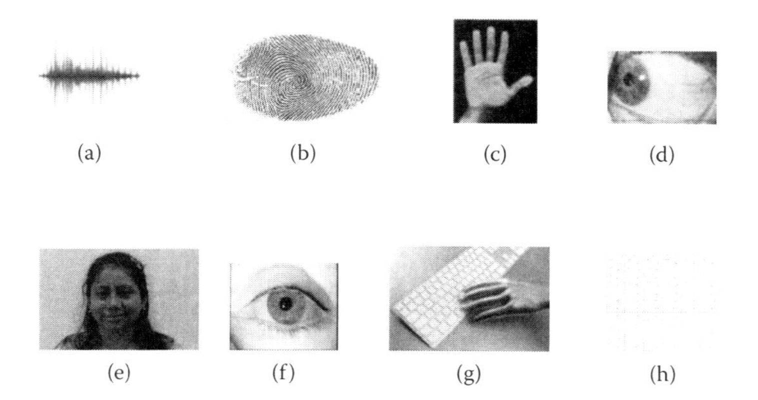

FIGURE 1.2: Examples of biometric characteristics: (a) voice, (b) fingerprint, (c) hand geometry, (d) sclera, (e) face, (f) iris, (g) keystroke, (h) handwriting.

1.2.7 Types of writers

On the literature survey, the field of writer verification generally considers two different types of writers. First is the genuine writers or original writer; second is the forged writer.

1.2.8 Genuine writer or original writer

When a handwritten document is produced by an authentic writer under normal conditions, the writer is called the genuine writer or original writer. The interwriter stability and variability of the original writer are affected by many factors, such as age, mental state, habit, and practical conditions [53].

1.2.9 Forged writer

There are usually two different types of forgeries available. The first one is the unskilled forgery, which is written by a person who does not know the pattern of the original or may be without much practice. The second type is a skilled forgery where a copy of the document is represented as an original or written by the genuine writer. Various skill levels of forgeries can be found in Nguyen et al. [55].

1.2.10 Database

A handwriting database has great importance in our study of writer verification area. All the research works into the writer verification field as they have been conducted for several years, there have some different languages of standard offline databases are created. Bulacu et al. [38] created a system on writer verification and identification on Roman databases like the IAM database [56], Unipen database [57], and Firemaker set [58]. They propose to combine allographic and textural features for text-independent writer identification with Arabic script. But due to the lack of availability of a Bangla script database, the development of a Bengali writer verification system has been negatively affected. Only Halder and Roy

[10] used Bangla characters (covering all alphabets, vowel modifiers, vowel modifiers) and Bangla numerals in [58] for writer identification in their research works.

1.3 Introduction to Bangla

1.3.1 Bangla script

Bangla is the seventh most popular language in the world and the second most popular language in India [59]. Almost 200 million people in the eastern part of the Indian subcontinent speak this language [59]. Bangla alphabets are used in texts of Bangla, Manipuri, and Assamese languages. Bangla script has more than 250 characters (consonants, vowels, and compound characters). A lot of research has already been done in this direction for several scripts [1,3]. The works have been applied successfully to various commercial applications, such as question document examination, bank check verification, etc. [7,60].

The biometric-based authentication system is used for two purposes: identification and verification.

Bangla has 11 vowels and 40 consonants and these are called the basic Bangla characters. It has lots of numbers, compound shapes, and complex letters. Figure 1.3 shows an example of Bangla alphabets. In Figure 1.3, we show the Bangla character shapes which are more complex than the Roman scripts. There exists a horizontal line (called a matra) which is

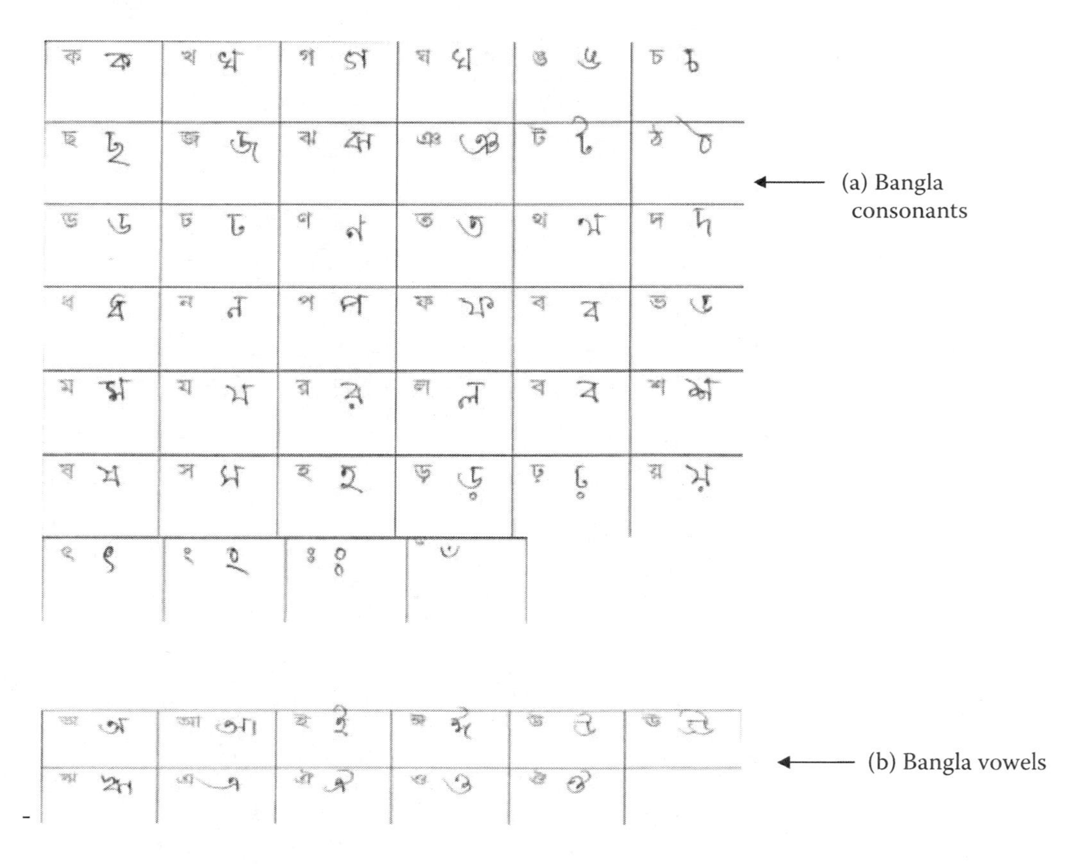

FIGURE 1.3: Examples of (a) Bangla consonants and (b) Bangla vowels.

sometimes absent for certain characters. In Bangla script, a vowel following a consonant takes a modified shape. This modified shape is placed at the left, right, and bottom of the consonant, or both at left and right. These modified shapes are called the modified characters.

The handwritten character verification system which is proposed in this article consists of four major phases: preprocessing phase, feature extraction techniques, dissimilarity measurement in the process calculating Kullback–Leibler (KL) divergence distances on the questioned document examinations between same-writer and different-writer samples, and finally, the classification phase.

In order to construct a more precise verification system, the following four methods are combined in this work:

1. DT-CWT feature extraction.
2. GLCM feature extraction.
3. KL divergence distance measure on writers.
4. Different benchmark classification methods.

1.4 Method

In this section, the overall idea of the proposed method has been described. For the present work, we have contemplated a sample database of 150 writers, consisting of different Bangla Isolated characters: consonants, vowels, numericals, and vowel modifiers are collected. The different- and same-writer variations can be seen in the distance value (dissimilarity measure) among handwriting from different writers. The experts analyze and verify the writers on the basis of these distance values. We can see the difference in Figure 1.4. In this work, we have used combined textural features of DT-CWT and GLCM to obtain that difference for writer verification. To verify unknown writers comparing a database of known writers, first textural features are calculated, then the dissimilarity feature vector is calculated and finally using them, the verification result is obtained.

Figure 1.5 depicts the proposed diagram of the writer verification process. At first, the collected data are preprocessed and the textural feature extraction methods of

FIGURE 1.4: Sample of four different isolated Bangla characters from three different writers.

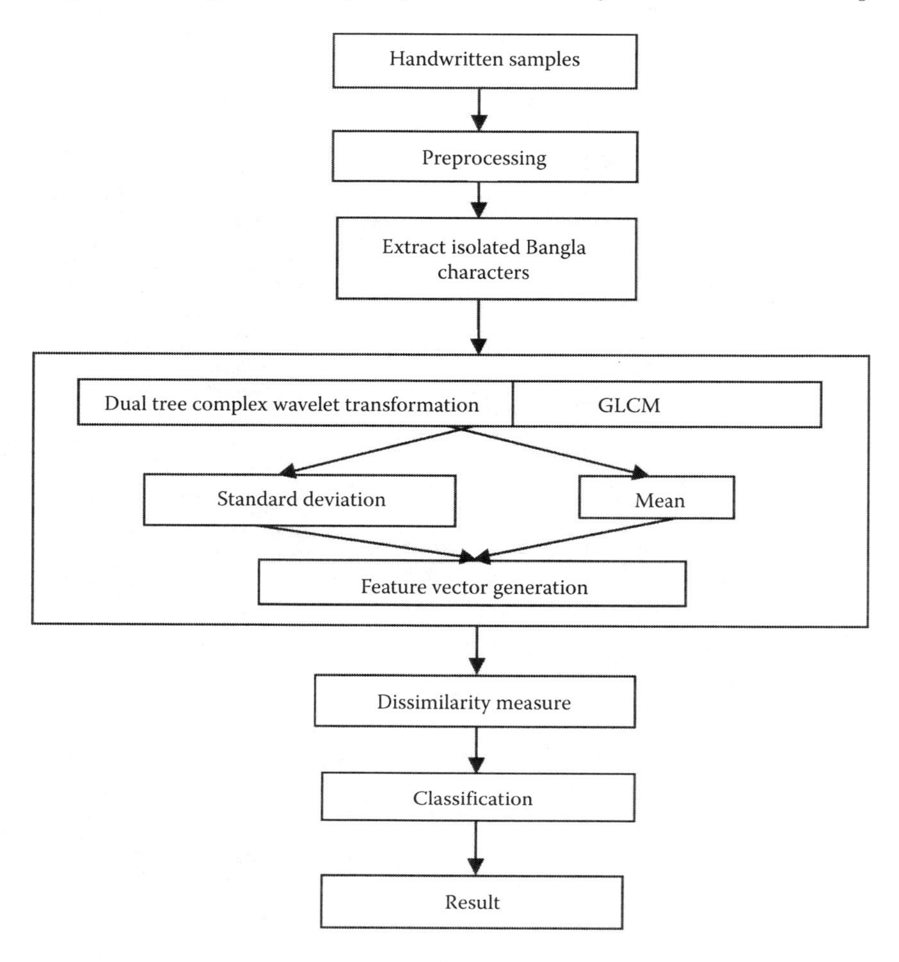

FIGURE 1.5: Diagram of the proposed writer verification system.

DT-CWT and GLCM are applied separately and then as a combined feature extraction method on the chosen dataset. The classifiers are then applied on the computed distances to calculate the dissimilarity measures between the questioned and the reference handwriting.

1.5 Data Collection and PreProcessing

Here we designed a simple data collection form [58] consisting of all Bangla numerals and alphabets. The total number of writer is 150 in the current work. Each writer has five sets of the same copy. For the present work, data collection has been done by different writers in [61]. Most of the writers of the database are right-handed. Out of 150 writers, now we have to consider full five sets of data from 123 writers (used for current purpose work). Here we show an example of our sample collection document form in Figure 1.6. The details about the data collection and the preprocessing method on the data can be found in [9].

FIGURE 1.6: Sample data collection of isolated Bangla handwritten characters.

1.6 Feature Extraction Techniques

1.6.1 Feature extraction based on dual-tree complex wavelet transformation

The proposed methodology is to use a modified DT-CWT [62–64] where multiscale analysis and direction-oriented feature extractions are possible.

We use the decomposition level six for the wavelet transform. In this work, first we normalize 128×128 dimensional image data and then the feature vectors have been computed over it. After using a Gaussian filter, we get 128-dimension image features. Subsequently, we get 64×64-dimensional feature vectors, which have been computed using the DT-CWT algorithm.

Complex-valued wavelet is defined in Equation 1.1:

$$\Psi_c(t) = \Psi_r eal(t) + \Psi_i mg(t) \tag{1.1}$$

Here $\Psi_c(t)$ is the complex wavelet filter, $\Psi_r eal(t)$ is the real part, and $\Psi_i mg(t)$ is the imaginary part of the wavelet. DT-CWT functions $h(p,q)$ have the following equation:

$$h(p,q) = d(p,q)e^{(j(w_p p + w_q q))} \tag{1.2}$$

where $h(p,q)$ is the DT-CWT functions, $d(p,q)$is a Gaussian function centered at $(0,0)$ and the center frequency is (w_p, w_q) of the corresponding subband. The complex coefficients of the k-th subband of the l-th level are as follows:

$$x_k^l = y_k^l + j z_k^l \tag{1.3}$$

In Equation 1.4 we calculate the magnitude of each subband:

$$x_k^l = \sqrt{((y_k^l)^2 + (z_k^l)^2)} \tag{1.4}$$

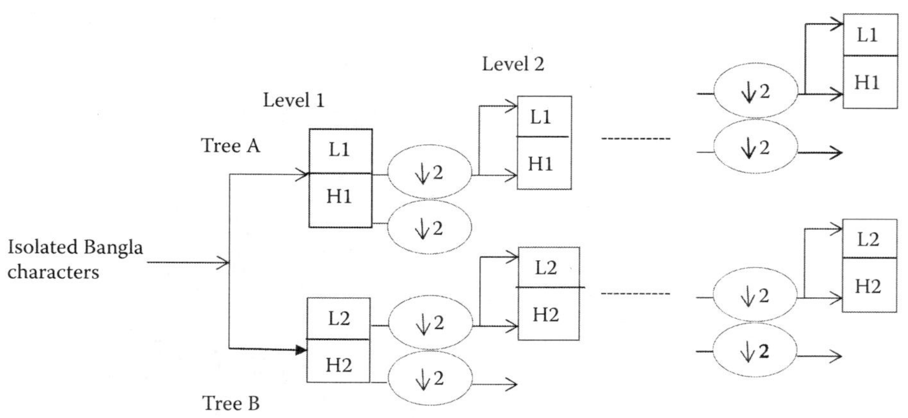

FIGURE 1.7: The schematic diagram of n level dual tree complex wavelet transforms (DT-CWT).

The magnitude is inconsiderate because $d(p, q)$ is slowly varying. At each level, complex coefficients angles are $\pm 15°$, $\pm 45°$, $\pm 75°$. These two properties are very useful for writer verification.

Figure 1.7 shows a DT-CWT feature extraction diagram for the verification system. DT-CWT can be use as a dual-tree formation [65]. The up sampling and down sampling before filtering and after filtering are needed to support the shift invariance. Here we use six levels of the dual-tree complex wavelet. We collected outputs of subband images into complex wavelet coefficients. Getting more information is referred to in [66].

1.6.2 Feature extraction based on GLCM

The GLCM is a texture analysis of statistical method and different combination of grayscale pixel value occur either horizontally, vertically, or diagonally of an image. It is a common textual analysis of images, we get it from Haralick et al. [67].

In Equation 1.5 the co-occurrence matrix G of an image P with size $M \times M$ is defined:

$$G(p, q) = \Sigma_{(}^{M} a = 1) \Sigma_{(}^{M} b = 1) \begin{cases} 1, if P(a, b) = s \, and \, P(a + \Delta_a, b + \Delta_b) = q \\ 0, otherwise. \end{cases} \tag{1.5}$$

Here, in Equation 1.5, the offset $(\Delta a, \Delta b)$ parameterization makes the co-occurrence matrix sensitive to rotation. The offset $(\Delta a, \Delta b)$ is specifying the distance between the pixel-of-interest and its adjacent pixel. Choosing an offset vector, the rotation of the image is not equal to 180°, will result in a different co-occurrence matrix for the same image. To achieve a degree of rotational invariance (i.e., 0°, 45°, 90°, 135°) using a set of offsets sweeping through 180 degrees at the same distance parameter Δ by forming the co-occurrence matrix. In this approach, GLCM is calculated with Energy, Contrast, Correlation, and Homogeneity in four directions, considering the pairs like $G[s, q]$ and $G[s, p]$. Following this approach, we obtain 22 feature vectors using the modified GLCM feature extraction method over our chosen dataset. Here, we use a maximum of 22 dimension features. Twenty-two dimensional feature vectors have been computed using GLCM method in MATLAB®, which is different from the work of Haralick [67]. Then the first-order statistics of standard deviation and mean are computed over the features.

1.6.3 Edge direction information

In handwriting verification, edge information is very important to enlarge the dimension of writing features. Here we use first-order statistics which capture the information in 172 dimensions. First-order statistics of feature depend on the pixel value of the histogram. In current work, two first-order statistics are calculated:

$$(Mean)\mu = \frac{1}{M}\Sigma_{(i=1)}^{M} E_i \tag{1.6}$$

$$(Standard\,deviation)S = \sqrt{\frac{1}{(M-1)}\Sigma_{(i=1)}^{M}|E_i - \mu|^2} \tag{1.7}$$

Here, vector I_i is made up of M scalar observations. Therefore, finally 172 feature vectors are obtained.

1.7 Dissimilarity Measure

DT-CWT generates $64X64$ dimensional vectors of the first-order statistics. In the next phase, we use the distance metric based on the dissimilarity representation in [68] using KL_{dist} (Kullback–Leibler divergence) method. To define the Kullback–Leibler divergence distance, let A and B be two discrete probability distributions. Then KL of B from A [69] is defined as follows:

$$KL_{dist}(A||B) = \Sigma_i A(i) \log \frac{A(i)}{B(i)} \tag{1.8}$$

In this method, Gaussian distribution is applied on each writer set to find out the KL_{dist} of same-writer sets. Different writer KL_{dist} distances are calculated between one writer from each set with different writers from all sets.

1.8 Classification Techniques

1.8.1 Multi-layer perceptrons networks

We can define an MLP by joining some neurons with functional neurons and gives one numerical output. The first hidden layer of the neurons network consists of functional neurons, whereas subsequent layers are created with numerical neurons. For example, a single hidden layer functional output of MLP enumerates the following function:

$$H(g) = \Sigma_{i=0}^{n} a_i T(b_i + \int w_i g d\mu) \tag{1.9}$$

where w_i is the function of the weight space.

Figure 1.8 shows the architectural graph of MLP with two hidden layers and one output. Here the network shown is of a fully connected type.

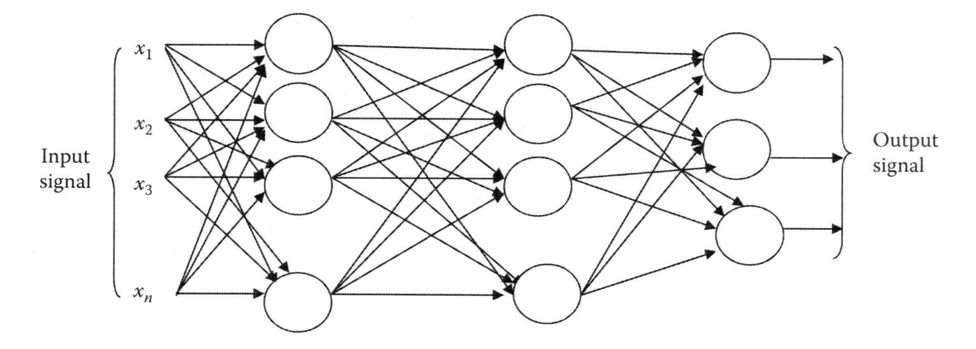

FIGURE 1.8: Architecture of MLP.

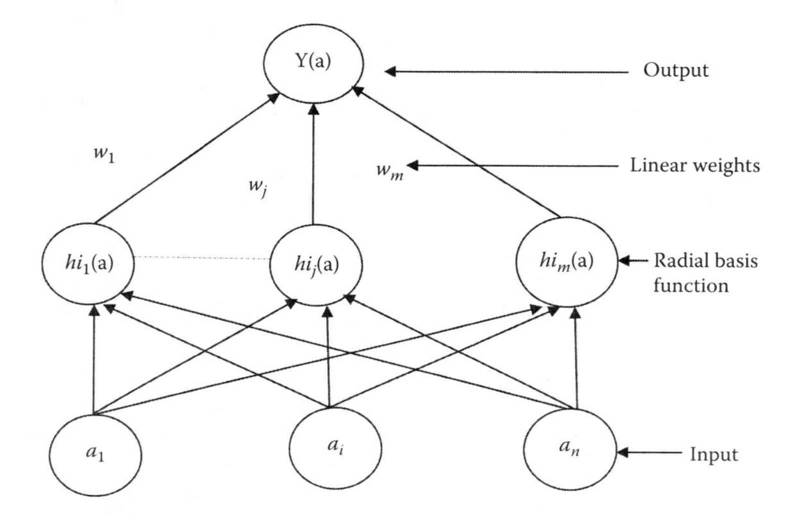

FIGURE 1.9: Architecture of traditional radial basis function network.

1.8.2 Simple logistic

Logistic Model Trees (LMTs) are the combination of two complementary classification schemes, namely, tree induction and linear logistic regression. It has been shown that LMTs perform ruthlessly with other cutting edge classifiers [70]. The drawback of LMTs is the time consuming method. The fixed number of iterations are repeatedly called in the LogitBoost algorithm [71]. In this work, we use SimpleLogistic as a classifier.

In the LogitBoost approach [71] on all data, we use the number of iterations that gives the lowest squared error on the test set from the average values over the cross-validations. We will refer to this method as SimpleLogistic [70].

1.8.3 Radial basis function networks

Radial basis function (RBF) networks are the supervised learning algorithms. In theory they could be recruited in any kind of model (linear or nonlinear) and any class of network (single layer or multilayer). However, since Broomhead and Lowes' work in 1988 [72], the RBF networks have traditionally been incorporated with radial functions in a single layer network as shown in Figure 1.9. The aim of RBF is to approximate the target function through linear combinations of radial kernels, such as Gaussian. Thus the output of an RBF network learning algorithm typically consists of a set of centers and weights for these functions.

1.9 Experimental Results

Standard datasets of Bangla handwriting are not available. So, we have to use Bangla handwriting samples from [58]. To enrich the result of our proposed system, we have used Bangla handwriting samples containing 71 Bangla characters, including vowel modifiers and numerals. Figure 1.3 shows the sample of the collected handwriting. From our datasets, 123 writers are selected and each writer has five sets. Each document consists of 71 characters. Each writer has 355 Bangla isolated characters. So, 123 writers have 43,665 characters.

Here we use one of the most popular tools, Weka [73] for executing machine learning algorithms for data mining tasks. For the purposes of our research work purpose, we have used the SimpleLogistic, RBF network (RBF) and multilayer perceptron (MLP) classifiers. We have used these classifiers as those have given some supportive results for our work than other classifiers of Weka. We have used five-fold scheme above to mention three different classifiers of Weka tool for verification of the writer.

1.9.1 Performance evaluation criteria

The geo-mean [33] metric used to evaluate our classifers results in highly complexed datasets. Therefore, the geo-mean [33] metric is acquired to evaluate the performance of our experiments over this chosen dataset. It is defined as follows:

$$Sensitivity = \frac{tp}{tp + fn} \tag{1.10}$$

$$Specificity = \frac{tn}{tn + fp} \tag{1.11}$$

$$Geo\text{-}mean = \sqrt{sensitivity \times specificity} \tag{1.12}$$

Here in Equations 1.11 and 1.12, tp, fn, tn, and fp are "true positive," "false negative," "true negative," and "false positive," respectively from receiver-operator characteristics (ROC). Finally, the g-mean-based error rate (%) [73] is calculated as follows:

$$g\text{-}mean(error\,rate) = (1 - (geo\text{-}mean))X100. \tag{1.13}$$

Error rates of verification are defined in [74] [Error rates in fault diagnostics or biometric verification/identification] as follows:

$$False\ acceptance\ rate\ (FAR) = fp \tag{1.14}$$

$$False\ rejection\ rate\ (FRR) = fn \tag{1.15}$$

where false acceptance rate (FAR) denotes the genuine writer rate rejected by the system. False rejection rate (FRR) denotes different writer rates incorrectly accepted by the system.

1.9.2 Dissimilarity vector calculation

The interclass dissimilarity vector length between samples from the same writer is 1,230 ($= {}^{5references}C_2$ X 123 writers).

About 3,050 (123 writers $\times$ ${}^{5}C_1$ $\times$ 5 references) dissimilarity vectors are calculated between samples from different writers as interclass dataset.

1.10 Verification Results

Figures 1.10 and 1.11 show the ultimate verification result when the verification approach is done by different classifiers, both on GLCM features alone and again on GLCM combined with DT-CWT features vectors, to verify a writer. The x axis in Figures 1.10 and 1.11 show the different classifiers and the y axis shows the percentages of the geo-mean error rate. Here we obtain higher accuracy results in Table 1.1 in comparison with single GLCM features, after combining these two texture features. But when we use only GLCM features on our isolated Bangla handwritten characters as shown in Figure 1.10, the accuracy becomes very low. However, in Figure 1.11 the accuracy for the proposed combined GLCM and DT-CWT features is high. Hence, it demonstrates that single-feature GLCM is enriched when it is combined with the DT-CWT method for feature extraction. With the proposed automated verification system, the innovative result of the isolated Bangla

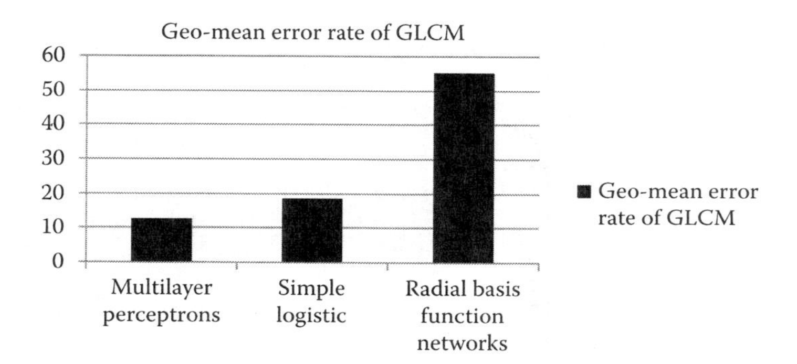

FIGURE 1.10: Comparative performances of the different geo-mean error rates on GLCM features using different classifiers.

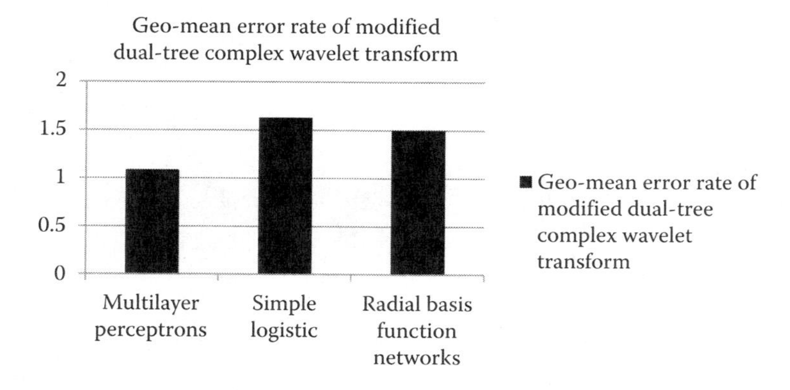

FIGURE 1.11: Comparative performances of the different geo-mean error rates on modified DT-CWT features using different classifiers.

TABLE 1.1: Geo-mean-error rate of modified dual tree complex wavelet transform and GLCM of different classifiers

Classifier name	Geo-mean error rate of modified dual tree complex wavelet transform	Geo-mean error rate of GLCM
MLP	1.0880	12.5646
SimpleLogistic	1.6304	18.77
RBF	1.4965	55.3039

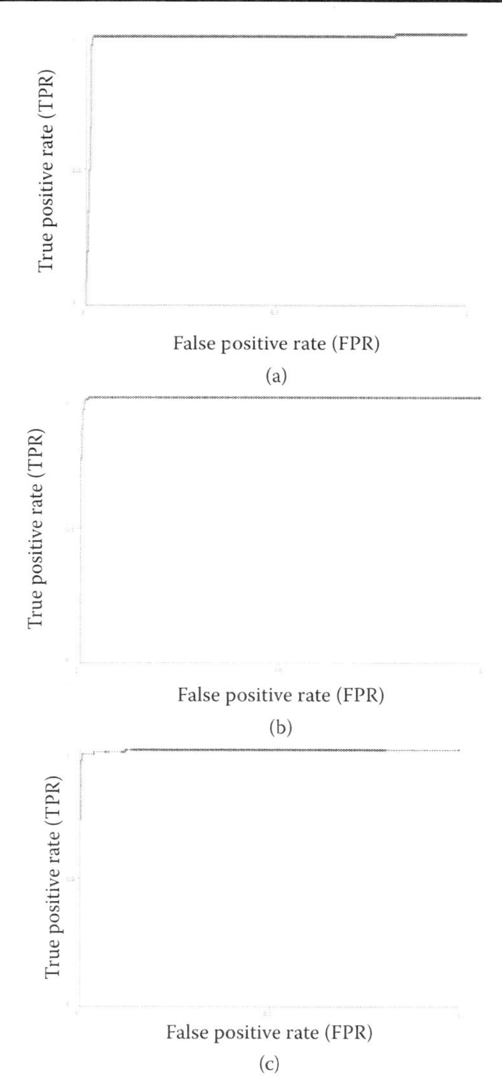

FIGURE 1.12: (a), (b), and (c) are different ROC curves using combined features with different classifiers.

characters reaches to 98.0695% value in accuracy, which is the highest percentage of verification obtained so far.

We have analyzed the ROC characteristics of the performances of the chosen classifiers over our isolated Bangla characters dataset. Figures 1.12 and 1.13 show the ROC curves

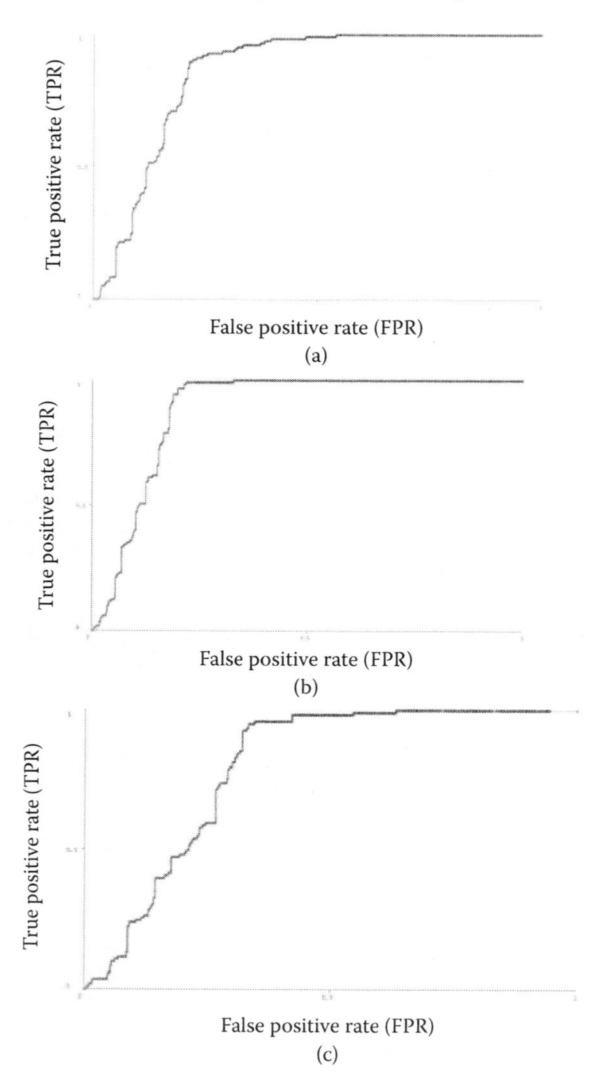

FIGURE 1.13: (a), (b), and (c) are different ROC curves using GLCM features with different classifiers.

obtained for the solutions of SimpleLogistic, RBF, and MLP classifiers in our experiments. The Precision, Recall, F-Measure, and ROC area values obtained for the classifiers using GLCM features have been shown in Table 1.2 for both same-writer and different-writer classes.

A higher value for ROC area denotes better classification results. For same-writer and different-writer classes, SimpleLogistic, RBF, and MLP classifiers obtain 0.858, 0.794 and 0.884 values respectively. Therefore, MLP classifier shows the highest efficiency in detecting same writer and different-writer classes in our chosen dataset.

Similarly, using combined GLCM and DT-CWT features, the parameters obtained from ROC characteristics have are in Table 1.3. The ROC area obtained using this proposed combined feature set by SimpleLogistic, RBF, and MLP classifiers are, respectively, 0.985, 0.998, and 0.996. Therefore, the RBF classifier shows the best performance among chosen classifiers with combined features.

TABLE 1.2: Class-wise evaluation parameters on solutions obtained from different classifiers using GLCM features

Classifier name	Precision		Precision		F-measure		ROC area	
	Same writer	Different writer	Same writer	Different writer	Same writer	Different writer	Same writer	Different writer
SimpleLogistic with GLCM	0.576	0.933	0.829	0.796	0.829	0.859	0.858	0.858
RBF with GLCM	0.450	0.777	0.220	0.910	0.295	0.838	0.794	0.794
MLP with GLCM	0.607	0.986	0.967	0.790	0.746	0.877	0.884	0.884

TABLE 1.3: Class-wise evaluation parameters on solutions obtained from different classifiers using combined GLCM and DT-CWT

Classifier name	Precision		Precision		F-measure		ROC area	
	Same writer	Different writer	Same writer	Different writer	Same writer	Different writer	Same writer	Different writer
SimpleLogistic with GLCM	0.953	0.994	0.984	0.984	0.968	0.989	0.985	0.985
RBF with combined GLCM and DT-CWT	0.984	0.992	0.976	0.995	0.980	0.993	0.998	0.998
MLP with combined GLCM and DT-CWT	0.961	0.997	0.992	0.986	0.976	0.992	0.996	0.996

Moreover, all classifiers show significant superior performances on our chosen dataset after using the combination of DT-CWT features with GLCM features. This signifies the efficiency of our proposed combined feature extraction method for the writer verification problem.

1.11 Conclusions

The automatic writer verification method is a very challenging task in the domain of finance and forensic analysis. In this chapter, to enhance the verification accuracy over isolated Bengali characters samples, we have proposed a hybrid model which is a combination of well-known GLCM and new DT-CWT–based feature extraction methods to compute dissimilarity between same-writer and different-writer classes. For comparitive study, a similar kind of work was done by some different algorithms using Bengal character dataset [75]. Furthermore, we have utilized Kullback–Leibler divergence measure as the dissimilarity norm over the extracted features. The advantages of our present method are as follows:

1. In our writer-dependent approaches we calculated dissimilarity vector, the writer specific Kullback–Leibler divergence dissimilarity measure is used to improved verification accuracy.

2. The individual GLCM based and combined GLCM–DT-CWT based features are extracted from samples obtained by multiband image scanners and those features are used to provide superior writer verification results.

3. Our experimental results show that the combined feature extraction method raises the accuracy from 73.67% to 98.9%, using the RBF network classifier.

4. Furthermore, when comparing using three different classifiers, the g-mean error rate has been reduced to 1.6-1.0% using our proposed combining methods.

Therefore, in our experimental results, our proposed combined feature extraction method has shown significant improvement over well-known GLCM based features for the automated writer verification problem on isolated Bengali characters. As for the future research works, the authors will utilize this improved feature extraction method for forgery detection.

References

1. S. H. Cha and S. Srihari. Multiple feature integration for writer verification. In *Proceedings of 7th International Workshop on Frontiers in Handwriting Recognition*, Amsterdam, Netherlands, September 11–13, 2000, pp. 333–342, 2000.

2. R. N. Morris. *Forensic Handwriting Identification.* San Diego, CA: Academic Press, 2000.

3. H. E. S. Said, T. N. Tan, and K. D. Baker. Personal identification based on handwriting. In *Fourteenth International Conference on Pattern Recognition*, pp. 149–160, 2000.

4. S. Srihari, S. H. Cha, H. Arora, and S. Lee. Individuality of handwriting: A validity study. In *Proceedings of 6th International Conference on Document Analysis and Recognition'01*, pp. 106–109. IEEE, 2001.

5. A. Dutta and S. Chaudhury. Bengali alpha-numeric character recognition using curvature features. *Pattern Recognition*, 26(12): 1757–1770, 1993.

6. Y. Tang and S. N. Srihari. Likelihood ratio estimation in forensic identification using similarity and rarity. *Pattern Recognition*, 47(3): 945–958, 2014.

7. S. Srihari, M. Beal, K. Bandi, V. Shah, and P. Krishnamurthy. A statistical model for writer verification. In *Proceedings of International Conference on Document Analysis and Recognition*, pp. 1105–1109, 2005.

8. H. Srinivasan, S. Kabra, C. Huang, and S. Srihariand. On computing strength of evidence for writer verification. In *Ninth International Conference on Document Analysis and Recognition (ICDAR 2007)*, vol 2, pp. 844–848. IEEE, 2007.

9. L. Wan, B. Wan, and Z.-C. Lin. On-line signature verification with two-stage statistical models. In *Proceedings of Eighth International Conference on Document Analysis and Recognition*, pp. 282–286. IEEE, 2005.

10. C. Halder and K. Roy. Individuality of isolated Bangla numerals. *Journal of Network and Innovative Computing*, 1: 33–42, 2013.

11. I. Tomai, B. Zhang, and S. N. Srihari. Discriminatory power of handwritten words for writer recognition. In *Proceedings of the 17th International Conference on Pattern Recognition*, pp. 638–641. IEEE, 2004.

12. A. Fornes, J. Llados, G. Sanchez, and H. Bunke. Writer identification in old handwritten music scores. In *Eighth IAPR International Workshop on Document Analysis Systems (DAS '08)*, pp. 638–641. IEEE, 2008.

13. R. Chaudhry and S. K. Pant. Identification of authorship using lateral palm print a new concept. *Journal of Forensic Science*, 141(1): 49–57, 2004.

14. L. Schomaker. Advances in writer identification and verification. In *Ninth International Conference on Document Analysis and Recognition (ICDAR 2007)*, pp. 1268–1273, IEEE, 2007.

15. R. Plamondon and S. N. Srihari. On-line and off-line hand writing recognition: A comprehensive survey. *IEEE Transactions on Pattern Analysis and Machine Intelligence*, 22(1): 63–84. IEEE, 2000.

16. D. Impedovo and G. Pirlo. Automatic signature verification: The state of the art. *IEEE Transactions on Systems, Man, and Cybernetics Part C: Applications and Reviews*, 38(5): 609–635, 2008.

17. H. Chayan, Sk. Md. Obaidullah, J. Paul, and K. Roy. Writer verification on bangla handwritten characters. In *Advanced Computing and Systems for Security*, pp. 53–68 New Delhi, India: Springer, 2016.

18. M. S. Shirdhonkar and M. Kokare. Off-line handwritten signature identification using rotated complex wavelet filters. *IJCSI International Journal of Computer Science Issues*, 8(1): 478-482, 2011.

19. D. Bertolini, L. S. Oliveira, E. Justino, and R. Sabourin. Texture-based descriptors for writer identification and verification. *Expert Systems with Applications*, 40: 2069–2080, 2013.

20. R. K. Hanusiak, L. S. Oliveira, E. Justino, and R. Sabourin. Writer verification using texture-based features. *International Journal on Document Analysis and Recognition (IJDAR)*, 15: 213–226, 2012.

21. V. Nguyen and M. Blumenstein. Techniques for static handwriting trajectory recovery: A survey. In *DAS '10 Proceedings of the 9th IAPR International Workshop on Document Analysis Systems*, pp. 463–470, 2010.

22. R. Plamondon and C. M. Privitera. The segmentation of cursive handwriting: An approach based on off-line recovery of the motor-temporal information. *IEEE Transactions on Image Processing*, 8: 80–91, 1999.

23. M. A. Ferrer, J. Francisco Vargas, A. Morales, and A. Ordonez. Robustness of offline signature verification based on gray level features. *IEEE Transactions on Information Forensics and Security*, 7: 966–977, 2012.

24. D. Ellen. *Scientific Examination of Documents: Methods and Techniques*, 3rd ed. Boca Raton, FL: CRC Press, 2005.

25. L. Jane. *Forensic Document Examination: Fundamentals and Current Trends*, 1st ed. Norwell: Academic Press, 2014.

26. U.-V. Marti, R. Messerli, and H. Bunke. Writer identification using text line based features. In *Proceedings of Sixth International Conference on Document Analysis and Recognition*, pp. 463–470. IEEE, 2001.

27. H. Arora, S. Lee, S. N. Srihari, and S. H. Cha. Individuality of handwriting. *Journal of Forensic Science*, 47: 1–17, 2002.

28. M. Bulacu, L. Schomaker, and L. Vuurpijl. Writer identification using edge-based directional features. In *Proceedings of Seventh International Conference on Document Analysis and Recognition*, pp. 937. IEEE, 2003.

29. A. Bensefia, T. Paquet, and L. Heutte. A writer identification and verification system. *Pattern Recognition Letters*, 26: 2080–2092, 2005.

30. I. Siddiqi and N. Vincent. Text independent writer recognition using redundant writing patterns with contour-based orientation and curvature features. *Pattern Recognition*, 43: 3853–3865, 2010.

31. D. Pavelec, E. Justino, L. V. Batista, and L. S. Oliveira. Author identification using writer-dependent and writer-independent strategies. In *Proceedings of the 2008 ACM Symposium on Applied Computing*, pp. 414–418, 2008.

32. E. Pkalska and R. P. Duin. Dissimilarity representations allow for building good classifiers. *Pattern Recognition Letters*, 23: 943–956, 2002.

33. T. Matsumoto and D. Muramatsu. Online signature verification algorithm with a user-specific global-parameter fusion model. In *IEEE International Conference on Systems, Man and Cybernetics (SMC 2009)*, pp. 486–491. IEEE, 2009.

34. S. N. Srihari, S. H. Cha, H. Arora, and S. Lee. Individuality of handwriting. *Journal of Forensic Science*, 47: 856–872, 2002.

35. R. Jain and D. Doermann. Writer identification using an alphabet of contour gradient descriptors. In *12th International Conference on Document Analysis and Recognition (ICDAR)*, pp. 550–554. IEEE, 2013.

36. B. Zhang, S. N. Srihari, and S. Lee. Individuality of handwritten characters. In *Proceedings of Seventh International Conference on Document Analysis and Recognition*, pp. 1086–1090. IEEE, 2003.

37. S.-H Cha and S. N. Srihari. On measuring the distance between histograms. *Pattern Recognition*, 35: 1355–1370, 2002.

38. M. Bulacu and L. Schomaker. Text-independent writer identification and verification using textural and allographic features. *IEEE Transactions on Pattern Analysis and Machine Intelligence*, 29(4): 701–717, 2007.

39. A. Schlapbach. *Writer Identification and Verification.* Clifton, VA: IOS Press, 2007.

40. I. Siddiqi and N. Vincent. A set of chain code based features for writer recognition. In *10th International Conference on Document Analysis and Recognition (ICDAR '09)*, pp. 981–985. IEEE, 2009.

41. S. N. Srihari and G. Leedham. A survey of computer methods in forensic document examination. In *Proceedings of 11th Conference International on Graphonomics Society (IGS 2003)*, pp. 278–281, 2003.

42. D. Maltoni, R. M. Bolle, D. Zhang, and J. P. Campbell. Guest editorial special issue on biometric systems. *IEEE Transactions on Systems, Man, and Cybernetics, Part C: Applications and Reviews*, 35: 273–275, 2005.

43. J. L. Wayman, A. Jain, D. Maltoni, and D. Maio (Eds.). *Biometric Systems Technology, Design and Performance Evaluation*. London, UK: Springer, 2005.

44. D. V. Klein. Foiling the cracker: A survey of, and improvements to, password security. In *Proceedings of the 2nd USENIX Security Workshop*, pp. 5–14, 1990.

45. D. Maltoni, D. Maio, A. Jain, and S. Prabhakar. *Handbook of Fingerprint Recognition*. London, UK: Springer Science and Business Media, 2009.

46. R. Sanchez-Reillo, C. Sanchez-Avila, and A. Gonzalez-Marcos. Biometric identification through hand geometry measurements. *IEEE Transactions on Pattern Analysis and Machine Intelligence*, 35: 1168–1171, 2000.

47. J. Daugman. The importance of being random: Statistical principles of iris recognition. *Pattern Recognition*, 36: 279–291, 2003.

48. A. Jain, R. Bolle, and S. Pankanti (Eds.). *Biometrics: Personal Identification in Networked Society*. New York, NY: Springer Science and Business Media, 2006.

49. J. P. Campbell. Speaker recognition: A tutorial. In *Proceedings of the IEEE*, 85(9): 1437–1462, 1997.

50. Z. Zhou, E. Y. Du, N. Luke Thomas, and E. J. Delp. Multi-angle sclera recognition system. In *Computational Intelligence in Biometrics and Identity Management (CIBIM)*, IEEE, 2011.

51. F. Monrose and A. Rubin. The importance of being random: Statistical principles of iris recognition. In *Proceedings of the 4th ACM Conference on Computer and Communications Security*, pp. 48–56, 1997.

52. W. Stone Henry and A. C. Sanderson. A prototype arm signature identification system. In *Proceedings of IEEE International Conference on Robotics and Automation*, vol 4, pp. 107–131, IEEE, 1987.

53. R. Plamondon and G. Lorette. Automatic signature verification and writer identification 17 the state of the art. *Pattern Recognition*, 22: 107–131, 1989.

54. R. Plamondon. A kinematic theory of rapid human movements: Part III. kinetic outcomes. *Biological Cybernetics*, 78: 133–145, 1998.

55. V. Nguyen, M. Blumenstein, V. Muthukkumarasamy, and G. Leedham. Off-line signature verification using enhanced modified direction features in conjunction with neural classifiers and support vector machines. In *Ninth International Conference on Document Analysis and Recognition (ICDAR 2007)*, vol. 2, pp. 734–738, IEEE, 2007.

56. U. V. Marti and H. Bunke. The iam-database: An English sentence database for off-line handwriting recognition. *International Journal on Document Analysis and Recognition*, 5: 39–46, 2002.

57. G. L. Schomaker, R. Plamondon, R. Liberman, and S. Janet. Unipen project of online data exchange and recognizer benchmarks. In *Proceedings of the 12th IAPR International Conference on Pattern Recognition, Computer Vision and Image Processing*, vol. 2, pp. 29–33. IEEE, 1994.

58. C. Halder and K. Roy. Individuality of isolated Bangla characters. In *Proceedings of ICDCCom*, pp. 1–6. IEEE, 2014.

59. G. F. Simons Lewis, M. Paul and C. D. Fennig. *Ethnologue: Languages of the World.* Texas: SIL International, 2016.

60. N. Gorski, V. Anisimov, E. Augustin, O. Baret, and S. Maximov. Industrial bank check processing: The A2iA Check-reader^{TM}. *International Journal on Document Analysis and Recognition*, 3: 196–206, 2001.

61. C. Halder, J. Paul, and K. Roy. Individuality of Bangla numerals. In *2012 12th International Conference on Intelligent Systems Design and Applications (ISDA)*. IEEE, 2012.

62. A. Eleyan, H. Ozkaramanli, and H. Demirel. Dual-tree and single-tree complex wavelet transform based face recognition. In *IEEE 17th Conference on Signal Processing and Communications Applications (SIU 2009)*, vol. 3, pp. 610–621. IEEE, 2009.

63. G.-Y. Zhang, S.-Y. Peng, and H-M. Li. Combination of dual-tree complex wavelet and svm for face recognition. In *2008 International Conference on Machine Learning and Cybernetics*, vol. 5. IEEE, 2008.

64. Y. Peng, X. Xie, W. Xu, and Q. Dai. Face recognition using anisotropic dual-tree complex wavelet packets. In *19th International Conference on Pattern Recognition (ICPR 2008)*. IEEE, 2008.

65. P. Zhang, T. Bui, and C. Suen. Wavelet feature extraction for the recognition and verification of handwritten numerals. *Wavelet Analysis and Active Media Technology*, 3: 3–9, 2005.

66. N. Kingsbury. Image processing with complex wavelets. *Philosophical Transactions of the Royal Society of London A: Mathematical, Physical and Engineering Sciences*, 357(1760): 2543–2560, 1999.

67. R. M. Haralick, K. Shanmugan, and I. Dinstein. Textural features for image classification. *IEEE Transactions on Systems, Man, and Cybernetics*, 3: 610–621, 1973.

68. S. Kullback and R. A. Leibler. On information and sufficiency. *The Annals of Mathematical Statistics*, 22: 79–86, 1951.

69. M. Kubat and S. Matwin. Combination of dual-tree complex wavelet and SVM for face recognition. In *Proceedings of the Fourteenth International Conference on Machine Learning*, vol. 97. Citeseerx, 1997.

70. N. Landwehr, M. Hall, and E. Frank. Logistic model trees. *Machine Learning*, 59: 161–205, 2005.

71. J. Friedman, T. Hasti, and R. Tibshirani. Additive logistic regression: A statistical view of boosting. *The Annals of Statistics*, 28: 337–407, 2000.

72. D. S. Broomhead and D. Lowe. Multivariable functional interpolation and adaptative networks. *Complex Systems*, 11: 321–355, 1988.

73. M. Hall et al. The WEKA data mining software: An update. *ACM SIGKDD Explorations Newsletter*, 11: 10–18, 2009.

74. S. Pankanti, N. K. Ratha, A. W. Senior, R. M. Bolle, and J. H. Connell. *Guide to Biometrics*. New York, NY: Springer, 2004.

75. J. Paul, C. Halder, Sk. Md Obaidullah, and K. Roy. Writer verification on Bangla handwritten characters. In *Advanced Computing and Systems for Security*, pp. 53–68. New Delhi, India: Springer, 2015.

Chapter 2

Hybrid Intelligence Techniques for Handwritten Digit Recognition

Hemant Jain, Ryan Serrao, and B.K. Tripathy

2.1 Introduction

The search for patterns in data has eluded most for a long time in history and recent advancements have allowed us to relish success in this field. The study of pattern recognition primarily focuses on automating the detection of regularities in data [1]. The use of these regularities to perform data classification and prediction has gained popularity and has led to rapid advancements in the field of pattern recognition. In the last two decades, the application of machine learning has proven to be very successful and in this chapter we will investigate several hybrid machine learning approaches for handwriting recognition.

The complexity of pattern recognition is often understated. In real-world data, the variability of the input (in which we must detect the pattern) is very high. Thus, generalization is one of the greatest hurdles of pattern recognition. Even though machine learning helps in generalization, unprocessed image data become hard for the model to understand. Due to this, the data are often pre-processed before giving them as an input to the model (sometimes in the first layer of the model itself). An important step in preprocessing is rescaling and translation of images. For example, in the problem of digit recognition, images of the digits are usually translated and scaled such that all digits can be stored in a rectangular boxes of equal dimensions. This method largely reduces the variability because location and size of digits in the box are similar, allowing us to accurately label these boxes.

The preprocessing stage for all kinds of data, including image, text, and video, is commonly referred to as feature extraction. The process of optimally performing feature extraction is called feature engineering. It is important to note that the same steps for preprocessing must be applied on both test and training data. The methods used and extent of preprocessing rely on noise and variability of data along with the computational power of the system being used. Large images take exponentially longer computational overhead than smaller images and this is often a deciding factor while selection of size of the preprocessed image. The challenge is to strike a balance between time taken to compute and the loss of information in image size reduction. There are several other methods such as image mean subtraction that accelerate the computation time but the motive of this chapter is to provide a broader overview. A reminder, however, is that during image size reduction, some of the information is often discarded and while smaller images are easier to learn patterns from, they may not contain the complex patterns in the original image. This often results in the loss of the overall accuracy of the system.

Soft computing consists of several computing techniques, including fuzzy logic, neural networks, and genetic algorithms that can be used to produce powerful hybrid intelligent systems for digit recognition. Handwritten digits recognition (HDR) has been a very popular application of computer vision in the field of character recognition. The use of digits just as any other universal symbol is everywhere, especially in the field of technology, banks, OCR, postal service, analysis of digits in engineering, recognition of numbers on number plates etc. Despite the design of several image processing techniques [2], the fact that the handwritten digits have no fixed image recognition pattern in each of its digits makes it all the more difficult and challenging to design an optimal recognition system. Recognizing numbers is clear to the human eye, but not so for the machines, especially when there are ambiguities on different classes (e.g., digit '1' and '7'). Figure 2.1 shows 15 different handwritten digits related to these issues.

HDR deals with two important steps. The first one is feature extraction and the second is the classification method used. Features are the basis for all classification techniques. It is based on the features that the classification algorithm learns to classify the digits accurately. The better the feature selection, the higher the classification accuracy.

Classification is another important step in digit recognition. Classifiers such as support vector machines (SVMs), neural networks, and convolutional networks have proven to be very accurate in the past.

The recognition system is applied in either online or offline mode. In offline mode, the digits are generated first followed by processing and feature extraction. While in online mode, digits are processed while being generated. In this kind of recognition, factors such as the speed of writing and pressure involved have a large influence on the system.

FIGURE 2.1: Different samples of handwritten digits in MNIST.

Hence, accuracy in recognition of handwritten digits is still limited due to large variation in shape, style, orientation, etc.

Hybrid intelligent systems can have different architectures, which have an impact on the efficiency and accuracy of digit recognition systems. For this reason, it is very important to optimize architecture design. The architectures can combine neural networks [3], fuzzy logic, and genetic algorithms in different ways to achieve the ultimate goal of pattern recognition. Handwritten digit recognition involves the following steps: image acquisition, preprocessing which involves noise removal [4], normalization, thinning and skeletonization, feature extraction, classification and recognition. Figure 2.2 depicts the proposed methodology.

In this chapter, we will be providing the detailed methodologies and the hybrid intelligent techniques applied so far in every step of handwritten digit recognition.

2.2 Literature Review

A vast amount of research has been done with respect to pattern analysis and recognition, especially with reference to recognition of handwritten digits. Certain standard databases have been used for benchmark comparison of various algorithms innovated. The earliest database to be used was created by AT&T, which consisted of 1200 digits. Recognition of these digits proved to be very easy having a high accuracy rate and hence had to be discarded. The US Postal Service introduced a new database known as the Zip Code database. It consisted of 7064 training examples and 2007 testing digits. All digits were machine segmented. But due to the extraneous ink and also omission of critical fragments, a limited accuracy was obtained. To overcome the limitations, the NIST database was introduced which consisted of 58,527 digits written by 500 writers. The training set represented the digits written by the US census workers and the testing sets were written by high school

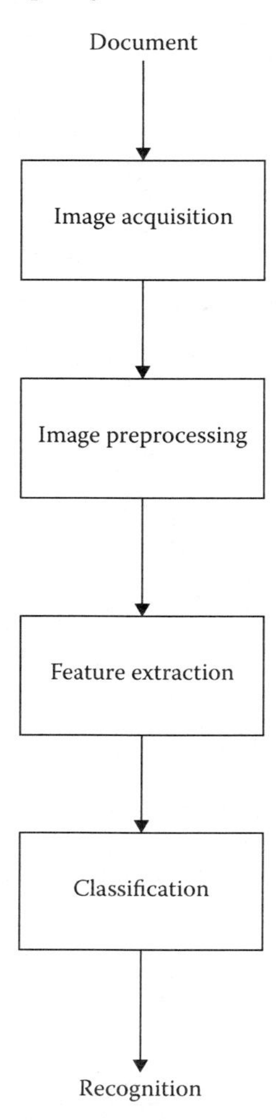

FIGURE 2.2: Flowchart for digit recognition.

students. There was a huge difference in the distribution of digits in the training and testing sets, because of which the obtained accuracy was low. Hence, a modified version of NIST was introduced called MNIST consisting of 60,000 digits [5]. All images were normalized to fit a 20×20 pixel image which was then transformed to center in a 28×28 image. The pixel intensities were gray-scaled to decrease aliasing.

A lot of classifiers have been tested for recognizing handwritten digits. One such classifier was the linear classifier where the output is determined by the weighted sum of the inputs. Nearest neighbors is another classifier which takes into account the Euclidean distance between input image pixels for mapping. Yann Le Cunn introduced LeNet1 [6] which proved to be very efficient for the recognition of handwritten digits. It used alternate layers of convolution and subsampling to extract essential features for digit recognition. An extension of LeNet1 is LeNet4, which consists of more feature maps and an additional hidden layer that is intended to improve the accuracy of the digit classifier. Another model is the boosted LeNet4, having the highest accuracy among the family. LeNet4 combined with KNN showed better performance.

Hybrid algorithms combine the strengths of individual algorithms for better performance and accuracy [7]. Principal component analysis was used on the MNIST dataset to reduce the dimensionality. A set of 40 features were extracted which were further classified using a polynomial classifier [8]. Another hybrid algorithm used Gaussian radial basis function (RBF) followed by a linear classifier giving high accuracy rates.

Convolutional neural networks (CNN) have proven to be excellent feature extractors. They are now being used in almost every domain by all researchers to extract the most efficient feature for further processing. CNNs have proven to be very efficient in digit recognition. Hence, combining CNNs with efficient classifiers have given tremendously positive results. One such example is the hybridization of CNNs with support vector machines (SVMs). Using this approach, high accuracy rates have been obtained [9]. Extreme learning machines (ELM) have proven to be very efficient classifiers in the recent years. Known for their fast learning rate, a combination of CNNs with ELMs has given very robust and accurate classifiers [10].

SVMs are excellent classifiers and have been used in diverse pattern recognition applications. Many efficient hybrid classifiers have been built using SVMs to improve classification accuracy. A. Bellili [11] introduced a hybrid classifier using SVM with multilayer perceptrons (MLP). M. Govindarajan [12] introduced an ensemble classifier by combining RBFs with SVM. Ayaaz et al. [13] showed an increase in the accuracy of the hybrid model as compared to individual models by a huge difference. SVMs have also been used with hybrid feature extractors for classifying highly efficient features [14].

Other hybrid models include the combination of neural networks (NNs) with hidden Markov models (HMMs) [15]. Another interesting application is the use of neuro-fuzzy techniques for recognizing digits adopted by M. Pinzolas et al. [16]. An extension is the utilization of the adaptive neuro-fuzzy method for pattern recognition [17,18] introduces hybrid tree classifiers for classification, which follows a two-step classification, namely hierarchical coarse and fine classification which combines artificial neural networks (ANN) with decision trees.

2.3 Conventional Models

2.3.1 Convolutional neural network

A convolutional neural network is a special deep learning architecture. It is essentially a multilayer neural network that is a combination of two operations: automatic feature extraction and classification. Feature extraction is the first operation where the network generates a feature map for the image/input. These filters or maps highlight the discriminating features in the raw images/input via two operations: convolutional filtering and down-sampling. This architecture is inspired from that of the visual cortex.

In the feature extractor or the convolutional layer, there are a set of filters that are modified as new images are encountered by the system. These filters extend through the full depth of the input image. In the forward pass, each of these filters is 'convolved' across the height and width of the input image and produce a two-dimensional activation map for each filter. This causes certain filters to learn to distinguish or highlight one image from the other. Through this iterative process, the network learns a set of filters (features) that activate upon seeing similar features at an arbitrary spatial position in the input image.

The performance and working of this model can be controlled by changing the hyper-parameters like depth/dimension of image, stride and padding.

For the fully connected layer, if l denotes the present layer,

$$x^l = f(u^l), \quad u^l = W^l x^{l-1} + b^l \tag{2.1}$$

where $f(.)$ is an activation function (usually sigmoidal) that outputs values in the range between 0 and 1.

The Error function E is calculated as:

$$E^n = \frac{1}{2}\sum_{k=1}^{c}(t_k^n - y_k^n)^2 = \frac{1}{2}||t_k^n - y_k^n||_2^2 \tag{2.2}$$

Finally, the weights are updated by using the formula:

$$\frac{\partial E}{\partial b} = x^{l-1}(\delta)^T \tag{2.3}$$

The three hyper parameters of the convolutional layer are the depth of the output volume, the strike S, or distance from the slide to the filter, and whether or not zero-padding is to be done. The depth signifies the number of unique filters we desire. If the stride is one, the filter moves one step at a time. The larger the stride, the smaller the output volumes of the layer. Due to variable size of the input volume and the stride, to avoid loss of data, zero-padding is done. The size of the zero-padding P depends on the size F of the receptive field of the convolution layer and the stride.

Thus the number of neurons that fit in volume is given by:

$$n = (W - F + 2P)/S + 1 \tag{2.4}$$

Therefore, we get an output volume of size $n \times n$ and this is why the value n should always be an integer. For this reason, it is recommended that the padding P always be equal to $(F - 1)/2$ when $S = 1$. It has empirically been proven that no advantages exist of having volumes with unequal height and width.

While dealing with CNN, we must also perform pooling—a nonlinear down sampling technique. The most popular of all pooling approaches is max pooling. The concept works on the assumption that if a feature is found, its precise location in the image is not as important as its relative position to that of the other features. Max pooling partitions the image or matrix into sets of smaller rectangles or region and outputs the maximum value for each of the subsections. It significantly reduces the possibility of over-fitting (Figure 2.3).

2.3.2 Extreme learning machines

Traditional neural networks that employed the feedforward technique to learn from data required a large amount of time to train the model. This problem was magnified when using

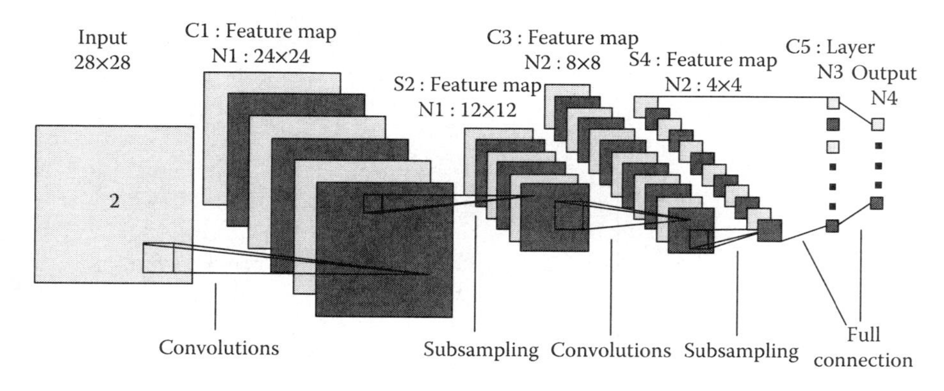

FIGURE 2.3: Structure of a convolutional neural network.

different types of neural networks that relied on low learning rate and slow gradient-descent-based optimization. The bottleneck, due to the requirement of several iterations before the model could actually learn, led to the development of a single layer feedforward neural network (SLFN) that would soon be called the extreme learning machine (ELM). This algorithm drastically reduced training time while providing a significant generalization to the data. An ELM is nothing but a SLFN with random hidden nodes whose values would be arbitrarily determined by making changes to the weights assigned to each of these nodes [19,20].

To understand the internal workings of ELM, let us assume there exists an ELM with n input neurons, l hidden layer neurons and m output neurons with weight between the input and hidden layer denoted by w_i and weight between hidden and output layer denoted by β_i. If we have N samples to learn from, in the format (x, t) (where x represented the attributes and t the label), we can express the network as:

$$\sum_{i=1}^{l} \beta_i g(w_i, b_i, x_j) = y_i, j = 1, 2, ..., N. \tag{2.5}$$

where the function $g(.)$ is the activation function on the hidden layer, β represent the weights between a pair of output and hidden neuron and b is the bias of the corresponding hidden neuron.

In SLFN, we must approximate the N samples with minimum or zero errors. Thus we calculate the hidden layer output matrix H using the weight matrix β and the label matrix T as:

$$H\beta = T \tag{2.6}$$

Therefore, ELMs are basically two-layered neural networks, consisting of a randomly organized input layer with a trained second layer. Such a model has proven to produce great results, especially with respect to pattern recognition, and are also trained many times faster than a general neural network with a back-propagation (BP) algorithm [21].

Below is the algorithm for updating β during training of ELM.

Step 1. Calculate the number of hidden layer neurons.

Step 2. Randomly initialize bias b of hidden layer neurons.

Step 3. Randomly assign values to the weights w between the input and hidden layers.

Step 4. Select an infinitely differential function as the activation function (like the sigmoid function).

Step 5. Calculate the hidden layer output matrix H.

Step 6. Lastly, calculate the new weight matrix β as $\beta = H^T H$ for getting the internal weights

2.3.3 Support vector machines

Support vector machines (SVMs) are supervised learning algorithms that are non-probabilistic linear classifiers [22]. By using different kernel functions, an SVM can use what is referred to as the kernel trick to perform nonlinear classification. The kernel function works by transforming nonlinear separable problems into linear separable problems by using a method of projection that projects the data into a higher dimensional feature space, after which it easily finds the optimal separating hyperplane [23]. The SVM method initially was proposed to solve a binary classification problem but some strategies were proposed that made it applicable to multiclass problems as well.

Currently, two prominent approaches are being used to solve the multiclass problem. The first works as follows: it constructs multiple two-class SVM classifiers for each combination of classes that are then combined using one of three strategies to create the final multiclass classifier. These three combination strategies are (i) one-against-all, (ii) one-against-one, and (iii) directed acyclic graph (DAG).The second approach considers all the multiclass data in one optimization problem and is often harder to optimize than the former approach.

The most popular version of SVM is called LIBSVM; it uses the one-against-one method for combining the binary SVM classifier. It constructs $k(k-1)/2$ classifiers for k-class problems and uses the following mathematical expressions, Equations 2.7 and 2.8, for optimizing a combination of them to create the multiclass classifier. For a k class problem, there are l training samples: $\{x_1, y_1\}, ..., \{x_l, y_l\}$, where x_i is the feature vector and y_i is the class label corresponding to it for each value of i from 1 to l.

$$min : P(w^{ij}, b^{ij}, \xi^{ij}) = \frac{1}{2}(w^{ij})^T.w^{ij} + C\sum_n \xi_n^{ij} \tag{2.7}$$

$$\begin{aligned} \text{S.T.} \quad & (w^{ij})^T\phi(x_n) + b^{ij} \geq 1 - \xi_n^{ij}, \quad y_n = i, \\ & (w^{ij})^T\phi(x_n) + b^{ij} \leq -1 + \xi_n^{ij}, \quad y_n \neq i \\ & \xi_n^{ij} \geq 0, n = 1, \ldots, k(k-1)/2 \end{aligned} \tag{2.8}$$

Each of the sub-classifiers are given a single vote for the class determined by it. For the final decision, LIBSVM applies a 'max win' algorithm to the vote from each of the sub-classifiers and selects the class with the maximum votes.

2.3.4 Hidden Markov model

Hidden Markov models (HMMs) were originally used in speech recognition, and after their success in the same, there was speculation regarding their application on handwriting recognition. The main reason behind the HMM's success was attributed to the mechanism it follows, whereby it segments and recognizes the input data simultaneously and was thus seen as suitable for handwriting recognition systems.

HMMs are statistical models that are optimally designed for analysis of sequential data and their working largely is based on the statistical model that analyzes the input data. The algorithm tries to find a finite sequence W that maximizes the value of the posterior probability $P(W|X)$, where X is the data given as input and W is the sequence of symbols/words present in the language model.

Therefore, from an input sequence X, the HMM is required to find the finite sequence of words, W, that maximize the value of $P(W) \times P(X|W)$ [Note: $P(X)$ is ignored by applying naive Bayes theorem]. In the following equation, $P(W)$ denotes the prior probability of the occurrence of word W, which can be estimated from the language model, and $P(X|W)$ is the observed probability derived from the language model.

$$W = arg\ max\ P(W|X) = arg\ max\ \frac{P(W)P(X|W)}{P(X)} = arg\ max\ P(W)P(X|W) \tag{2.9}$$

2.4 Hybrid Models

There are several Hybrid models for handwriting recognition [24]. Many such models are discussed in this section.

2.4.1 Convolutional neural network–support vector machine (CNN-SVM)

Convolutional neural networks are designed to act as tools for feature engineering automation and while they create optimal feature vectors for the input data, the final classification based on these features is not as sophisticated. Therefore, the idea to combine them with a more complex classifier by replacing the CNN's classification layer with other sophisticated classification models is an area worth exploring. This method of creating a hybrid model, using a CNN-SVM combination, has been applied to the problem of handwriting recognition and outperformed either of the two algorithms.

There are thus many reasons to combine a CNN with an SVM classifier. Firstly, hand-designed feature extraction is a highly tedious and time-demanding task. It requires skilled workers and even then fails to process raw images effectively. Methods like CNN automate feature extraction methods and extract features directly from the raw images (Figure 2.4).

Figure 2.4 shows how the two algorithms can be combined. It is important to note that all images should be normalized and resized before giving them as input to the CNN. The images are fed into the CNN without any modifications and the process is repeated for a certain number of epochs until the training process achieves convergence (maximum stable accuracy after several epochs/iterations of training). Then the last (output) layer of the CNN is replaced with a SVM which accepts the feature vectors generated by the preceding CNN layers.

Case Study:

Xiao-Xiao Niu et al. [9] proposed a similar hybrid CNN-SVM model whose architecture is shown in Figure 2.4. The images were normalized to a size of 28 × 28 pixels before being fed into the model. The experimental values of the size of different layers N1, N2, and N3 are 25, 50, and 100 respectively. The researchers found that a value smaller than 25 led to a decrease in performance and greater than 25 to an increase in performance. The value of

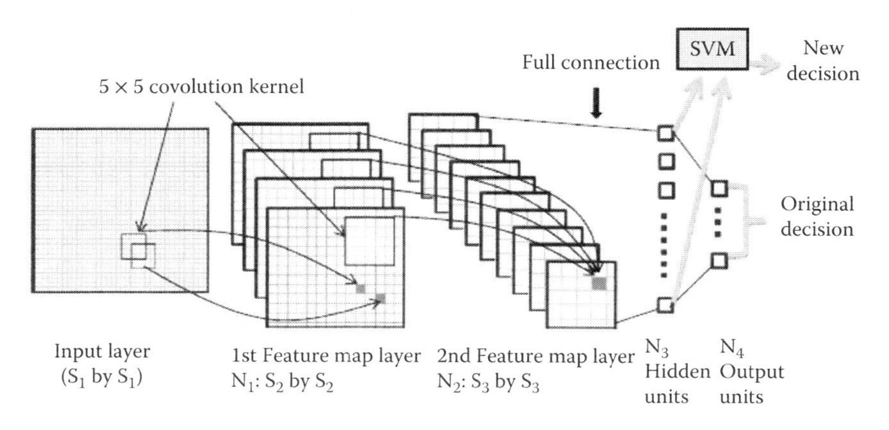

FIGURE 2.4: Structure of the CNN-SVM model.

TABLE 2.1: Performance comparison between CNN and CNN-SVM on MNIST

Error rate (%)	Simple CNN	Hybrid CNN-SVM
Training	0.28	0.11
Testing	0.59	0.19

TABLE 2.2: Comparison of results of CNN-SVM with other algorithms on MNIST

Reference	Method	Distortion	Error rate (%)
Lauer et al.	TFE-SVM	Affine	0.54
Simard et al.	Convolutional NN	Elastic	0.40
Ranzato et al.	Convolutional NN	Elastic	0.39
LeCun	Boosted LetNet-4	Affine, scaling, squeezing	0.7
Ciresan et al.	6-layer	NN elastic	0.35
Mizukami et al.	KNN	Displacement computation	0.57
Keysers	KNN	Nonlinear deformation	0.52
Xiao-Xiao et al.	Hybrid CNN–SVM	Elastic, scaling, rotation	0.19

N1 was set to 25 after considering a trade-off between accuracy and time cost. Finally, the hybrid model was compared with a simple CNN as shown in Table 2.1. The error rate of the hybrid CNN-SVM model was lower (0.19 as compared to 0.59 during testing), suggesting the hybrid approach had a higher accuracy. The researchers went on to compare their proposed CNN-SVM model with other models as well, as shown in Table 2.2.

2.4.2 Convolutional neural network–extreme learning machine (CNN-ELM)

After the success of other hybrid models in combining CNN with other classifier algorithms, by replacing the last layer of a CNN (the classification layer) with a classifier, there was a large number of researchers who tried to find the optimal classifier to replace the last layer of CNN, like [9] and [25]. While the empirically proven structure of a CNN where a C-layer (convolutional) and S-layer (subsampling) were alternated and the last or output layer of the CNN was replaced by advanced classifiers like SVM and random forest classifier.

On one such occasion, an ELM was combined with a CNN is such a fashion [26]. Like many neural-network-based models, it contained a feedforward neural network that was composed of a single hidden layer. This hidden layer in turn comprised of an input, hidden, and output layer. Traditional neural networks worked on the principle of tuning the internal parameters primarily by means of an iterative approach and this approach was time-demanding.

There are diverse opinions of the applicability of ELM to more complex tasks. The reason this algorithm was widely accepted was that it tends to have a low training time. The algorithm is still widely used all around China and was often called 'fast machine learning' due to its ability of learn from a small amount of labeled data. The patterns that could be learned were restricted in terms of complexity and although there was significant success when the patterns needed to be learned were simple, the same did not translate to success with data having highly, nonlinear, complex functions (Figure 2.5).

This was one of the major reasons to combine ELM with a deeper model that could express and learn such complexities in data while negligibly increasing the training time of the original model. Several reasons, including the surprising success of various hybrid ELM models like the CNN-ELM model, suggested that while standalone ELM was not very

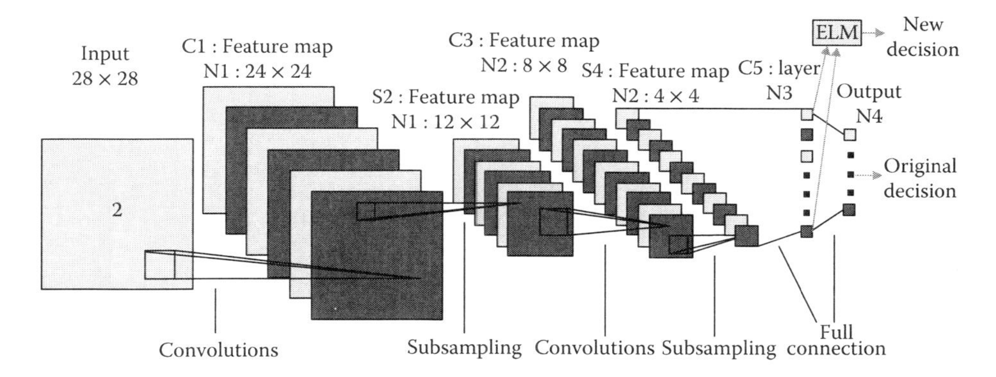

FIGURE 2.5: Structure of the CNN-ELM model.

TABLE 2.3: Performance comparison between CNN, ELM, and Hybrid CNN-ELM on MNIST

Error rate (%)	CNN	ELM	Hybrid CNN-ELM
Training	0.3	0.66	0.07
Testing	0.8	2.47	0.67

TABLE 2.4: Comparison of results of CNN-ELM with other algorithms on MNIST

Reference	Method	Error rate (%)
LeCun et al.	CNN LeNet5	0.95
R. Salakhutdinov et al.	DBM	0.95
L.L.C. Kasun et al.	ML-ELM	0.97
Proposed hybrid	CNN-ELM	0.67

effective on real-world data, combinations of ELM with deeper models could prove to be very effective.

Table 2.3 shows how a combination of CNN and ELM outperformed either of the two by a considerable margin or error rate. Thus we can say that CNN-ELM integrated the advantages of both ELM and CNN. This is also seen in Table 2.4 where CNN-ELM is compared with other successful implementations like (CNN) LeNet5, DBM, and ML-ELM and is seen to have the lowest error rate of 0.67% in testing data.

2.4.3 Neuro-fuzzy

While solving the problem of handwriting recognition, the problem of OCR needs to be solved. For this, two fundamental approaches are used: feature classification and template matching. Both these approaches have their drawbacks. To counter these drawbacks, size normalization and fuzzy similarity measures are used.

This method reduces the size of the scanned digits to a fixed level/dimension by computing the density of pixels in each region. Each of these normalized images are treated as fuzzy sets that are then compared to the templates by means of a fuzzy similarity measure. Fuzziness handles the problems of variability, noise, and ambiguity in data. These fuzzy similarity measures are then fed to a neural network which is used to classify it based on the similarity values input into it. Due to this, the size of the normalized image needs to be constant or fixed.

Case Study:

In one such hybrid neuro-fuzzy approach, the images fed to the model were first converted to black-and-white. The next step involves calculation of black pixel density (for which black pixels are treated as 1 and white as 0). The size needs to remain constant for the normalized image because of the need for the density to be independent of image size. If the normalized, resized image is $m \times n$, the pixel density of the zones are calculated as:

$$M[i,j] = \text{number of pixels}(i,j)/\text{total number of pixels} \tag{2.10}$$

For which a 2D fuzzy set can be generated using the formula:

$$x = \{([i,j], M[i,j]; 1 \leq i,j \leq m, M[i,j] \in [0,1]\} \tag{2.11}$$

With these generated fuzzy sets, a fuzzy comparison with the templates is performed and the similarity E and dissimilarity D scores are calculated as:

$$E(A,B) = \frac{A \cap B}{A \cup B} \tag{2.12}$$

$$D(A,B) = 1 - E(A,B) \tag{2.13}$$

where $\cap$ and $\cup$ are fuzzy union and intersection respectively of the two fuzzy sets (here A and B). Here, $\mu_c(x)$ represents the membership value of x in the fuzzy set C.

In the work under review, fuzzy intersections and fuzzy unions are performed using t-norms and t-conorms as:

$$\mu_{A \cap B}(x) = t[\mu_A(x), \mu_B(x)] \tag{2.14}$$

$$\mu_{A \cup B}(x) = s[\mu_A(x), \mu_B(x)] \tag{2.15}$$

When D_i^j is the dissimilarity between the input and template T_i^j, w_{ij}^N is the weight of the neuron S^N and bias is denoted by b^N, U^N is net input to the Nth neuron in the layer, given by:

$$U^N = \sum_{j=0}^{9} \sum_{i=0} k_j D_i^j w_{ij}^N + b^N \tag{2.16}$$

Using these scores we may take a final decision on the class but instead, we feed these scores into an ANN like a multilayer feedforward perceptron that outputs a value between 0 and 1 for each possible character (10 since there are 10 digits for MNIST). The one with the highest value is selected and is labeled as the digit associated with the input image to the overall system. The paper concluded with the result that by using a 6×6 normalization, an accuracy of about 90% for training data and 93% for test/validation data was obtained.

2.4.4 Particle swarm optimization–back-propagation with momentum (PSO-BPM)

This hybrid algorithm tends to combine the strengths of the two individual algorithms. The back-propagation with momentum (BPM) algorithm has a strong tendency to find the locally optimistic result, while on the other hand, particle swarm optimization (PSO) has a stronger tendency to find the globally optimistic result [27]. This hybridization helps find the most optimized results. The BPM algorithm mainly involves two stages, the first being the propagation of the activation values from the input to the output layer. Let zin_j be the input received to the hidden layer and b the bias. $X_i, i = 1, 2, 3, ..., n$ forms the input layer and z_j is the output value obtained when passed through the activation function.

$$zin_j = b_{0j} + \sum_{i=1}^{n} x_i.v_{ij} \tag{2.17}$$

$$z_j = f(zin_j) \tag{2.18}$$

The value propagated from the input layer along with the bias w is then used to calculate the output units $y_k, k = 1, 2, 3, 4, ..m$ by finding the sum of the weighted inputs yin_j and output unit y_j:

$$yin_j = w_{0k} + \sum_{j=1}^{p} z_j.w_{jk} \tag{2.19}$$

$$y_k = f(yin_k) \tag{2.20}$$

The second stage consists of the error correction for improvement of the classification. Each output unit finds the error between the actual value $t_k, k = 1, 2, 3, 4, ...m$ and the observed value $y_k, k = 1, 2, 3, 4, ...m$. This error is then propagated backwards through the network to improve the causal weights. The following shows the calculation of the error function:

$$\delta_k = (t_k - y_k)f'(yin_k) \tag{2.21}$$

The change in weight $\triangle w_{jk}$ and the bias correction term $\triangle w_{0k}$ can be calculated as follows:

$$\triangle w_{jk} = \alpha.\delta_k.z_j \tag{2.22}$$

$$\triangle w_{0k} = \alpha.\delta_k \tag{2.23}$$

Each layer sums the change in the error across each output unit and multiplies it with the derivative of the activation function to get the error information term across the weights and bias between the particular layer and the previous layer.

$$\delta_{inj} = \sum_{k=1}^{m} \delta_k.w_{jk} \tag{2.24}$$

$$\delta_j = \delta_{inj}.f'(zin_j) \tag{2.25}$$

$$\triangle w_{ij} = \alpha.\delta_j.x_i \tag{2.26}$$

$$\triangle w_{0j} = \alpha.\delta_j \tag{2.27}$$

Weights are then updated according to the following rule:

$$w_{jk}(new) = w_{jk}(old) + \triangle w_{jk} \tag{2.28}$$

$$b_{0k}(new) = b_{0ok}(old) + \triangle b_{ik} \tag{2.29}$$

The PSO is a global optimization algorithm which tries to find the best possible solution for a problem. The solution begins by defining a population. Each particle in the population keeps track of the coordinates of the locally best solution (fitness value) and the globally best solution. All particles then update their position based on the formula shown below:

$$V_i^{k+1} = w.V_i^k + c_1.rand_1().(pbest_i - s_i^k) + c_2.rand_2().(gbest_i - s_i^k) \tag{2.30}$$

where,

V_i^k: velocity of particle i at iteration k
w: weight
c_j: causal value
$rand$: random number between 0 and 1
s_i^k: position of particle i at iteration k
$pbest_i$: pbest of particle i
$gbest_i$: gbest of the group

$$w = w_{Max} - [(w_{Max} - w_{Min})iter]/maxiter \tag{2.31}$$

where,

w_{Max}: initial weight
w_{Min}: final weight
$maxiter$: max iteration number
$iter$: current iteration number

The current position of the particle can be updated as follows:

$$S_i^{k+1} = S_i^k + V_i^{k+1} \tag{2.32}$$

The hybrid PSO-BPM algorithm uses the initial particles as the weight vectors to train the network. The propagation of the activation value is done using the BPM algorithm. The output is then compared with the actual desired result and the difference is calculated. This difference forms the error and is considered as the fitness value for changing the particle's position. The updated particles are then used further to calculate the next set of error functions. This procedure repeats until the globally optimized result is obtained. The following shows the hybrid algorithm in detail:

Step 1: Initialize the positions (weights and bias) and velocities of the particles.

Step 2: Calculate the forward propagation value through the network.

Step 3: Calculate the error and correct the position of the particle by updating the weights and bias.

Step 4: Update the velocity of the particle.

Step 5: Repeat steps 2 to 4 to train the network.

2.4.5 Multilayer perceptron–support vector machine (MLP-SVM)

MLPs are the most common method used for recognizing handwritten digits. Due to its fast learning and easy training it is highly preferred by researchers. But MLP also has certain limitations when it comes to classification and designing of the system [28]. It is a black box and hence displays no relation between the hidden layers and the output nodes. Another limitation is that in representation of the feature space for classification purposes, it obtains a hyperplane separated surface which is not optimally separated in terms of area difference between two classes. Hence, it becomes difficult to classify the first two maximum classes, leading to error on terms of classification of digits, especially with digits like 1 and 7 or 4 and 9.

To overcome this limitation, MLP is combined with SVM to increase the accuracy rate, especially when it comes to classifying highly similar patterns. SVM basically transforms the feature space into a high-dimensional space where all patterns are almost linearly separable. This helps especially when used with MLP as it has a limitation of of deriving a small marginalized feature space leading to erroneous results.

Case Study:

An experiment was conducted to test the efficiency of the hybrid MLP-SVM algorithm. For this, segmented images from the mail Zip Codes were considered. The dataset was divided into three sets, namely set1, set2, and set3. The MLP network was trained on the set1 consisting of 44,081 digits with 138 feature nodes. Optimized SVMs were derived from the set2 consisting of 44,081 digits.

The entire hybrid model is trained using a two-step process. Three different datasets (set1, set2, set3) are used for the design and validation of the model. $Label(x_k)$ and $Class(x_k)$ denote the true label and the recognized class of the pattern, and MLP_{max1} and MLP_{max2} denote the first and the second maximum of the output from the MLP layer. Algorithm 2.1 explains the procedure for the design of the hybrid model:

Algorithm 2.1

Step 1: Training process

(a) An optimized MLP is trained on the dataset1, and the majority class is grouped into pairs of (C_i, C_j) called $S_{MajSubPairs}$.

(b) For each pair (C_i, C_j) in $S_{MajSubPairs}$, extract all patterns x_k from set2, such that $Label(x_k) = C_i$ or $Label(x_k) = C_j$ having $MLP_{max1}(x_k) = C_i$ and $MLP_{max2}(x_k) = C_j$ or vice versa. The subset formed is denoted by $Subset)(C_i, C_j)$.

(c) A two-class classification problem is created for which an SVM is used to train the model on the $Subset_{(C_i,C_j)}$ where an output of $+1$ denotes C_i and an output of -1 gives C_j.

Step 2: Decision process

Here labeling of unknown pattern x_k takes place.

(a) Find the MLP output for every pattern x_k.

(b) If $MLP_{max1}(x_k) = C_i$ and $MLP_{max2}(x_k) = C_j$ for pairs of (C_i, C_j) then, $Class(x_k) = SVM(C_i, C_j)(x_k) = C_i$ for an output of $+1$ else output is C_j else $Class(x_k) = MLP_{max2}(x_k)$.

The performance of the hybrid model was tested on the dataset3. Results of the testing show the superior performance of the hybrid model when compared to the use of the individual MLP model. Table 2.5 compares the results of the MLP model with the hybrid model showing the actual recognition and the theoretical recognition rate.

TABLE 2.5: Performance comparison between MLP and MLP-SVM

Classifier	Recog. (%)	Th. Recog. (%)
MLP	97.45	99.00
MLP + 5 local SVMs	9.71	98.10
MLP + 10 local SVMs	97.90	98.41
MLP + 15 local SVMs	98.01	98.60

2.4.6 Radial basis function network–support vector machine (RBF-SVM)

A large number of ensemble classifiers have, empirically, been proven to outperform the classifiers of which they are comprised. One such combination is that of RBF networks with an SVM. An RBF network is comprised of three layers, with the outer two being the input and output layer and the intermediate layer being the hidden layer and works similar to a MLP network. This model bases its decision on the center and sharpness of the Gaussian functions in the data. The vectors derived from the training data are used to calculate the centers and standard deviations. The RBF network has feedforward connections between the input-hidden and output-hidden layers. There are also connections between a bias node and each of the output nodes.

While training, each hidden node calculates the distance between its weight vector and the activation received as input from the input layer. The output nodes compute their activation value as a weighted sum of the hidden nodes. The activation of an output node is high when the input vector to the network is similar to the center of the basis function. In an RBF network the radius of the hyper-spherical cluster, formed around the center of the basis function, is given by the value of the radius parameter.

Case Study:

In this paper, we see the RBF-SVM algorithm implemented and tested on the US Zip Code dataset.

Algorithm for hybrid RBF-SVM using arcing classifier: Inputs: D, a set of d-tuples

k, the number of models in the ensemble.

Base classifiers are RBF, SVM.

Outputs: Hybrid RBF-SVM model (M).

Procedure:

Step 1. for $i = 1$ to k repeat Steps 2, 3, 4, and 5.

Step 2. By using sampling with replacement, create a training dataset D_i.

Step 3. Using D_i, derive a model M_i.

Step 4. Classify each element in the training data D_i.

Step 5. Based on accuracies of correctly classified examples in D_i, assign the model M_i weights to initialize it.

Step 6. To make predictions on a tuple using the hybrid model, take into account classifications of all k models and return the one with the maximum votes.

TABLE 2.6: Performance analysis of RBF-SVM on MNIST

Algorithm	Accuracy rate (%)
RBF	86.46
SVM	93.98
Hybrid RBF-SVM	99.13
Rajib et al.	98.26
Om Prakash et al.	98.50
Moncef et al.	98.00
Xin Wang et al.	95.00

In Table 2.6 we can see how the performance of the hybrid SVM is compared with the performance of RBF, SVM, and some popular algorithms. We see that the proposed hybrid RBF-SVM model has the highest accuracy rate of 99.13%. Thus, we see how the integration of RBF with SVM has significantly higher performance over either of the two models.

2.4.7 Multi-Layer perceptron–hidden Markov model (MLP-HMM)

HMM models were successful with many pattern recognition datasets due to their ability to segment and recognize human written script. Neural networks in some ways outperformed HMM in cases where an isolated character needed to be recognized. A combination of these two seemed a potentially wholesome solution to the problem where nonlinear patterns and sequential patterns could be effectively understood and leveraged for classification.

Case Study:

A hybrid hidden Markov model–MLP neural network, HMM-MLPNN, has been used to solve the problem of Arabic handwriting recognition. For the images in the dataset, preprocessing and normalization were done. Input images were resized to a fixed dimension and a low-pass filter was applied on the resized image. To deal with the inherent complexities of Arabic script, an enhanced version of the beta-elliptical strategy was used wherein the image/signal was segmented and each character in the continuous stroke was separated from the previous and succeeding character.

The proposed model uses MLP-NN to extract features by converting class probabilities to posterior probabilities and feeding them to a subsequent HMMs recognizer. The proposed MLP-NN system is comprised of OCONs (one class one network) where each class or handwritten character corresponds to a particular OCON. In the work being reviewed, each neural network was trained by a standard back-propagation algorithm with training parameters (the rate: $l = 0.01$; the momentum factor: $a = 25$; and the iterative number for training: $epochs = 4000$). It is important to note that the number of hand strokes are not equal throughout and varies from two to seven.

The MLP posterior probabilities $P(S|X)$ are divided by the prior state probabilities $P(S)$ in order to approximate the observation probabilities of our HMM.

$$P(X|S) \approx \frac{P(S|X)}{P(S)} \tag{2.33}$$

$$where, \quad -log\ P(X|S) = -log\ P(S|X) + \alpha\ log\ P(S) \tag{2.34}$$

Here α is the priori scaling value. Therefore, the posterior probabilities that are obtained are related to each other. The Viterbi algorithm is applied on the discrete HMMs after training the MLP-NNs. The HMMs are then made to work on the input signal after vector quantization is used to transform the dense vectors into discrete symbols (Figure 2.6).

The algorithm was applied on segmented characters and not isolated characters. The 6000 segmented words were sampled from the three sets of ADAB. 378,950 segment strokes were obtained, which were input to the MLP-NN and character skeletons/templates were generated to be used by the HMM for final character recognition. The performance of the hybrid MLP-HMM model was compared with the basic MLP, continuous MLP, and discrete MLP and shown in Table 2.7. It was seen that the hybrid system has the highest accuracy rate of 97.58%.

2.4.8 Hybrid tree classifier

This classifier combines artificial neural networks (ANNs) with decision trees for recognition of digits. The methodology involves two types of classification, namely hierarchical coarse classification and fine classification. Six subsets are classified {0}, {6}, {8}, {1,7}, {2,3,5}, {4,9} using a coarse classifier which is a three-layer neural network trained using the BP algorithm. The classification is done based on the similarity of character features. A decision tree is used to classify digits in the subsets consisting of more than one digit while {0}, {6}, {8} are classified rightly by the ANN. Figure 2.7 demonstrates the methodology involved in using the hybrid tree classifier to classify digits.

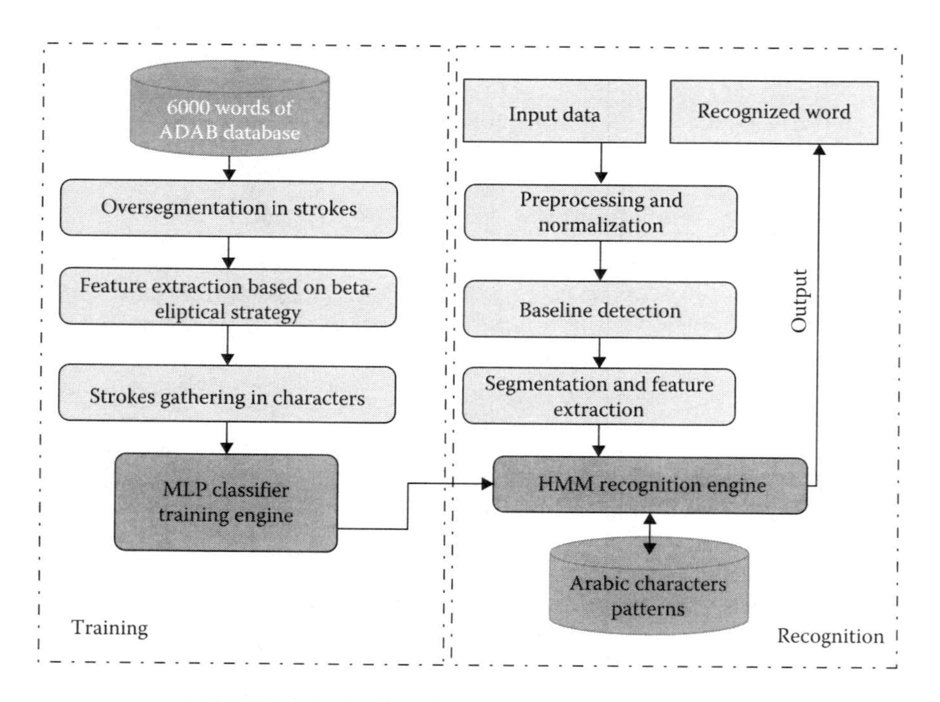

FIGURE 2.6: Flowchart of hybrid MLP-HMM.

TABLE 2.7: Performance analysis of MLP-HMM on MNIST

Algorithm	Top 1 acc.(%)	Top 5 acc.(%)	Top 10 acc.(%)
Basic MLP system	71.14	71.14	71.14
Continuous HMM system	90.15	91.89	91.89
Discrete HMM system	91.26	93.25	93.25
Hybrid MLP-HMM system	96.45	97.08	97.58

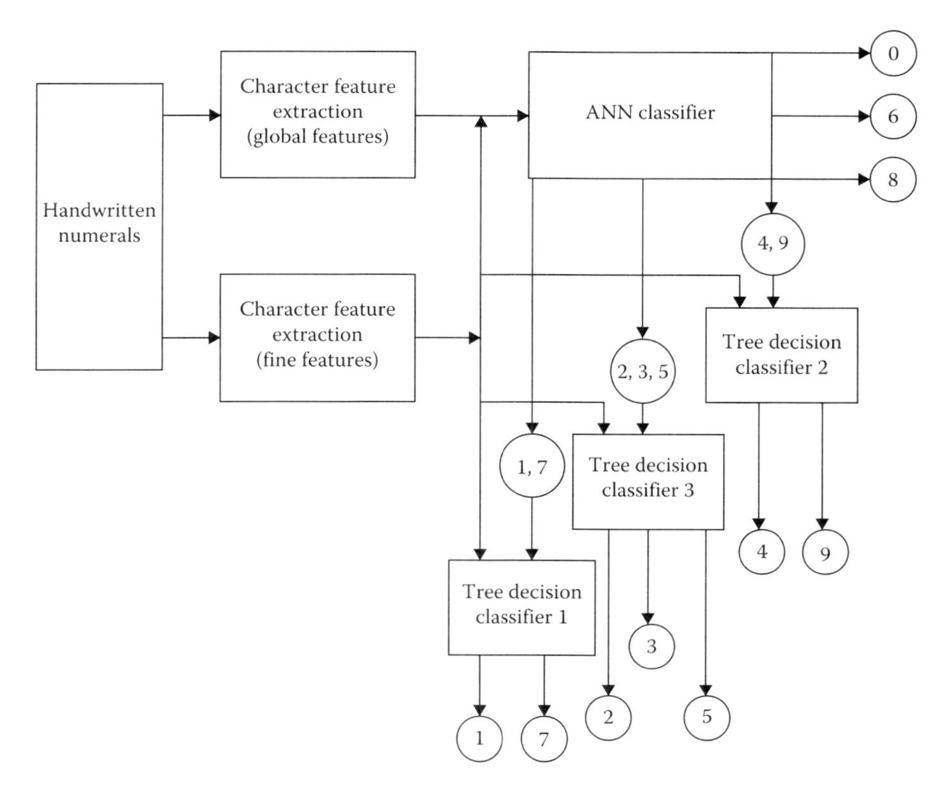

FIGURE 2.7: Methodology for hybrid decision tree classifier.

Filtering, segmentation, and normalization of data is done before feature extraction. Additional broken stroke connection and the character slant correction algorithm are used for further preprocessing. The image is then converted to a 32×24 sized image.

Two types of feature extraction are mainly performed on the image dataset. The first technique focuses on extracting global features of the digits such as middle line feature, concave features, width features, and point features. The other type focuses on extracting local features, which makes use of a floating type detector to identify certain types of segments present in the digit based image (Figure 2.7).

Case Study:

A study was conducted based on two sets of experiments. Two sets of data were used for the following experiment. Data 1 consisted of 10000 free handwritten numerals written by 200 different people out of which 5000 were set as training and the rest as testing dataset. Data 2 consisted of the NIST database out of which 2500 were chosen as training and 2500 as testing. Both experiments were trained first on Data 1 and tested on both the datasets and the next time trained on Data 1 and Data 2.

For experiment 1, a three-layer feed forward network was used for recognition. A total of 106 features (98 global + 8 local) were fed into the ANN. The network had ten output classes for the different digits. For experiment 2, a coarse classifier was used to represent digits into the six classes, shown as {0}, {6}, {8}, {1,7}, {2,3,5}, {4,9}. The performance rate improves as the ANN has only six classes to recognize. The same data is then further tested using the hybrid algorithm to further classify the digits into ten different classifier based on the coarse classifier. The results in Table 2.8 show the superior performance of the hybrid model when compared to ANNs shown in Table 2.9.

TABLE 2.8: Performance of hybrid classifier on handwritten digits

Training samples	Testing samples	Recognition rate (%)	Rejection rate (%)	Misrecognition rate (%)
Data 1(5000)	Data 1(5000)	97.80	1.35	0.85
Data 1(5000)	Data 2(2500)	97.60	1.25	1.15
Data 1(5000) + Data 2(2500)	Data 1(5000)	98.10	1.15	0.75
Data 1(5000) + Data 2(2500)	Data 2(2500)	97.90	1.20	0.90

TABLE 2.9: Performance of ANN on handwritten digits

Training samples	Testing samples	Recognition rate (%)	Rejection rate (%)	Misrecognition rate (%)
Data 1(5000)	Data 1(5000)	96.70	1.95	1.35
Data 1(5000)	Data 2(2500)	95.50	2.50	2.00
Data 1(5000) + Data 2(2500)	Data 1(5000)	97.60	1.30	1.10
Data 1(5000) + Data 2(2500)	Data 2(2500)	96.10	1.60	2.30

2.4.9 Adaptive neuro-fuzzy

This method involves utilization of fuzzy rules for recognition of handwritten digits. It involves finding a radius vector for optimum results. The output of the system is represented as a linear combination of the input values along with constants multiplied. Finding the optimized radius vector plays a very crucial role in the recognizing of digits.

Optimization methods are used for the purpose of improving the accuracy rate of various solutions by comparing the actual output with the desired output. In optimization algorithms, a fitness function is mainly used to evaluate the strength of the solution. Based on the fitness value, a solution is decided to be optimal. For example, in the recognition of digits, one of the most common fitness functions to compute error between the target output t_i and actual output y_i used by neural networks is the mean square error (MSE) function shown by:

$$MSE = \frac{\sum_{i=1}^{n}(t_i - y_i)^2}{n} \tag{2.35}$$

Based on this value the optimal weights of the system is decided. The lower the value of the fitness function, the better the set of weights for the system [29].

Evolutionary methods have been used for a long time by researchers for finding the most optimized solution for a problem. Such algorithms are mainly inspired by nature. A genetic algorithm (GA) is one such algorithm, which consists of biologically inspired processes such as selection, mutation, and crossover. Genetic algorithms have been used in a varied set of problems ranging from finding the most accurate weights for a neural network to scheduling problems to feature selection and many more. It starts with defining a random population and evaluates the fitness function for all in the population. The top particles, known as the elite group, are retained for the next generation while the remaining are selected using mutation and crossover of the elite particles. This way a new generation is created and the same procedure is applied until the optimized solution is obtained.

Other optimization algorithms involve particle swarm optimization (PSO) which starts with an initial population and tries to change position of every particle to the locally best and the globally best solution. Many other biologically inspired algorithms include ant colony optimization, fish swarm optimization, bacterial foraging optimization algorithm, firefly algorithm, artificial bee colony optimization algorithm (BA), etc.

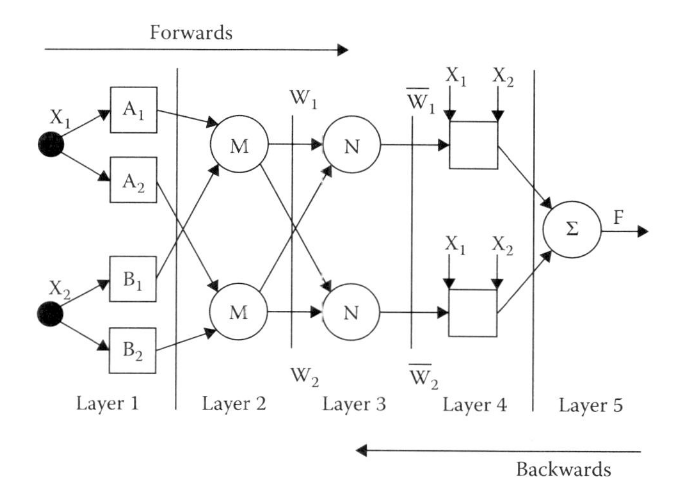

FIGURE 2.8: Structure of ANIS.

TABLE 2.10: Performance analysis of ANIS and IBA-ANIS

Classifier	Accuracy rate (%)
ANIS	97.74
GA-ANIS	99.17
ICA-ANIS	99.26
PSO-ANIS	99.31
BA-ANIS	99.31
IBA-ANIS	99.52

Figure 2.8 shows the structure of the adaptive neuro-fuzzy inference system (ANIS).

Various algorithms have been used to optimize the vector radius for better accuracy and performance of which the improved bees algorithm (IBA) gave the best performance. Table 2.10 shows the performance of the ANIS model and the optimized ANIS using various optimization algorithms such as PSO, IBA, BA, GA, and independent component analysis (ICA).

2.5 Results and Discussion

In this section we discuss the performance of various conventional models, hybrid classifiers, and also the state-of-the-art techniques presented in [30]. Many conventional models such as LeNet1, LeNet4, boosted LeNet4, deep belief network (DBN), deep Boltzmann machines (DBMs), SVMs, and RBF have proven to be very efficient classifiers and their hybrid systems have given even better performances. In Table 2.11, we present the performances of all conventional models on the MNIST dataset tested for digit recognition.

From the above description, all classifiers perform well with respect to classification of handwritten digits on the MNIST dataset. Boosted LeNet4 gives the highest accuracy. Below a comparative analysis is done of the various hybrid models tested on the MNIST dataset.

TABLE 2.11: Performance analysis of conventional models on handwritten digits (MNIST dataset)

Classifier	Error rate (%)
Linear classifier	8.4
RBF	3.6
Linear vector quantization	2.79
KNN	2.4
ANIS	2.26
MLP	1.91
LeNet1	1.7
SVC-poly	1.69
Large fully connected MLP	1.6
SVC-rbf	1.41
Deep Boltzmann machine	1.13
LeNet4	1.1
Deep belief network	0.95
LeNet5	0.80
Boosted LeNet4	0.70

TABLE 2.12: Performance analysis of hybrid models on handwritten digits (MNIST dataset)

Hybrid classifier	Error rate (%)
Principal component analysis–Poly	3.3
LeNet4-KNN	1.1
MLP-ELM	0.97
IBA-ANIS	0.48
CNN-ELM	0.67
CNN-SVM	0.19

Other hybrid algorithms are discussed above in their respective sections. The various hybrid algorithms represented in Table 2.12 show how hybridization help in better accuracy rates of the classifiers when compared to their individual performances.

On comparing the various hybrid algorithms with the conventional ones—LeNet1, LeNet4, LeNet5, and boosted LeNet4, which are basically CNNs—all give very high accuracy rates of above 99% but hybridized performs even better. The following can be observed when CNNs are combined with ELMs and SVMs which give an error rate of only 0.67% and 0.19% respectively.

KNN gives a high error rate of 2.4% but when combined with CNN, the error falls down to 1.1%. The same applies with the ANIS which individually gives an error of 2.26% but performs better by giving an error of 0.48% when combined with IBA to optimize its accuracy. This shows the superior performance of hybrid models.

2.6 Conclusion

This chapter presents an in-depth explanation on the various hybrid methods used for classifying patterns with respect to features obtained from handwritten digits. A huge amount of models have been developed by researchers in the past 50 years, and an elaborate explanation is provided about the major classifying models such as CNN, ELM, HMM, and SVM. Nine different hybrid models are explained to show the different methodologies involved in pattern classification such as the CNN-SVM, CNN-ELM, neuro-fuzzy, PSO-BPM, MLP-SVM, RBF-SVM, NN-HMM, hybrid tree classifier, and the adaptive neuro-fuzzy system. Case studies are shown to depict the superior performances of hybrid models with respect to conventional models.

To further demonstrate the efficiency of the hybrid models, a comparative analysis of all algorithms is presented showing their performance on the MNIST dataset. Even though conventional models show high accuracy, combining models help increase individual performances as the strengths of both the algorithms are captured to help with better accuracy. Hence, hybrid systems show better performance on digit recognition with the best classification accuracy demonstrated by CNN-SVM on MNIST dataset.

2.7 Future Work

This chapter has presented many different hybrid algorithms to recognize handwritten digits. Further enhancement can be done by incorporating various models and trying to combine them for better classification accuracy. Ensemble classifiers can be used to make use of the multiple classifiers and various different hybrid ensemble classifiers can be generated. To further improve, bagging and boosting can be applied for more optimization.

Optimization is another area which can be explored. Different optimization algorithms can be made use of to train neural networks. The fuzzy cognitive Map is yet another pattern classification method which can be applied and used in digit recognition. Also, there are various pattern classification problems, and the methods explained in this chapter can be extended to various other challenging pattern analysis problems to improve performance and thereby increase accuracy.

References

1. Bishop, C. M. (2007). *Pattern Recognition and Machine Learning*, p. 128. New York: Springer.

2. Shah, F. T. and Yousaf, K. (2007). Handwritten digit recognition using image processing and neural networks. In World Congress on Engineering (Vol. 2, pp. 648–651).

3. Denker, J. S., Henderson, D., Howard, R. E., and Hubbard, W. (1990). Handwritten character recognition using neural network architectures. In Proceedings of the 4th USPS Advanced Technology Conference, Washington D.C., pp. 1003–101.

4. Caldern, A., Roa, S., and Victorino, J. (2003). Handwritten digit recognition using convolutional neural networks and gabor filters. In Proceedings of International Congress on Computational Intelligence.

5. El Kessab, B., Daoui, C., Bouikhalene, B., Fakir, M., and Moro, K. (2013). Extraction method of handwritten digit recognition tested on the mnist database. *International Journal of Advanced Science & Technology*, 50, 99–110.

6. Bottou, L., Cortes, C., Denker, J. S et al. (1994). Comparison of classifier methods: A case study in handwritten digit recognition. In *Pattern Recognition*, Vol. 2-Conference B: Computer Vision & Image Processing., Proceedings of the 12th IAPR International. Conference on (Vol. 2, pp. 77–82). IEEE.

7. Bishnoi, D. K. and Lakhwani, K. Advanced approaches of handwritten digit recognition using hybrid algorithm. International Journal of Communication and Computer Technologies, 1(57), 186–191.

8. LeCun, Y., Jackel, L. D., Bottou, L et al. (1995). Learning algorithms for classification: A comparison on handwritten digit recognition. Neural networks: The statistical mechanics perspective, p. 261, 276.

9. Niu, X. X. and Suen, C. Y. (2012). A novel hybrid CNNSVM classifier for recognizing handwritten digits. *Pattern Recognition*, 45(4), 1318–1325.

10. Guo, L. and Ding, S. (2015). A hybrid deep learning CNN-ELM model and its application in handwritten numeral recognition. *Journal of Computational Information Systems*, 11(7), 2673–2680.

11. Bellili, A., Gilloux, M., and Gallinari, P. (2001). An hybrid MLP-SVM handwritten digit recognizer. In Proceedings of Sixth International Conference on Document Analysis and Recognition (pp. 28–32). IEEE.

12. Govindarajan, M. (2016). Recognition of handwritten numerals using RBF-SVM hybrid model. *Int. Arab J. Inf. Technol.*, 13(6B), 1092–1098.

13. Ayyaz, M. N., Javed, I., and Mahmood, W. (2012). Handwritten character recognition using multiclass svm classification with hybrid feature extraction. *Pakistan Journal of Engineering & Applied Science*, 10, 57–67.

14. Tuba, E., Tuba, M., Simian, D., and Street, I. R. Handwritten digit recognition by support vector machine optimized by bat algorithm. In 24th International Conference in Central Europe on Computer Graphics, Visualization and Computer Vision, (WSCG 2016) (pp. 369–376).

15. Bengio, Y., LeCun, Y., Nohl, C., and Burges, C. (1995). LeRec: A NN/HMM hybrid for on-line handwriting recognition. *Neural Computation*, 7(6), 1289–1303.

16. Pinzolas, M., Villadangos Alonso, J., and Gonzlez de Mendvil, J. R. (2001). A neuro-fuzzy system for isolated hand-written digit recognition. *Mathware & Soft Computing* 8(3), 291–301.

17. Bayat, A. B. (2013). Recognition of handwritten digits using optimized adaptive neuro-fuzzy inference systems and effective features. *Journal of Pattern Recognition and Intelligent Systems*, 1(2), 25–37.

18. Ping, Z. and Lihui, C. (2002). A novel feature extraction method and hybrid tree classification for handwritten numeral recognition. *Pattern Recognition Letters*, 23(1), 45–56.

19. Cambria, E., Huang, G. B., Kasun, L. L. C et al. (2013). Extreme learning machines [trends & controversies]. *IEEE Intelligent Systems*, 28(6), 30–59.

20. Huang, G. B., Zhu, Q. Y., and Siew, C. K. (2006). Extreme learning machine: Theory and applications. *Neurocomputing*, 70(1), 489–501.

21. Le Cun, B. B., Denker, J. S., Henderson, D., Howard, R. E., Hubbard, W., and Jackel, L. D. (1990). Handwritten digit recognition with a back-propagation network. In Advances in Neural Information Processing Systems Citeseer.

22. Milgram, J., Cheriet, M., and Sabourin, R. (2006). "One against one" or "one against all": Which one is better for handwriting recognition with SVMs? In Tenth International Workshop on Frontiers in Handwriting Recognition. Suvisoft.

23. Bahlmann, C., Haasdonk, B., and Burkhardt, H. (2002). Online handwriting recognition with support vector machines—a kernel approach. In Proceedings of Eighth International Workshop on Frontiers in Handwriting Recognition (pp. 49–54). IEEE.

24. Katiyar, G. and Mehfuz, S. (2016). A hybrid recognition system for off-line handwritten characters. *SpringerPlus*, 5(1), 1.

25. Guo, Q., Tu, D., Lei, J., and Li, G. (2014). Hybrid CNN-HMM model for street view house number recognition. In Asian Conference on Computer Vision (pp. 303–315). Springer International Publishing.

26. Huang, G. B., Zhou, H., Ding, X., and Zhang, R. (2012). Extreme learning machine for regression and multiclass classification. *IEEE Transactions on Systems, Man, and Cybernetics, Part B (Cybernetics)*, 42(2), 513–529.

27. Lagudu, S. and Sarma, C. V. (2013). Hand writing recognition using hybrid particle swarm optimization & back propagation algorithm. *International Journal of Application or Innovation in Engineering & Management (IJAIEM)*, 2(1), 75–81.

28. Hsu, C. W. and Lin, C. J. (2002). A comparison of methods for multiclass support vector machines. *IEEE Transactions on Neural Networks*, 13(2), 415–425.

29. Man, Z., Lee, K., Wang, D., Cao, Z., and Khoo, S. (2013). An optimal weight learning machine for handwritten digit image recognition. *Signal Processing*, 93(6), 1624–1638.

30. Liu, C. L., Nakashima, K., Sako, H., and Fujisawa, H. (2002). Handwritten digit recognition using state-of-the-art techniques. In Proceedings of Eighth International Workshop on Frontiers in Handwriting Recognition (pp. 320–325). IEEE.

Chapter 3

Artificial Immune Recognition System for Offline Handwritten Signature Verification

Yasmine Serdouk, Hassiba Nemmour, and Youcef Chibani

3.1 Introduction

Handwritten signature is widely used to authenticate official documents, such as contracts and checks. As the signing process depends on the physical and psychological conditions of the writer, the signature is highly variable. Therefore, signature verification systems should be as robust as possible to deal with such variability. In this respect, researchers try to develop systems which reduce the intrapersons variability while enlarging interpersons variations. Roughly, the training of a signature verification system is achieved by learning multiple signature samples in order to be able to authenticate a questioned (unknown) signature in the test stage. As shown in Figure 3.1, in the two stages, the process is composed of four main modules: data acquisition, preprocessing, feature generation, and classification that is generally named "verification" in the signature verification field.

According to the acquisition mode, two approaches are conceivable for such a system: the online verification and the offline verification [1]. Using an electronic device like a tablet, the online verification takes into account the dynamic information within signatures, such as speed, time, and pressure. Conversely, dynamic information is not available in the offline approach since the signature is written over a sheet of paper and acquired through a scanner. Nevertheless, although offline systems are less accurate than online systems, they have more practical application areas like to validate a financial transaction, to sign a bank check or an official contract, etc. This is why this chapter is focused on the development of an offline signature verification system. Next, the preprocessing is useful to improve the signature image quality before the feature generation. For instance, noise removal, size normalization and binarization are typical preprocessing techniques [2].

After the signature have been acquired and preprocessed, the next step is to extract discriminant features from each treated signature. In the past years, for an effective characterization, various global and local features were used in the literature. As global features, one can mention the mathematical transforms, such as wavelets, ridgelets, curvelets, and

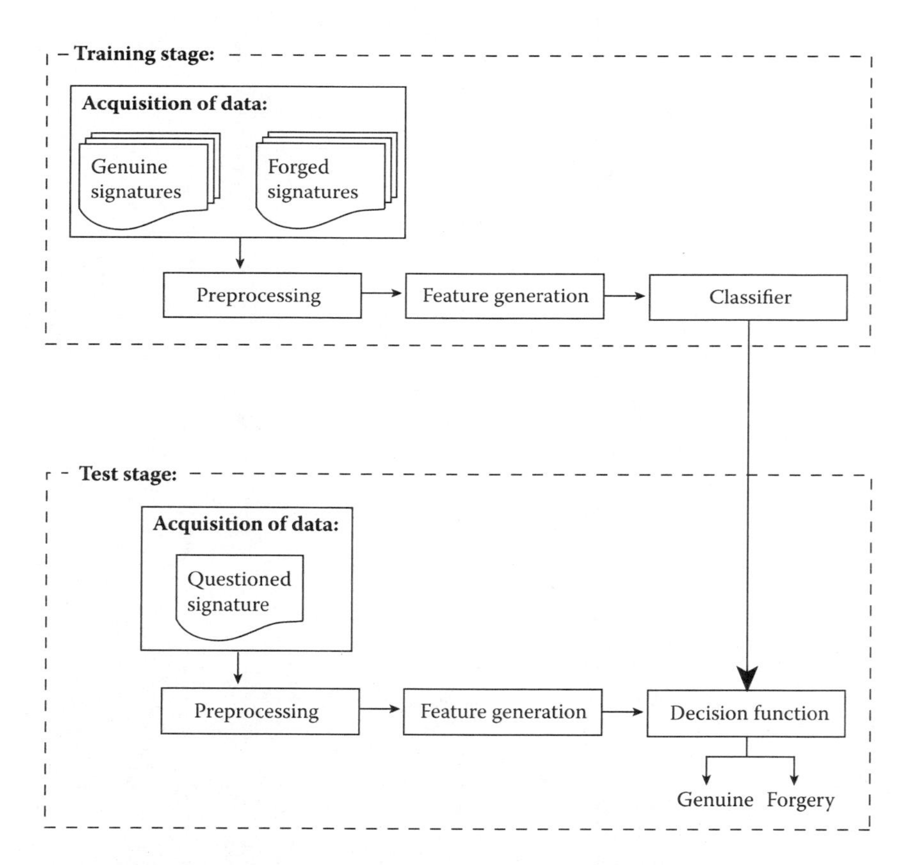

FIGURE 3.1: Flowchart of a signature verification system.

contourlets [3–6]. Recently, local features showed promising performances to describe specific parts of signature images by dividing the image into several parts and then, extracting features from each part of the image [7]. In this respect, we note topological features, such as pixel density and pixel distribution, curvature features, orientation features, and gradient features [7–9]. Also, the local binary patterns (LBPs) provide interesting results by describing gray level variations within signature images [10].

Finally, the last module has the ability to decide about the signature authenticity. For that, a set of labeled signatures are enrolled in the training stage. Then, a questioned signature is classified as genuine or forged using the decision function of the learned model. Several classifiers are used in the literature. The most popular are dynamic time warping, neural networks, hidden Markov models, and support vector machines (SVMs) [11,12]. Currently, the verification using SVMs gives the most promising results [13,14]. However, the scores reported in literature are not optimal and still need improvement. Recently, many interesting mechanisms inspired from natural immune systems allowed the development of artificial immune systems (AISs) that tackle with several pattern analysis applications, such as diagnosis tests [15] and watermarking [16].

Presently, to further improve the classification of handwritten signatures, a novel bio-inspired method, called the artificial immune recognition system (AIRS) [17] is employed. The AIRS classifier adopts a supervised learning process to develop new representative data for each class, called memory cells. These memory cells are used to separate the two classes according to the k-nearest neighbor (kNN) rule. For feature generation, different descriptors that characterize topological, textural, and gradient information are employed to assess the performance of the AIRS classification. Experiments are carried out on the Center of Excellence for Document Analysis and Recognition (CEDAR) dataset and the results in terms of error rates reveal promising performances and often comfortably outperform various literature results.

The rest of this chapter is organized as follows. In the next section, the two types of dependent/independent protocols of signature verification systems are described, followed by an overall presentation of the natural versus artificial immune systems. Section 3.4 introduces the AIRS algorithm and its applicability in the signature verification field. Feature generation methods are described in Section 3.5. Experimental results are given in Section 3.6. Finally, the main conclusions are reported and discussed in the last section.

3.2 Signature Verification Strategies (Dependent/Independent)

There are two protocols to perform automatic offline signature verification [18]. The first protocol allows a writer-dependent verification since it develops a specific system for each user. The second protocol is writer-independent because it builds a global system for all users.

3.2.1 The writer-dependent approach

The writer-dependent strategy is the standard approach for signature verification [11, 19,20]. As illustrated in Figure 3.2, for each signer, a specific model is built to separate between its genuine and forged signatures. This approach needs a large dataset to get an efficient system and each time a new writer is included in the system, the training process must be repeated. This is why researchers proposed an alternative solution using the writer-independent approach.

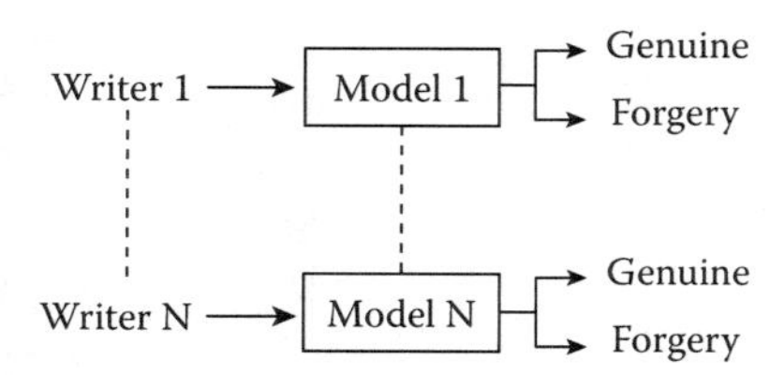

FIGURE 3.2: Signature-verification-based writer-dependent strategy.

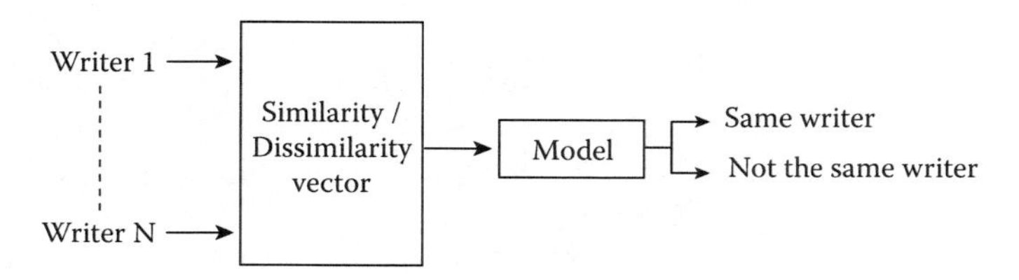

FIGURE 3.3: Signature-verification–based writer-independent strategy.

3.2.2 The writer-independent approach

Based on a similarity/dissimilarity representation, the writer-independent strategy employs one model to train all writers in a dataset (see Figure 3.3), which leads to a single system that can process any number of new users. In the literature, several implementations of the writer-independent approach are investigated [8,21,22]. However, the writer-independent results still need to be improved and as the majority of the works presented in the literature provide better scores using the writer-dependent strategy, the latter is employed in this chapter.

3.3 Natural Immune System versus Artificial Immune System

In the same way that artificial neural networks (ANNs) are inspired from the human brain, artificial immune systems (AISs) are founded on mechanisms inspired from the mammalians' natural immune system (NIS). The NIS is composed of numerous lymphocyte cells, which interact with the others to provide an immune response in order to attack a foreign presence (also called antigens). Among these lymphocyte cells, AIS for machine learning tends to employ the B cells, which mature in the bone marrow, because the NIS memory is largely occupied by these cells. It should be noted that each B cell embodies a population of antibodies and then, based on the degree of affinity between these antibodies and antigens, a B cell becomes stimulated. Thereafter, through a cloning and mutation process, stimulated B cells will produce offspring that are used to attack and destroy the foreign presence.

The AIS showed promising performances in various applications of feature extraction, data analysis, and machine learning [15]. More recently, an interesting concept of a resource limited artificial immune system based on artificial recognition balls (ARBs) was proposed by Timmis and Neal in [23]. Note that ARBs represent a collection of B cells, which provide the primary memory mechanism of the system. Based on this concept, a supervised learning classification using an artificial immune recognition system (AIRS) was proposed by Watkins [17]. In the AIRS algorithm, the concept of ARBs is maintained and a B cell could be simplified by having only a single antibody. The AIRS training consists of developing new

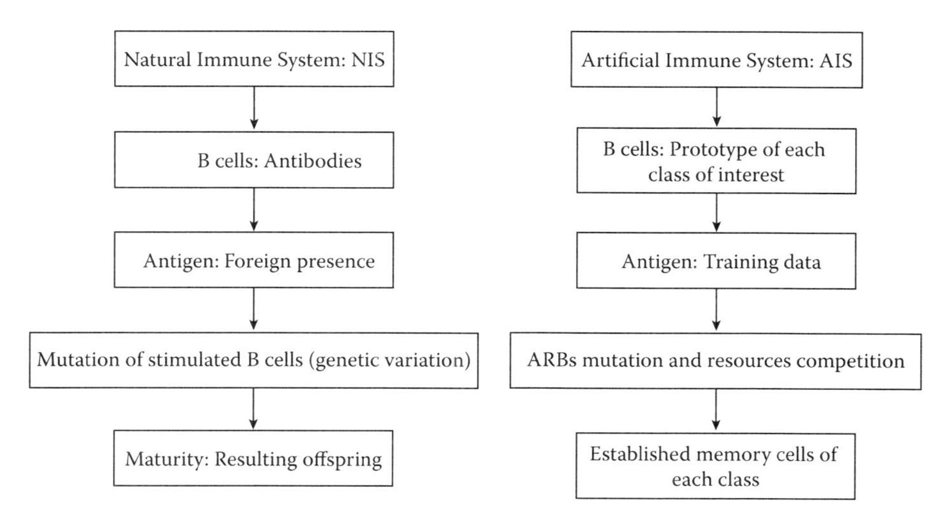

FIGURE 3.4: NIS versus AIS.

antibodies (or memory cells), which are representative data of each class, through a mutation and resources competition processes. The mutation process generates ARBs and each ARB acquires a resources number based on its stimulation level to the antigen. However, since the available resources are limited, the least stimulated antibodies are removed from the ARB set until the number of resources held by the cells equal the number of resources allowed in the system. This process, called resources competition, ensures the development of more representative cells (i.e., the strongest ARBs that allow the recognition of antigens) in order to constitute the established memory cells. The analogy between NIS and AIS is illustrated in Figure 3.4.

3.4 Artificial Immune Recognition System (AIRS)

The artificial immune recognition system (AIRS) is a resource limited artificial immune classifier [24]. For the good understanding of the AIRS algorithm, a description of some key terms followed by a definition of the AIRS parameters is given in what follows.

3.4.1 Key terms definitions

Antigen: a feature vector of a training signature image that is introduced into the AIRS training process. Precisely, it is used as a reference within the ARBs mutation and resources competition.

Antibody: a feature vector of a training signature image coupled with its associated class. This vector is referred to as an antibody when it is part of the ARBs population or memory cells.

Affinity: a value between 0 and 1 representing the antigen-antibody closeness calculated by an Euclidian distance. Thereby, the smaller the affinity measure, the more the antigen is similar to the antibody.

Artificial recognition ball (ARB): also known as an antibody. It represents a mutated clone provided with its resources number and its stimulation value.

Stimulation: a value between 0 and 1 that represents the response of an ARB to an antigen. It is inversely proportional to the Euclidian distance between the ARB and the antigen.

Affinity threshold (AT): an average affinity value that is calculated upon all antigens of the training set.

Memory cell match (MC-match): represent the most stimulated memory cell at the first exposure of a given antigen.

Memory Cell Candidate (MC-candidate): an antibody from the ARBs population, of the same class as the training antigen, which is the most stimulated to the antigen.

Memory Cells (MC): represent the ARBs that are the most stimulated to the antigen (a training signature). However, an MC can be replaced if a memory cell candidate (MC-candidate) is more stimulated to a given antigen than the memory cell match (MC-match) and the affinity between MC-candidate and MC-match is less than a threshold proportional to AT.

Established memory cells: antibodies from the ARBs population that survive in the resources competition process and are the most stimulated to the antigen.

3.4.2 AIRS parameters definition

Resources number: an integer value which defines the maximal number of ARBs in the system. It is compared to the resources number sum of each ARB. Indeed, to each ARB is assigned a resources number that corresponds to the multiplication of its stimulation value with the clonal rate.

Clonal rate: an integer value that controls the number of generated mutated clones that an ARB is allowed to produce. It is also used to assign resources to an ARB.

Affinity threshold scalar (ATS): a value between 0 and 1 that, when multiplied by the AT, provides a cut-off value for the memory cell replacement in the AIRS training process.

Stimulation threshold: a real ranged between 0 and 1 used as a stopping criterion in the training routine of an antigen. The training of that antigen does stop only when the average stimulation value of the ARBs of each class is above the stimulation threshold.

Mutation rate: a parameter between 0 and 1 that indicates the probability that any given feature of an ARB will be mutated.

Hyper-mutation rate: a parameter which determines the number of mutated clones that a given memory cell is allowed to inject into the cell population.

3.4.3 Overview of the AIRS algorithm

The training of AIRS is composed of three main steps. The first, called initialization, is a preliminary stage that is performed only once before the beginning of the training routine. Thereafter, the training process of each signature that is considered as an antigen is presented in Algorithm 3.1. It includes three steps: ARBs generation, ARBs resources competition, and the process of extending the set of memory cells. Finally, the last stage performs the classification of unknown data.

3.4.3.1 Initialization

The first stage begins with a normalization of data to scale the affinity values in the range [0–1]. Then, an affinity threshold (AT) that is used in the condition of a memory cell replacement is calculated upon all training signature images:

$$AT = \frac{\sum_{i=1}^{n-1} \sum_{j=i+1}^{n} \mathit{affinity}(ag_i, ag_j)}{n(n-1)/2}. \tag{3.1}$$

ag_i: feature vector of the ith antigen, which is a training signature.
affinity: Euclidian distance.
n: number of antigens.

Also, the MC and ARB pools need to be initialized. For that, a random initialization is carried out by randomly selecting a training sample from each class.

3.4.3.2 Training process

Algorithm 3.1 describes how the AIRS training process of each antigen is carried out.

Algorithm 3.1: Antigen training in AIRS

Step 1. ARBs generation: a set of mutated clones is generated by using MC-match and added to the initial ARBs pool. Listing 3.1 presents the source code used in the mutation routine, followed by Listing 3.2 that presents the ARBs generation process. It is worth to nothing that in this chapter, the verification system is achieved using MATLAB®.
Recall that MC-Match represents the most stimulated antibody between all antibodies within the MC pool. Then, the stimulation is calculated as follows:

$$ST(ag_i, MC) = 1 - \mathit{affinity}(ag_i, MC). \tag{3.2}$$

Step 2. ARBs resources competition: each ARB that belongs to the same class as the present antigen is allocated a resources number according to the product (stimulation(ARB, antigen) × clonal rate). Then, based on the predefined resources number parameter, a competition is performed in order to select and keep only ARBs with high stimulation. This step is repeated until the average stimulation of ARBs of each class reaches the predefined stimulation threshold.

Step 3. MC-candidate introduction: after selecting MC-candidate, one can compare its stimulation with that of MC-match. If the MC-candidate stimulation is higher than the MC-match stimulation, MC-candidate is added to the MC pool. In this case, if the Euclidian distance between MC-candidate and MC-match is lower than the product (AT × ATS), MC-match is removed from the MC pool.

3.4.3.3 Verification rule

After completing the training, the established MC pool is used to classify test samples. Precisely, the classification is achieved by a majority vote among the k closest MC to a given test antigen. This is done using the k-nearest neighbors (kNN) rule.

Listing 3.1: Mutation routine code.

```
% Description:
% This function is employed for the mutation routine within the
% AIRS training process.

% Function inputs:
   % mcMatch: Memory Cell Match.
   % mutation_rate: Mutation rate.
   % dim: Feature vector size.
   % ag_c: Class of the actual training antigen.
   % nbClass: Classes number (here nbClass=2, i.e. genuine and
   %          forged classes).

% Function outputs:
   % clone: A vector containing the MC-Match clone
   %        (after the mutation).
   % mutt: An index that is equal to 1 if a mutation just been done
   %       or 0 if not.
   % class: Class of the generated clone.

% CODE:
function [clone, mutt, class]=mutate(mcMatch, mutation_rate, dim, ...
                                     ag_c, nbClass)
   mutt=0;
   for i=1:dim,
      change=randi(1000)/1000;
      changeto=randi(1000)/1000;
      if(change < mutation_rate)
         % perform a mutation of an element from the
         % mcMatch vector.
          mcMatch(i)=changeto;
          mutt=1;
      end
   end

   if(mutt==1)
      change=randi(1000)/1000;
      chanto=randi(nbClass);
      if(change < mutation_rate)
        % assigning the class to the mutated clone.
          class=(chanto);
      else
          class=ag_c;
      end
      clone=mcMatch;
   end
```

Listing 3.2: ARBs generation code.

```
% Description:
% This function generates a set of ARBs from an input antigen
% within the AIRS training process.

% Function inputs:
    % mcMatch:  Memory Cell Match.
    % stimu: Stimulation(Mc_match,antigen).
    % hypMutRate: Hyper Mutation rate.
    % cloneRate: Clonal rate.
    % mutation_rate: Mutation rate.
    % dim: Feature vector size.
    % ag_c: Class of the actual training antigen.
    % nbClass: Classes number (here nbClass=2, i.e. genuine and
    %         forged classes).

% Function outputs:
    % MU: The set of generated ARBs.
    % class: A vector containing the class of each clone in MU.
    % index: Number of clones generated.


% CODE:
  function [MU, class, index]=arbgen(mcMatch,stim,hypMutRate, ...
                          cloneRate, mutation_rate, dim, ag_c, nbClass)

    nbClones=hypMutRate*cloneRate*stim;% the maximum of clones
                                       % allowed to generate.
    nbClones=floor(nbClones);
    index=1;
    i=1;
    MU(:,i)=mcMatch;
    class(i)=ag_c;
    while i < nbClones
        mcClone=mcMatch;
        % Call to the function "mutate" described in Listing 3.1
        [mcClone, mutation, class]=mutate(mcClone, mutation_rate, ...
                                 dim, ag_c, nbClass);
        if(mutation==1)
            MU(:,i+1)=mcClone;
            class(i+1)=class;
            index=index+1;
        end
        i=i+1;
    end
```

3.5 Feature Generation

The feature generation step has an important role in a signature verification system. In fact, to generate discriminant features that characterize effectively signature images, several kinds of descriptors can be used. In this work, topological, textural, and gradient features are employed to assess the performance of the AIRS classifier. Precisely, the pixel density and longest run features (LRF) extract the topological information while textural characteristics are provided by the orthogonal combination of local binary patterns (OC-LBP). Finally, the last descriptor, called gradient local binary patterns (GLBP), characterizes at the same time the gradient and textural information within signatures.

3.5.1 Pixel density

The pixel density (also called grid features) is a topological descriptor that is computed in local regions of images. Precisely, signature images are first divided into a uniform grid and then, for each cell in the grid, the pixel density is computed as in Equation 3.3. The final feature vector contains the densities of all cells.

$$dp = l/n \ . \tag{3.3}$$

Such as, l is the number of black pixels belonging to a signature image in the cell and n represents the size of the cell.

3.5.2 Longest run feature (LRF)

The longest run feature (LRF) aims to extract the largest string of text pixels according to the four main directions: horizontal, vertical, and the two main diagonals [9]. For each direction, one LRF value is captured along each line by accumulating the longest succession of pixels belonging to the signature. Thus, for each signature we obtain four LRF features that are normalized according to their directional length. An example of the LRF computation according to the horizontal direction is shown in Figure 3.5.

3.5.3 Description of orthogonal combination-based LBP (OC-LBP)

The local binary patterns (LBP) operator was introduced by Ojala [25] to describe the texture characteristics within images. It is calculated on a radius R through the comparison of the gray level values of a pixel with those of its neighbors. Figure 3.6 illustrates the computation of the LBP code according to eight neighbors ($P = 8$), which implies a radius $R = 1$.

First, a thresholding is applied to obtain a binary code as in Figure 3.6b, it is calculated as follows:

$$s(g_p) = \begin{cases} 1 & g_p \geq g_c \\ 0 & otherwise \end{cases} \tag{3.4}$$

g_c: the gray level value of the central pixel $I(x, y)$.
g_p: the gray level of the pth neighbor.

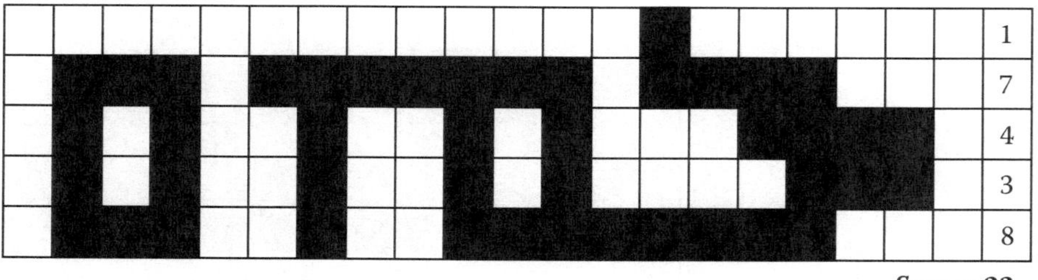

FIGURE 3.5: LRF computation according to the horizontal direction.

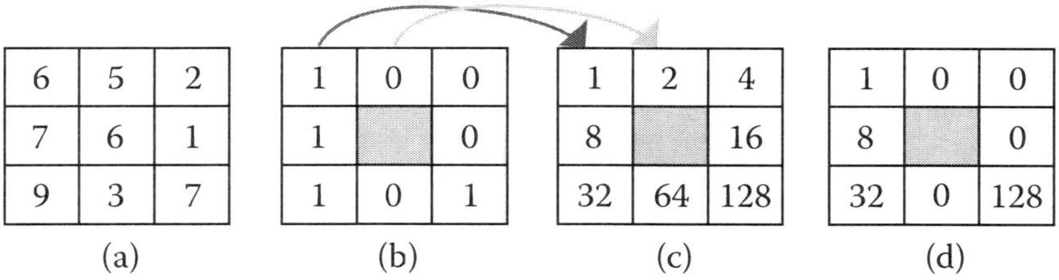

6	5	2
7	6	1
9	3	7

(a)

1	0	0
1		0
1	0	1

(b)

1	2	4
8		16
32	64	128

(c)

1	0	0
8		0
32	0	128

(d)

FIGURE 3.6: Original LBP calculation [25]: (a) 8×8 neighbors of a central pixel $= 6$; (b) thresholded neighborhood; (c) the corresponding weight; (d) multiplication between (b) and (c). The LBP code for the central pixel is $1 + 8 + 32 + 128 = 169$.

Then, the values of the pixels in the thresholded neighborhood are multiplied by the corresponding weight of Figure 3.6c. The result of this multiplication is summed to obtain the LBP code as:

$$LBP^{P,R}(x, y) = \sum_{p=0}^{P-1} s(g_p)2^p \ . \tag{3.5}$$

Furthermore, the pixel neighbors are collected on a circle to provide a rotation invariance. In such a case, the interpolated gray level of the pth neighbor is defined as follows:

$$g_p = I\left(x + R\ sin\left(\frac{2\pi p}{P}\right), y - R\ cos\left(\frac{2\pi p}{P}\right)\right) \ . \tag{3.6}$$

Finally, the histogram LBP constitutes the signature features. Despite the robustness of this descriptor, the size of LBP histogram can have a compounding effect on the LBP characterization. Precisely, for P neighbors, the histogram has 2^P components. So, 16 neighbors lead to 65,536 features. To overcome this issue, several LBP variants were developed with a reduced size of LBP. For example, the uniform LBP histogram contains $(P(P-1)+3)$ and the (LBP_{riu}) limits the histogram size to $(P+2)$ [26,27]. In this chapter, we employ the orthogonal combination of local binary patterns (OC-LBP), which has been previously introduced to highlight color information in images [26].

OC-LBP is a modified form of LBP, which reduces the histogram size to $(4 \times P)$. Considering P pixels situated on a circular neighborhood with a radius R, OC-LBP corresponds to the concatenated histograms computed according to the orthogonal directions. As depicted in Figure 3.7, two histograms are generated by using the orthogonal neighborhood. A first histogram is computed by considering horizontal and vertical neighboring pixels. Then, another histogram is calculated by considering diagonal neighboring pixels. This leads to two histograms that contain 16 elements each.

3.5.4 Gradient local binary patterns (GLBP)

This operator is obtained by combining textural and gradient characteristics. In various pattern recognition applications, LBP were proposed to perform statistical and structural analysis of textural patterns [28]. GLBP is a gradient feature that is computed by considering the LBP neighborhood of pixels. This operator is exploited here to characterize signature images (see Figure 3.8). It can be summarized as follows [29]:

1. Compute the thresholded LBP code.
2. Compute width and angle values from the uniform patterns, such as:

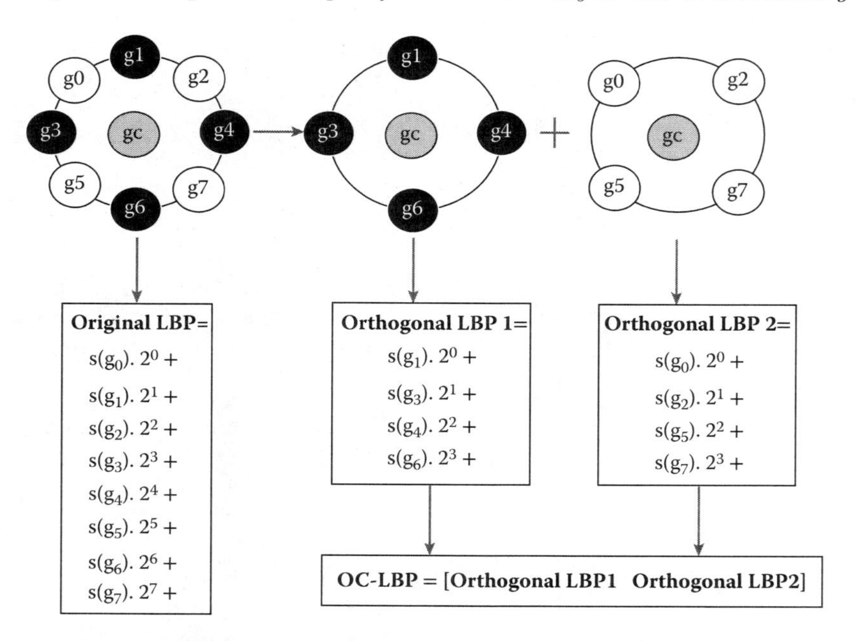

FIGURE 3.7: Original LBP and OC-LBP comparison. (From C. Zhu et al., *Pattern Recognition*, 46, 1949–1963, 2013.)

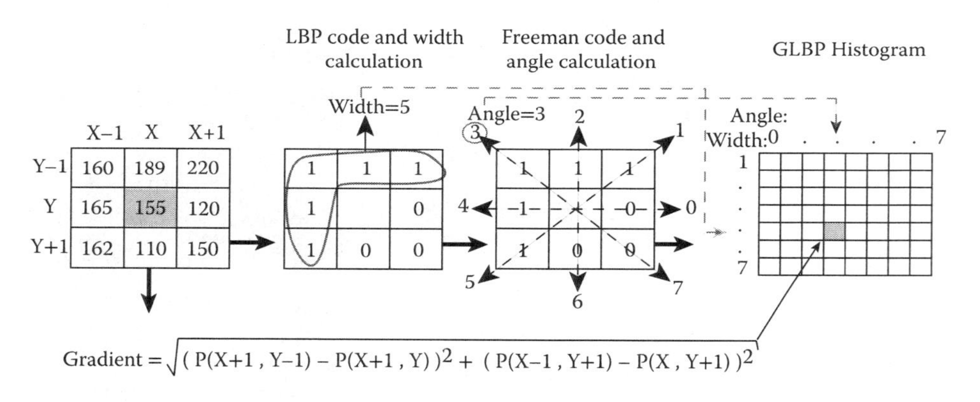

FIGURE 3.8: Flowchart for GLBP computation. (From Y. Serdouk et al., *Expert Systems with Applications*, 51, 186–194, 2016.).

- The width value corresponds to the number of "1" in LBP code.
- The angle value corresponds to the Freeman direction of the middle pixel within the "1" area in LBP code.

3. Compute the gradient at the 1 to 0 (or 0 to 1) transitions in uniform patterns.

4. Width and angle values define the position within the GLBP matrix, which is filled by the gradient information.

The angle can take values from zero to seven according to Freeman directions while width values can go from one to seven in uniform patterns. This means that the size of the GLBP matrix is 7×8, in which gradient features are accumulated.

3.6 Experimental Analysis

Using the writer-dependent approach, the signature verification system (SVS) is built for each writer of the CEDAR dataset to separate original signatures from forgeries. Performance evaluation is based on two types of errors: the false rejection rate (FRR, i.e., rejection of genuine signatures) and the false acceptance rate (FAR, i.e., acceptance of forged signatures). These two error rates are relative, which means that any parameter change in the classification that reduces the FAR will automatically provide an increase in the FRR and vice versa. In general, the type of error that is better to minimize depends on the application. Also, the average error rate (AER) that corresponds to the average sum of FAR and FRR is used to express the verification performance. Before introducing the experimental results, the CEDAR dataset is presented in the following subsection.

3.6.1 CEDAR dataset

The research Center of Excellence for Document Analysis and Recognition (CEDAR) developed an offline signature dataset that is originally presented in [30]. It contains signatures of 55 individuals, each with 24 genuine signatures and 24 forgeries forged by 20 skillful forgers. An example of five genuine and five forged signatures of an individual is shown in Figure 3.9. Each signature of this database was scanned at 300 dpi in 8-bit gray scale and stored in the portable network graphics (PNG) format. Presently, signature images are binarized to facilitate the LRF generation. This preprocessing is achieved using Otsu's method. To perform the proposed signature verification system, the CEDAR corpus is divided into two subsets (Training and Test). So, for each signer, 16 genuine and 16 forged signatures are used to train the model, while the remaining signatures (eight genuine + eight forgeries) are used to test the verification performance of the proposed system.

3.6.2 AIRS parameters selection

To perform the writer-dependent verification, user-defined parameters of AIRS should be experimentally tuned for each writer. However, among these parameters, the clonal rate, resources number, mutation rate and the hyper-mutation rate have a substantial effect on the verification accuracy. Thereby, several run-passes are carried out to separate effectively genuine signatures from forgeries. Figure 3.10 illustrates the AER variations according to

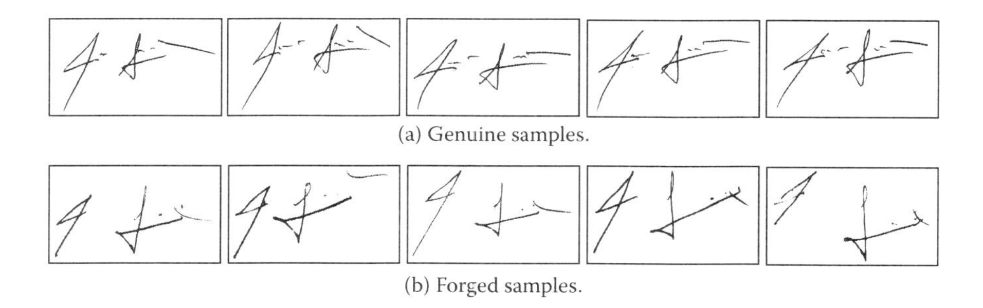

(a) Genuine samples.

(b) Forged samples.

FIGURE 3.9: Signatures of one writer from CEDAR dataset: (a) genuine samples and (b) forged samples.

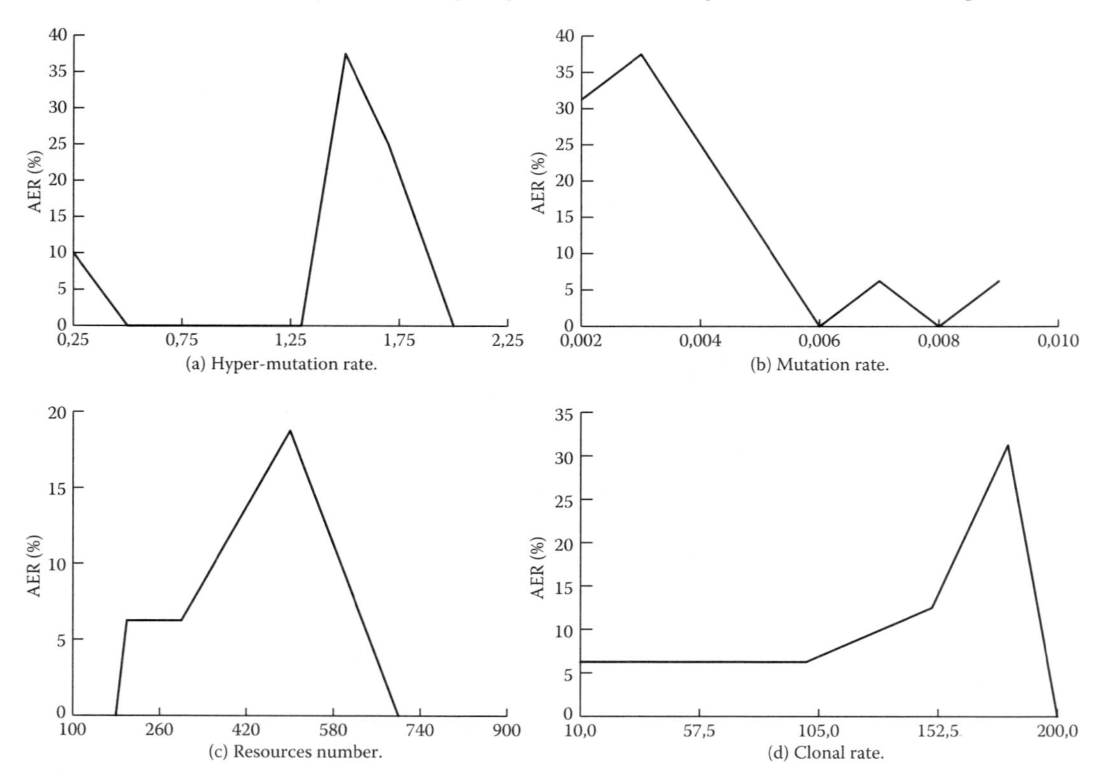

FIGURE 3.10: AER variations according to AIRS parameters: (a) hyper-mutation rate, (b) mutation rate, (c) resources number, and (d) clonal rate.

TABLE 3.1: Selected values of AIRS parameters

Parameters	Value
Clonal rate	200
Resources number	180
Mutation rate	0.006
Hyper-mutation rate	1.30
Stimulation threshold	0.88
ATS	0.50

these parameters for a randomly selected writer. The parameters are varied within the following ranges: the clonal rate in [10–200] with a step of 50, the resources number in [100–900] with a step of 100, the hyper-mutation rate in [0.25–2] with a step of 0.25 and in order to avoid large changes in the new ARBs, the mutation rate that expresses the mutation probability takes small values in [0.002–0.009] with a step of 0.001.

As can be seen, for each parameter there are several values that give the same AER. Therefore, despite using several setup parameters, we don't need extensive experiments to find the best values. Finally, the optimal parameters selection for the first writer in the CEDAR dataset is presented in Table 3.1.

3.6.3 Performance evaluation

The performance of the AIRS classifier is assessed using different types of characteristics that were presented earlier: pixel density, LRF, LBP, OC-LBP and GLBP. Thereby, Table 3.2 and Figure 3.11 resume the obtained results. Moreover, each feature vector size is reported in Table 3.2. Note that the regions partitioning of the pixel density features were experimentally tuned to (5×8) regions. Also, the LBP features are calculated using only four neighbors in a radius R=1. This choice is due to the complexity of performing the AIRS training with a large size of the feature vector. Indeed, with four neighbors, LBP size is equal to 16 while with eight neighbors, LBP size is equal to 256. However, as the OC-LBP allows a reduction of the original LBP size, we performed it using eight neighbors in a radius R=1.

From these outcomes, it is easy to see that the GLBP descriptor provides the best scores where all error rates are smaller than 3%. Indeed, the AIRS classifier performs well when gradient and textural characteristics are combined within 56 components in the GLBP feature vector. However, when textural features are applied individually using LBP and OC-LBP, an increase in terms of error rates is noted. On the other hand, concerning the topological features, LRF provides better performance with a feature vector of only four components than the pixel density with 40 components. This can be explained by the fact that LRF has less redundant information by calculating only the more representative density at the horizontal, vertical and diagonal directions. Moreover, compared to the state of the art reported in Table 3.3, our results provide a gain of more than 4% in all error rates.

TABLE 3.2: Verification results for the CEDAR dataset

Features	Classifier	Size	FRR (%)	FAR (%)
Pixel density		40	03.63	04.71
LRF		4	03.18	02.04
$LBP^{4,1}$	AIRS	16	07.50	09.09
$OC\text{-}LBP^{8,1}$		32	09.77	11.36
GLBP		**56**	**02.27**	**01.59**

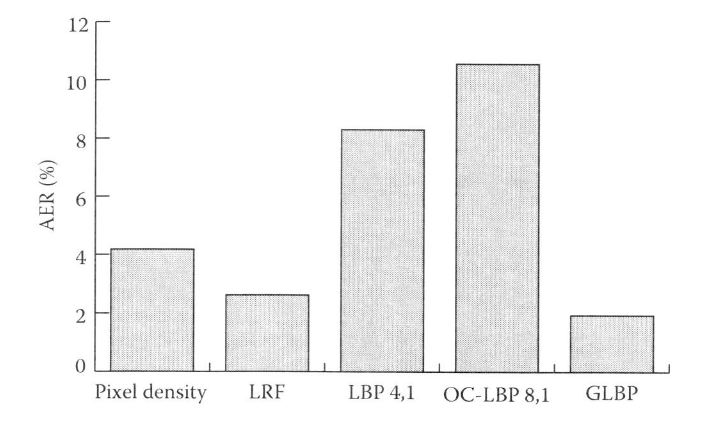

FIGURE 3.11: AER results for each descriptor.

TABLE 3.3: CEDAR state of the art

Ref.	Features	Classifier	Signatures per class	AER (%)
[22]	Surroundedness	MLP	24	08.33
[31]	Signature morphology	SVM	24	11.59
[32]	Gradient + concavity	Graph matching	16	07.90
[32]	Zernike moments	Harmonic distance	16	16.40
[33]	Gradient + equimass pyramid	Adaptive features thresholding	16	07.58
[34]	Chord moments	SVM	16	06.02

3.7 Conclusion

In this chapter, we investigated the use of the artificial immune recognition system (AIRS) for solving automatic handwritten signature verification. Inspired from the learning mechanisms of the natural immune system, AIRS tends to develop new representative data (memory cells) that characterize effectively each of the genuine and the forged classes. Thereafter, according to the k-nearest neighbors (kNN) among the established memory cells, a questioned signature is classified as being genuine or forged. Experiments were conducted on the CEDAR dataset, which is composed of 55 writers. Results showed that this bio-inspired classification method is very effective for verifying the authenticity of handwritten signatures, especially when gradient and textural features are combined to characterize signatures. Indeed, according to the writer-dependent approach, GLBP features provide a quite satisfactory average error rate of about 1.9% that considerably surpasses the state of the art methods. Nevertheless, in spite of the effectiveness and the simplicity of the AIRS classifier, it has two major drawbacks. The first is related to its six set-up parameters that require an experimental tuning for each writer to achieve a minimal training error rate. This implies several run-passes to carry out the training process. Also, the kNN classification constitutes the second limitation since it uses only the memory cells to classify test samples without any knowledge of the original training data. Thereby, in future works, one can substitute the kNN classification by a more reliable decision and then, globally tune the parameters of this classifier for all writers. Moreover, more improvements can be obtained by using features combination.

References

1. M. T. Das and L. C. Dulger. Signature verification (SV) toolbox: Application of PSO-NN. *Engineering Applications of Artificial Intelligence*, 22(4–5):688–694, 2009.

2. L. G. Hafemann, R. Sabourin, and L. S. Oliveira. Offline handwritten signature verification—Literature review. *arXiv preprint arXiv:1507.07909*. pp. 1–18, 2015.

3. P. S. Deng, H. M. Liao, C. W. Ho, and H. Tyan. Wavelet-based off-line handwritten signature verification. *Computer Vision and Image Understanding*, 76(3):173–190, 1999.

4. H. Nemmour and Y. Chibani. Off-line signature verification using artificial immune recognition system. In *International Conference on Electronics, Computer and Computation (ICECCO)*, pp. 164–167, Ankara, Turkey, 2013.

5. Y. Guerbai, Y. Chibani, and B. Hadjadji. The effective use of the one-class svm classifier for handwritten signature verification based on writer-independent parameters. *Pattern Recognition*, 48(1):103–113, 2015.

6. M. R. Pourshahabi, M. H. Sigari, and H. R. Pourreza. Offline handwritten signature identification and verification using contourlet transform. In *International Conference of Soft Computing and Pattern Recognition (SoCPaR)*, pp. 670–673, Malacca, 2009.

7. M. B. Yilmaz, B. Yanikoglu, C. Tirkaz, and A. Kholmatov. Offline signature verification using classifier combination of hog and lbp features. In *International Joint Conference on Biometrics (IJCB)*, pp. 1–7, Washington, DC, 2011.

8. D. Bertolini, L. S. Oliveira, E. Justino, and R. Sabourin. Reducing forgeries in writer-independent off-line signature verification through ensemble of classifiers. *Pattern Recognition*, 43(1):387–396, 2010.

9. Y. Serdouk, H. Nemmour, and Y. Chibani. New off-line handwritten signature verification method based on artificial immune recognition system. *Expert Systems with Applications*, 51:186–194, 2016.

10. S. Pal, A. Alaei, U. Pal, and M. Blumenstein. Performance of an off-line signature verification method based on texture features on a large indic-script signature dataset. In *12th IAPR Workshop on Document Analysis Systems (DAS)*, pp. 72–77, Santorini, Greece, 2016.

11. H. Baltzakis and N. Papamarkos. A new signature verification technique based on a two-stage neural network classifier. *Engineering Applications of Artificial Intelligence*, 14(1):95–103, 2001.

12. D. Impedovo, R. Modugno, G. Pirlo, and E. Stasolla. Handwritten signature verification by multiple reference set. In *International Conference on Frontiers in Handwriting Recognition (ICFHR)*, pp. 19–21, Montréal, Canada, 2008.

13. E. Frias-Martinez, A. Sanchez, and J. Velez. Support vector machines versus multi-layer perceptrons for efficient off-line signature recognition. *Engineering Applications of Artificial Intelligence*, 19(6):693–704, 2006.

14. E. J. R. Justino, F. Bortolozzi, and R. Sabourin. A comparison of SVM and HMM classifiers in the off-line signature verification. *Pattern Recognition Letters*, 26(9):1377–1385, 2005.

15. H. Kodaz, S. Özsen, A. Arslan, and S. Günes. Medical application of information gain based artificial immune recognition system (AIRS): Diagnosis of thyroid disease. *Expert Systems with Applications*, 36(2 Part 2):3086–3092, 2009.

16. O. Findik, I. Babaoglu, and E. Ülker. A color image watermarking scheme based on artificial immune recognition system. *Expert Systems with Applications*, 38(3):1942–1946, 2011.

17. A. Watkins. Airs: A resource limited artificial immune system. Master's thesis, Faculty of Mississippi State university, USA, 2001.

18. S. Srihari, A. Xu, and M. Kalera. Learning strategies and classification methods for offline signature verification. In *9th International Workshop on Frontiers in Handwriting Recognition (IWFHR)*, pp. 161–166, Tokyo, Japan, 2004.

19. E. J. R. Justino, F. Bortolozzi, and R. Sabourin. Off-line signature verification using hmm for random, simple and skilled forgeries. In *6th International Conference on Document Analysis and Recognition (ICDAR)*, pp. 1031–1034, Seattle, WA, 2001.

20. E. Ozgunduz, T. Senturk, and M. E. Karsligil. Off-line signature verification and recognition by support vector machine. In *13th European Signal Processing Conference*, pp. 1–4, Antalya, 2005.

21. C. Santos, E. J. R. Justino, F. Bortolozzi, and R. Sabourin. An off-line signature verification method based on document questioned experts approach and a neural network classifier. In *9th International Workshop on Frontiers in Handwriting Recognition (IWFHR)*, pp. 498–502, Tokyo, Japan, 2004.

22. R. Kumar, J. D. Sharma, and B. Chanda. Writer-independent off-line signature verification using surroundedness feature. *Pattern Recognition Letters*, 33:301–308, 2012.

23. J. Timmis and M. Neal. A resource limited artificial immune system for data analysis. *Knowledge-Based Systems*, 14:121–130, 2001.

24. A. B. Watkins, J. Timmis, and L. Bogess. Artificial immune recognition system (AIRS): An immune-inspired supervised learning algorithm. *Genetic Programming and Evolvable Machines*, 5:291–317, 2004.

25. T. Ojala, M. Pietikainen, and D. Harwood. A comparative study of texture measures with classification based on featured distributions. *Pattern Recognition*, 29:51–59, 1996.

26. C. Zhu, C.-E. Bichot, and L. Chen. Image region description using orthogonal combination of local binary patterns enhanced with color information. *Pattern Recognition*, 46:1949–1963, 2013.

27. M. Pietikainen, G. Zhao, A. Hadid, and T. Ahonen. *Computer Vision Using Local Binary Patterns*. Springer-Verlag, New York, 2011.

28. W. Zhang, S. Shan, W. Gao, X. Chen, and H. Zhang. Gabor binary pattern histogram sequence (lgbphs): A novel non-statistical model for face representation and recognition. In *10th International Conference on Computer Vision (ICCV)*, pp. 786–791, Beijing, 2005.

29. N. Jiang, J. Xu, W. Yu, and S. Goto. Gradient local binary patterns for human detection. In *International Symposium on Circuits and Systems (ISCAS)*, pp. 978–981, Beijing, 2013.

30. M. K. Kalera, S. Srihari, and A. Xu. Off-line signature verification and identification using distance statistics. *International Journal of Pattern Recognition and Artificial Intelligence*, 18:228–232, 2004.

31. R. Kumar, L. Kundu, B. Chanda, and J. D. Sharma. A writer-independent off-line signature verification system based on signature morphology. In *1st International Conference on Intelligent Interactive Technologies and Multimedia (IITM)*, pp. 261–265, Allahabad, India, 2010.

32. S. Chen and S. Srihari. A new off-line signature verification method based on graph matching. In *18th International Conference on Pattern Recognition (ICPR)*, pp. 869–872, Hong Kong, 2006.

33. R. Larkins and M. Mayo. Adaptive feature thresholding for off-line signature verification. In *23rd International Conference on Image and Vision Computing*, pp. 1–6, Christchurch, 2008.

34. M. M. Kumar and N. B. Puhan. Off-line signature verification: Upper and lower envelope shape analysis using chord moments. *IET Biometrics* 3:347–354, 2014.

Chapter 4

Parametric Object Detection Using Estimation of Distribution Algorithms

Ivan Cruz-Aceves, Jesus Guerrero-Turrubiates, and Juan Manuel Sierra-Hernandez

4.1 Introduction

In the pattern recognition field, detection of curves in images is a significant and challenging problem since relevant information about an object in the image is linked to the shape of its boundary. The Hough transform (HT) is a technique used in image analysis, computer vision, and digital image processing that finds imperfect instances of objects within a certain class of shapes by a voting procedure. This voting strategy is performed in a parameter space, where the candidate objects to be detected are obtained by taking the local maxima in a matrix that is known as accumulator [1]. This parameter space requires that the curve or object to be detected must have a predefined parametric equation.

The HT was first applied for the identification of lines in images. Then, it was extended to identify circles, where by this modification is known as the circle Hough transform (CHT). The HT, universally used today, was proposed by Richard Duda and Peter Hart in 1972, who called it the *generalized Hough transform.*

In the following sections, the HT for line, circle, and parabola detection are explained in detail. In addition, methods based on evolutionary computing for reducing the time-consumption of the HT strategy are introduced.

4.2 Hough Transform

In image processing, a classic problem is the identification of shapes. A feasible solution for this task is by means of parametric forms. The Hough transform (HT) represents a method widely used to find parametric forms, such as lines and circles. This section explains how the HT works to find three different parametric forms and exposes some disadvantages of the method.

4.2.1 Line detection

The generalized HT algorithm was first used to find lines in images. This is the most basic form of the HT, since it can represent the object by using only two parameters. Equation 4.1 represents a line in the Euclidean space as follows:

$$y = mx + b, \tag{4.1}$$

where b is the intersection with the y-axis and m is the slope of the line given by the following:

$$m = \frac{y_2 - y_1}{x_2 - x_1}, \tag{4.2}$$

where (x_i, y_i) are two points in the Euclidean space that belong to the line. Since Equation 4.1 cannot represent all the lines because m is undetermined when $x_2 = x_1$, another representation of the line is required. Duda and Hart [2] proposed to use the *Hesse normal form* to represent a line. This parametric form is described as follows:

$$r = xcos(\theta) + ysin(\theta), \tag{4.3}$$

where r represents the perpendicular distance from the line to the origin and θ the angle of the line, respectively. Consequently, one point of the true line is mapped at a time, and a number of lines can pass through this point, then several angles and distances must be tested. Figure 4.1 illustrates a set of (θ, r) of one single point, which forms a sine wave in the Hough space.

When multiple curves intersect in a single point in the HT space, this point will represent the parameters of a line in the Euclidean space as can be seen in Figure 4.2, where the example shows a line that can be defined by taking two of the three points on the Euclidean space. Each point will produce a sine wave on the Hough space, that will intersect the other sine waves produced by points that belong to the same line. If these intersections are added one by one, it will be possible to observe a peak with amplitude proportional to the number of curves passing through that point, this graph is known as the accumulator [1].

Figure 4.3 shows how the accumulator is computed with the number of intersections, then, the coordinate point with the maximum number of intersections will show a unimodal

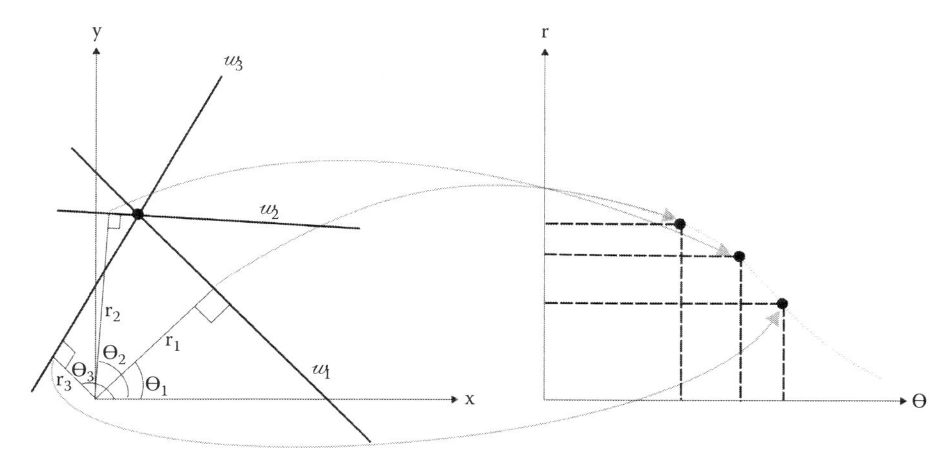

FIGURE 4.1: Mapping points from the Euclidean space to the Hough space.

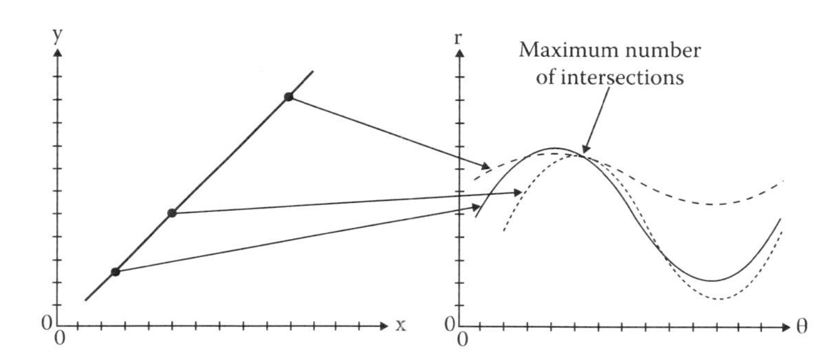

FIGURE 4.2: Relation between Euclidean and Hough spaces.

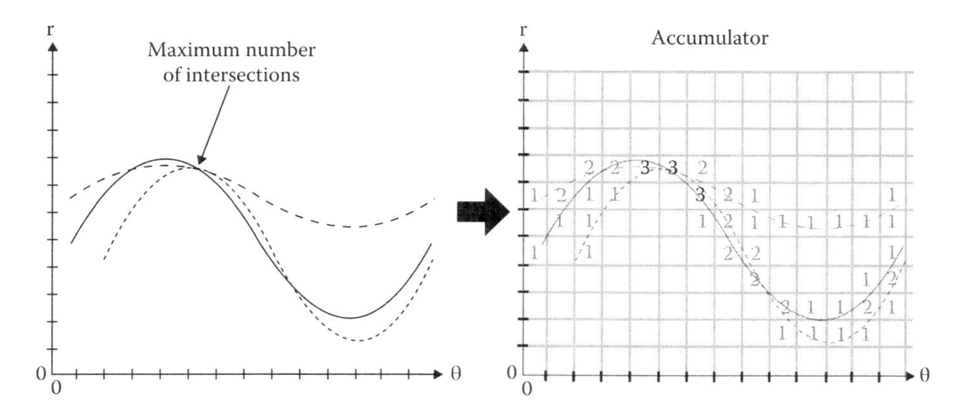

FIGURE 4.3: Hough space computing the accumulator.

peak in a 3D graph. It can be seen that there exist other peaks that do not belong to the real parameters of the line. These peaks are false positive parameter values, and must be avoided in order to obtain the real peak that represents the parameters of the true line.

To illustrate a real Hough accumulator, Figure 4.4 presents a synthetic image containing two lines with different slope angles. When the HT algorithm is applied to this image, a 3D accumulator is generated. Figure 4.5a shows the sine waves intersecting in two coordinates

FIGURE 4.4: Synthetic image for line detection.

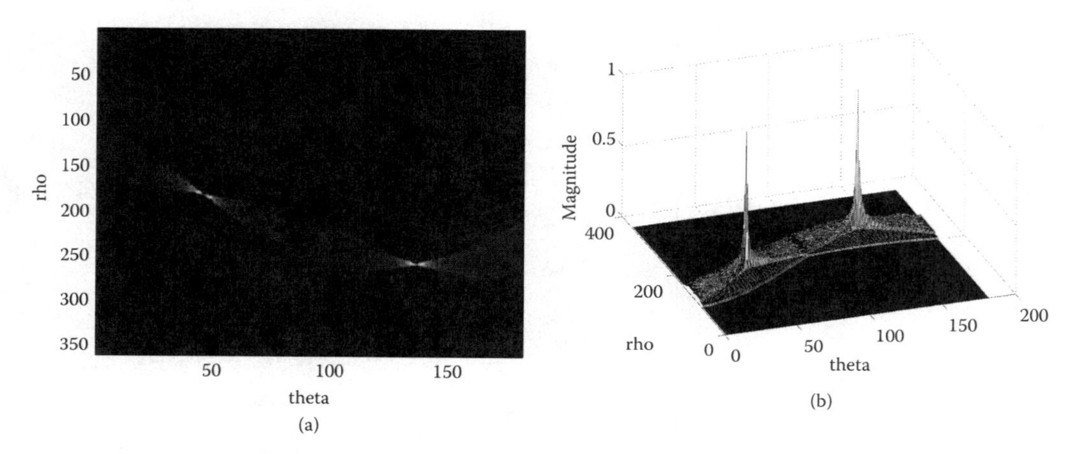

FIGURE 4.5: Accumulator for line detection: (a) visualization in a 2D space and (b) visualization in a 3-D space.

denoting the existence of two lines in the image, where it can be seen that the white areas represent the two peaks as they are shown in the 3D accumulator of Figure 4.5b.

A straight line is the simplest parametric form, the accumulator is easy to compute due to the simplicity of the parametric equations. However, a circle is more complex due to the number of parameters that represent this object. In the following section, the problem of detecting circular objects is explained in detail.

4.2.2 Circle detection

A further modification to the HT is the circular Hough transform (CHT) [3], which is used for circle detection. In this problem, three different parameters are required to define a circle (a, b, r), where (a, b) represent the coordinate values of the center of the circle and r is its radius. Using these three parameters, a circle in the Euclidean space can be determined as follows:

$$(x - a)^2 + (y - b)^2 = r^2. \tag{4.4}$$

Similarly to the HT for line detection, the most commonly strategy to observe the accumulator in the Hough space is by taking the trigonometric equations for the circle. These equations are defined as follows:

$$a = x - rcos(\theta) \tag{4.5}$$

$$b = y - rsin(\theta) \tag{4.6}$$

where the coordinate values (x, y) represent a single point in the Euclidean space, r is the radius to be tested, and θ is in the range $[0, 360°]$.

Since three different parameters are required to describe circles, each parameter represents a search dimension in the Hough space, obtaining a 3D accumulator. Figure 4.6 shows the accumulator for a fixed value of $r = d$, where it can be observed that for each point in Euclidean space is transformed into a circle in Hough space. Similar to HT for line detection, the point with the maximum number of intersections in the Hough space represents a circle detected in the Euclidean space.

Figure 4.7 shows a test image to apply the CHT. Two circles of different radius, $r_1 = d$ and $r_2 = D$, are introduced. Center positions (a_1, b_1) and (a_2, b_2) are also different for both circles.

Figure 4.8 shows an accumulator layer for $r = d$ in two and three dimensions. It can be seen as a common intersection point for the smaller circle in Figure 4.8a, and it is represented by the unimodal peak in Figure 4.8b. The detection of the circle with the maximum radius

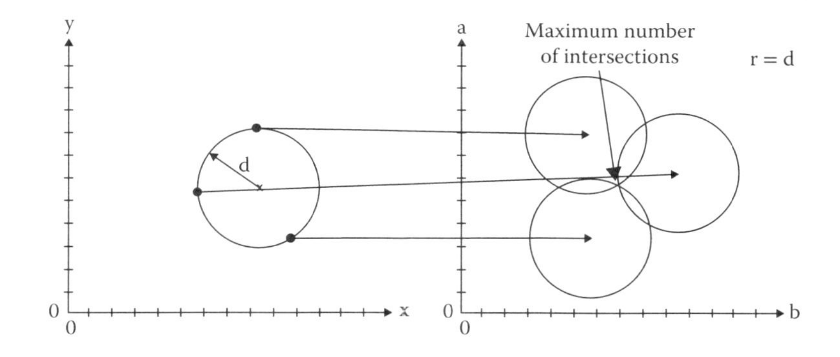

FIGURE 4.6: Euclidean space and Hough space for circle detection.

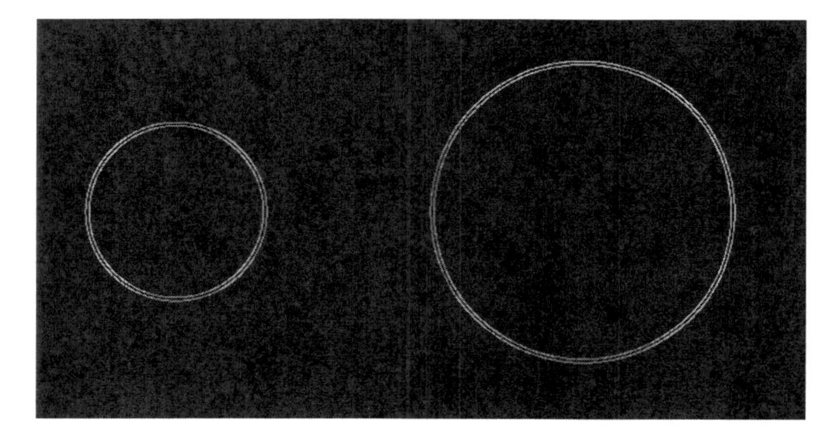

FIGURE 4.7: Test image for circular Hough transform.

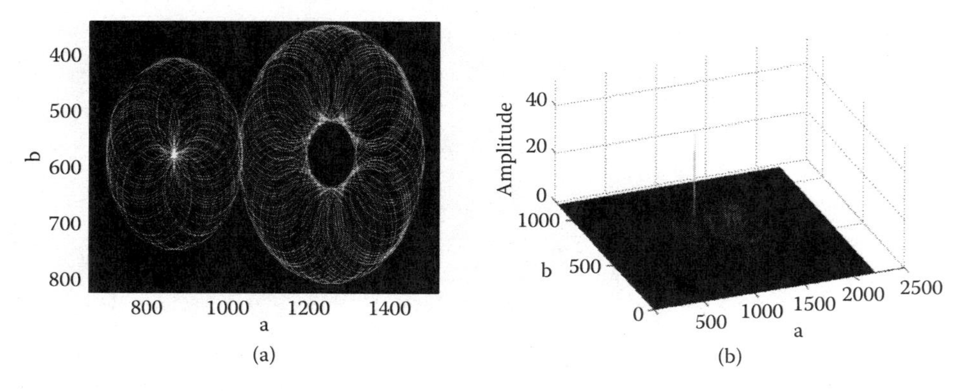

FIGURE 4.8: Accumulator for circle detection: (a) visualization in a 2D space and (b) visualization in a 3-D space.

presents disadvantages when the accumulator layer is analyzed. According to Figure 4.8a, each point belonging to the big circle in Euclidean space is transformed into a circle in Hough space. These circles are supposed to intersect in the center coordinates (a_2, b_2). However, most circles intersect along a circular perimeter from the real center and the correct location of the center cannot be located. In order to find the parameters of the big circle, a new accumulator must be computed. Then, the HT must be applied for every different radius of circles in the image. Hence, it can be concluded that the number of dimensions in the accumulator is proportional to the number of parameters needed to represent the object, increasing the computational complexity, and consequently, the difficulty to accurately find the object of interest.

4.2.3 Parabola detection

The HT can also be extended for parabola detection. However, there are four parameters to be considered if a rotation parameter is involved. These parameters are (x_0, y_0, a, θ), where (x_0, y_0) represent the coordinate values of the parabola vertex, a is the aperture, and θ is the rotation. There have been many approaches for finding non-rotated parabolas [4–6], in order to simplify the detection algorithm. In this chapter, non-rotated parabola detection is also presented. On the other hand, the parametric equation to define a parabola in the Euclidean space is described as follows:

$$(y - y_0)^2 = 4a(x - x_0); \tag{4.7}$$

in addition, when the directrix is parallel to the y-axis, and if the directrix is parallel to the x-axis, the equation is modified as follows:

$$(x - x_0)^2 = 4a(y - y_0). \tag{4.8}$$

Equations 4.7 and 4.8 can be represented by only three parameters, due to the fact that rotation parameter is not present. However, it is necessary to implement a 3D accumulator in order to detect parabolas, where a strategy to find the highest peak is required.

Figure 4.9 presents a synthetic image to test the algorithm for the parabola detection problem. The algorithm used in the present test can be found in MATLAB® [7] and was developed by Clara Sanchez.

Figure 4.10 shows the obtained accumulator for the previously mentioned synthetic image. From the accumulator graph it can be seen that there is not a single higher peak; consequently, the problem consists of finding the appropriate location and value of the accumulator.

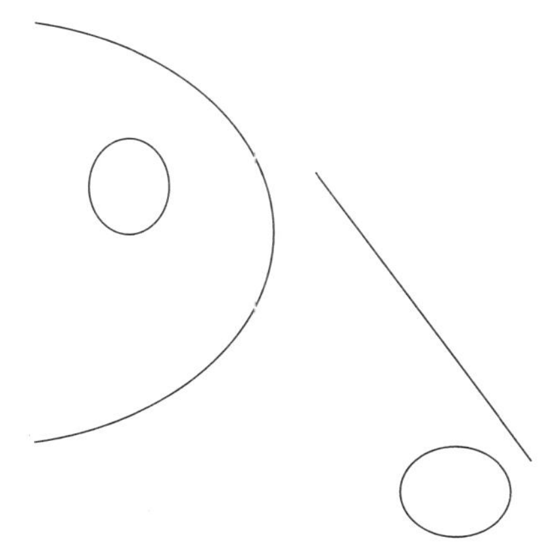

FIGURE 4.9: Synthetic test image for parabolic shapes.

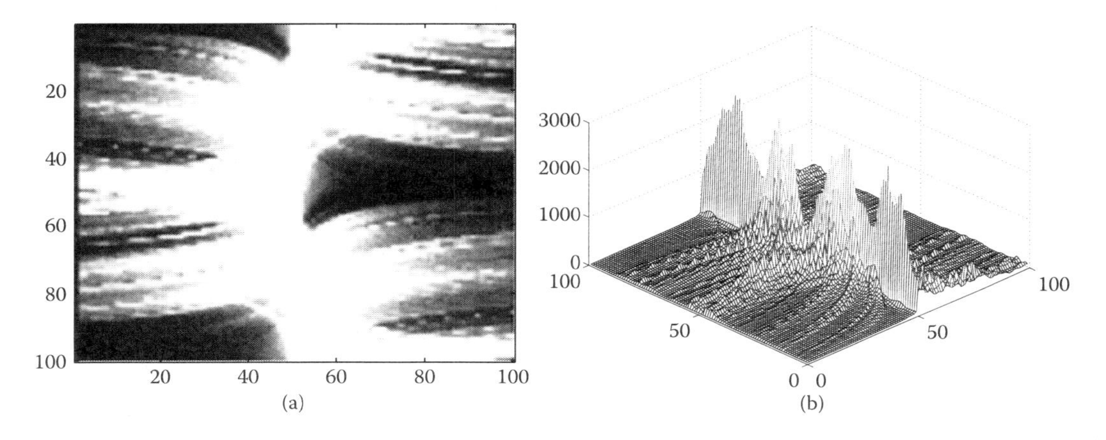

FIGURE 4.10: Accumulator for parabola detection: (a) visualization in a 2D space and (b) visualization in a 3-D space.

The aforementioned parametric equations for line, circle, and parabola forms can be found by using some techniques that do not need the accumulator strategy to determine the best set of parameters. In this chapter, estimation of distribution algorithms are used to reduce the computational complexity of the Hough transform strategy, as is shown in the following section.

4.3 Estimation of Distribution Algorithms

The estimation of distribution algorithms (EDAs) represent an extension to the field of evolutionary computation (EC). EDAs are useful to solve problems in the discrete and continuous domain by using some statistical information of potential solutions, also called individuals [8–10]. Similar to EC techniques, EDAs perform the optimization task using binary encoding and selection operators over a set of potential solutions called population. The main difference regarding classical EC techniques is that EDAs replace the crossover and mutation operators by building probabilistic models at each generation based on global statistical information of the best individuals.

In the present work, the univariate marginal distribution algorithm (UMDA) has been adopted as optimization strategy, because it works ideally for linear problems [11–14]. UMDA uses a binary codification for each possible solution, and it generates a probability vector $\mathbf{p} = (p_1, p_2, p_3, \ldots, p_n)^T$, where $\mathbb{P}$ is the marginal probability of the *ith* bit of each individual, to be one or zero in the next generation. Then, UMDA tries to approximate the probability distribution of the individuals in $\mathbb{P}$, which can be defined as follows:

$$\mathbb{P}(x) = \prod_{i=1}^{n} \mathbb{P}(X_i = x_i), \tag{4.9}$$

where $x = (x_1, x_2, \ldots, x_n)^T$ is the binary value of the *ith* bit in the individual, and X_i is the *ith* random value of the vector X. To select a subset of best individuals, an objective function is needed, which is used to determine the fitness of the current potential solution. From the subset of candidate solutions, a probability vector is computed in order to generate a new population based on its distribution. This process is iteratively performed until a stop condition is achieved, and the best solution is chosen to be the individual with the best fitness value along the evolutionary process. According to the above description, UMDA can be implemented as follows:

Univariate Marginal Distribution Algorithm

1. Initialize the number of generations t.
2. Initialize the number of individuals n into the predefined search space.
3. Select a subset of the best individuals S of $m \leq n$.
4. Calculate the univariate marginal probabilities $p_i^s(x_i, t)$ of S.
5. Generate n new individuals by applying $p(x, t+1) = \prod_{i=1}^{n} p_i^s(x_i, t)$.
6. Stop if convergence criterion is satisfied (e.g., number of generations), otherwise, repeat steps (3)–(5).

4.4 Parametric Object Detection Using EDAs

4.4.1 Line detection

The detection of straight lines in images is the first problem that was solved by the Hough transform. This algorithm finds the line even with lost information (incomplete lines); however, the process to find a single or multiple lines is highly demanding in terms of computational time.

The problem of finding straight lines has an important role in the engineering field with a number of applications, for example, object recognition, vehicle guidance, and camera calibration. In this section a strategy to find straight lines in images by using EDAs while reducing the computational time regarding the HT is introduced.

4.4.1.1 Solution representation

The detection of lines in images can be accomplished if a parametric equation that represents this form can be determined. The general equation for a line in the Euclidean space is defined as follows:

$$y - y_1 = m(x - x_1) \tag{4.10}$$

where m is the slope of the line and it is represented by Equation 4.2; then, it can be seen that only two points are required, as follows:

$$y - y_1 = \frac{y_2 - y_1}{x_2 - x_1}(x - x_1). \tag{4.11}$$

Since the search space is the entire binary image, the two points are the coordinate values of two pixels of the image. These pixels are used to form an individual (potential solution) for the UMDA algorithm.

To form an individual, all the pixels of interest are listed by their relative position of an origin and labeled with an index $ind = \{1, 2, 3 \ldots, N\}$, where N is the total number of pixels of interest in the image. Figure 4.11 represents how these pixels are labeled. The straight line is dashed to represent the individual pixels that form it, where the origin is placed at the upper left corner of the image.

The coordinate values (x_{ind}, y_{ind}) of the two pixels $\{i, j\}$ that form an individual are used in Equation 4.11 to compute a line. A single individual will be formed by concatenating the two indices $\{i, j\}$ in its binary form in a single vector. Table 4.1 shows an example of an individual formed with the indices {21,58} of two pixels of interest. Then, it must be taken into account that the number of required bits to represent the maximum index will depend on the total number of pixels of interest N on the entire image.

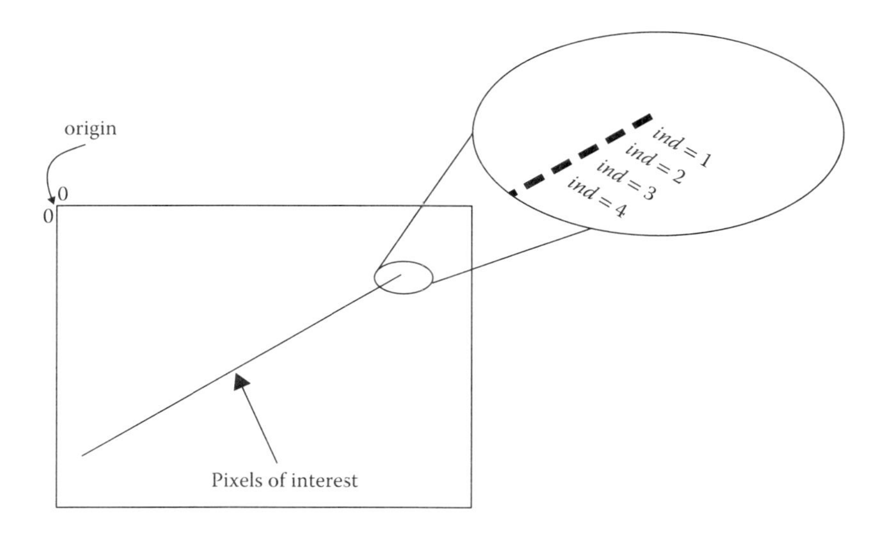

FIGURE 4.11: Labeling the pixels of interest in the image.

TABLE 4.1: Codification of individuals by concatenating the two indices

	Index i						Index j					
Individual	0	1	0	1	0	1	1	1	1	0	1	0

After the individual has been formed, it is necessary to evaluate the quality of this possible solution. Subsequently, a fitness function must be proposed to quantitatively evaluate the population of individuals.

4.4.1.2 Fitness function

To quantitatively assess the fitness of an individual, a binary image I_{VS} (virtual shape) of the same size of the input image is created. The value of all the pixels on the virtual image are initialized as zero value. Then, with the two indices stored in the individual, a line is computed by using Equation 4.11. Subsequently, all the pixels of interest in the line obtained by an individual, are stored in I_{VS} by setting to 1 the corresponding pixels. Finally, the Hadamard Product [15] between the binary form f_{binary} of the input image and the virtual image is calculated as follows:

$$Hd = f_{binary} \odot I_{VS}. \tag{4.12}$$

The resulting product Hd will only contain the number of coincident interest pixels between f_{binary} and I_{VS}. The quality of the individual can be measured by applying Equation 4.13. The fitness function $F(I)$ accumulates the number of coincident pixels between virtual shape and the input image as follows:

$$F(I) = \frac{\sum_{i=0}^{N_s-1} Hd(x_i, y_i)}{N} \tag{4.13}$$

where $Hd(x_i, y_i)$ contains all the pixels of interest obtained by Equation 4.12 and N is the total number of interest pixels in the input image.

Figure 4.12a shows an example of the Hadamard product between the virtual image formed by an individual and the binary image. It can be seen that the virtual image is only matching with a few pixels of the real image then, the function in Equation 4.13 will have a low fitness value. Figure 4.12b shows that the individual has two pixels that belongs to the same line, then the virtual image has all the pixels of interest of the line in the binary image. Consequently, the fitness value will be high.

4.4.1.3 Optimization process

The first solution of the UMDA algorithm may not have high fitness value, since this initial population is formed by taking 20 random pixels of interest and forming 10 individuals. Then, for each individual, a virtual image is created and the Hadamard product is performed for every one of them.

In the next step, the 10 fitness values generated are stored into a vector and placed from maximum to minimum value. The first individual that has the higher fitness value is taken as the best solution in the population. Then, to compute the next generation and ensure that the fitness value will be better, a probability vector with the best individuals of the current population is computed. A subset of individuals will be used to create a new population, as it is illustrated in Figure 4.13, where the first value p_n in the probability vector is obtained by the selected individuals that contain a 1 in that particular position.

The value p_n is computed as follows:

$$p_n = \frac{\sum_{i=1}^{s} Individual_s(bit)}{s} \tag{4.14}$$

where $bit = \{1, 2, 3, \ldots, n\}$ is the number of bits that represents an individual, and s is the total number of selected individuals to compute the probability vector. From this vector, the new population is generated by the assumption that for each position i, a 1 is generated

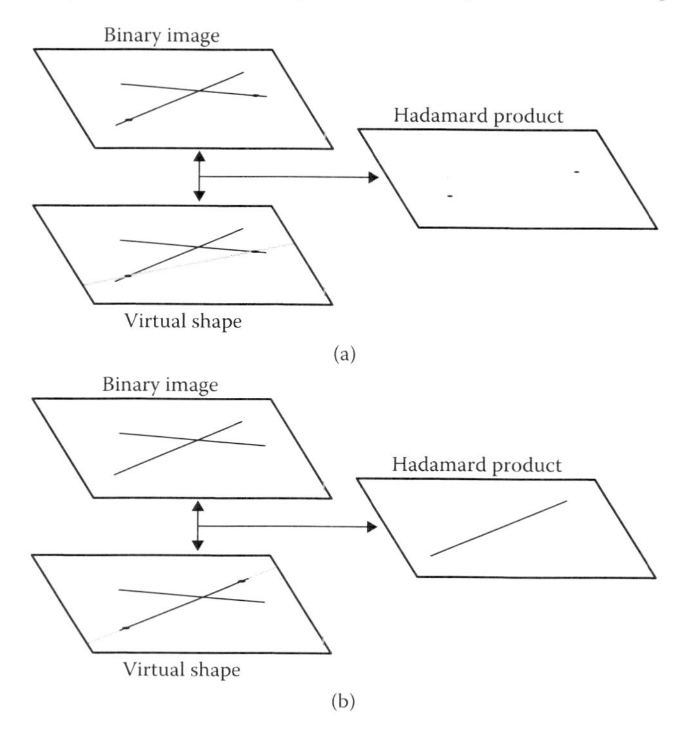

FIGURE 4.12: Hadamard product between virtual and input images for line object detection: (a) low detection rate and (b) high detection rate.

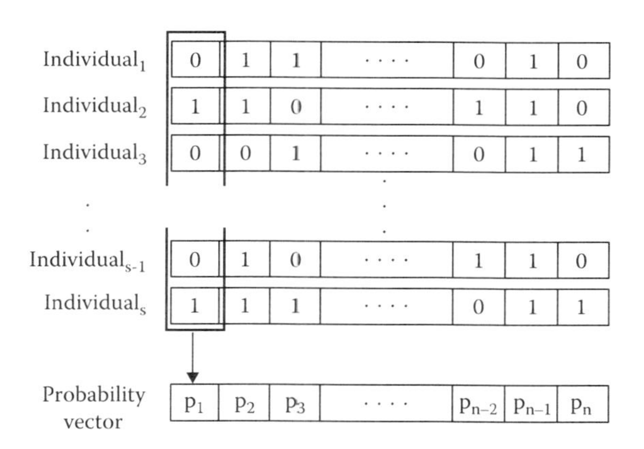

FIGURE 4.13: Calculation of probability vector.

with a probability of p_n. During the evolutionary process, it is necessary to store all the *elite* individuals, since UMDA represents a stochastic method.

4.4.2 Circle detection

This section presents the use of UMDA to find circles in images. The main advantage of applying a stochastic method instead of the HT is the computational time, which can be significantly reduced.

4.4.2.1 Solution representation

The target of this method is to find circles in images, then, a parametric equation that represents the circle is needed. The general equation of a circle is defined as follows:

$$(x - x_0)^2 + (y - y_0)^2 = r^2, \tag{4.15}$$

where (x_0, y_0) is the center of the circle and r is the radius. Since, these parameters are unknown, an optimization process to determine these parameters is required. A circle can be found if the coordinate values of at least three points that belong to the circle boundary are known. The three coordinate values of the pixels are used to find the center (x_0, y_0) and the radius. The x coordinate value of the circle can be determined as follows:

$$x_0 = \frac{\begin{vmatrix} x_j^2 + y_j^2 - (x_i^2 - y_i^2) & 2(y_j - y_i) \\ x_k^2 + y_k^2 - (x_i^2 - y_i^2) & 2(y_j - y_i) \end{vmatrix}}{4((x_j - x_i)(y_k - y_i) - (x_k - x_i)(y_j - y_i))} \tag{4.16}$$

where the three indices $\{i, j, k\}$ represent the coordinate values of three pixels of interest in the image. The y coordinate value can be defined as follows:

$$y_0 = \frac{\begin{vmatrix} 2(x_j - x_i) & x_j^2 + y_j^2 - (x_i^2 - y_i^2) \\ 2(x_k - x_i) & x_k^2 + y_k^2 - (x_i^2 - y_i^2) \end{vmatrix}}{4((x_j - x_i)(y_k - y_i) - (x_k - x_i)(y_j - y_i))}. \tag{4.17}$$

Finally, the radius of the circle is found by substituting x_0 and y_0 as follows:

$$r = \sqrt{(x - x_0)^2 + (y - y_0)^2}. \tag{4.18}$$

All the pixels of interest are listed by their relative position of an origin and labeled with an index $ind = \{1, 2, 3 \ldots, N\}$, where N is the total number of interest pixels in the image. The coordinate values (x_{ind}, y_{ind}) of the three pixels $\{i, j, k\}$ that form an individual are used in Equations 4.16 and 4.17, to compute the constant values.

Similar to Ayala et al. [16], the individual is formed by concatenating the three indices $\{i, j, k\}$ in a single vector. Table 4.2 shows an example of an individual formed with the indices {20,50,47} of three boundary pixels. The number of required bits to represent the indices will depend on the number of pixels of interest, N, in the entire image.

Since we are using the indices to form the individuals, the UMDA algorithm can easily eliminate unfeasible solutions by measuring the quality of the individual by using a fitness function.

4.4.2.2 Fitness function

To quantitatively evaluate the fitness of an individual, a binary image I_{VS} (virtual shape) of the same size of the input image is created and it is initialized with an intensity zero for all the pixels. In the second step, with the three indices stored in an individual, a circle is computed by using Equations 4.16 through 4.18, respectively. Subsequently, all the pixels of interest in the circle formed by the individual are stored in I_{VS} by setting to 1 the

TABLE 4.2: Codification of individuals by concatenating the three indices

	Index i						Index j						Index k					
Individual	0	1	0	1	0	0	1	1	0	0	1	0	1	0	1	1	1	1

corresponding pixels. Finally, the Hadamard product [15] between the binary form f_{binary} of the input image and the virtual image is calculated as follows:

$$Hd = f_{binary} \odot I_{VS}. \tag{4.19}$$

Figure 4.14a represents the boundary pixels in the binary image that form a circle in the virtual shape; since this circle does not contain points that belong to a single circle, the fitness value will be low. On the other hand, in Figure 4.14b, the three boundary pixels that form the virtual image belong to a single circle; this solution, then, will be the best in terms of the fitness function.

The first solution of UMDA is formed with random points over the image, this leads to a probability that the initial population has no optimal solutions. Then, for each individual, a virtual image is created and the Hadamard product is performed to compute the fitness value. Since UMDA is a stochastic algorithm, the best solution is chosen to be the individual with the best fitness value along the evolutionary process.

4.4.3 Parabola detection

The parabola detection has been a challenging problem in the pattern recognition area. In the literature, the recognition of this parametric object in different type of images has been commonly addressed by using the Hough transform. This section introduces the use of UMDA for detecting parabolic shapes in images.

The best way to describe a parabola is by using the general equation as follows:

$$Ay^2 + By + C = x, \tag{4.20}$$

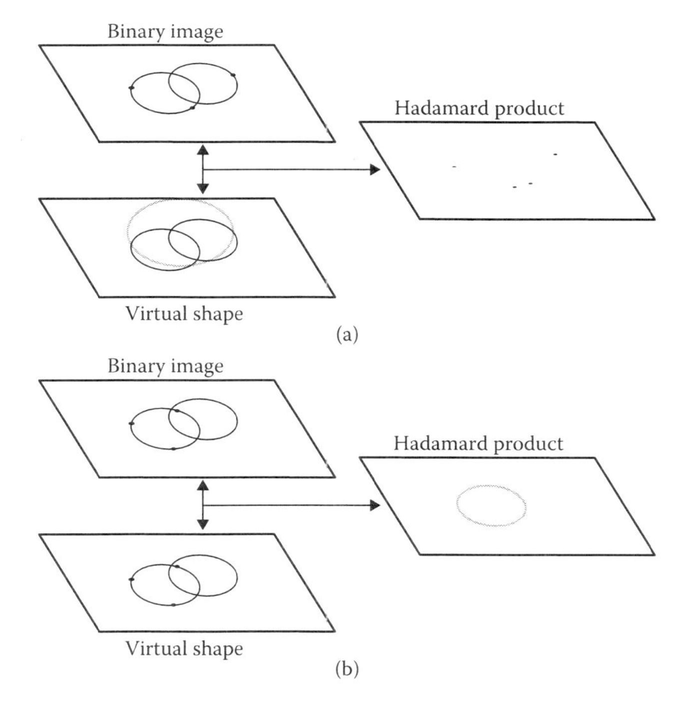

FIGURE 4.14: Hadamard product between virtual and input images for circle object detection: (a) low detection rate and (b) high detection rate.

where A, B, and C are constant values which need to be determined. To compute these three values, at least three pixels in the search space (binary image) are required. The coordinate values (x_{ind}, y_{ind}) of the three pixels $\{i, j, k\}$ that form an individual are used in Equations 4.21 through 4.23, to compute these constant values. The set of equations in terms of the indices of the three boundary pixels is represented as follows:

$$Ay_i^2 + By_i + C = x_i, \tag{4.21}$$

$$Ay_j^2 + By_j + C = x_j, \tag{4.22}$$

$$Ay_k^2 + By_k + C = x_k, \tag{4.23}$$

The first constant, A, can be found if Equation 4.22 is subtracted from Equations 4.21 and 4.23 from Equation 4.21 as follows:

$$Ay_i^2 + By_i + C - x_i - Ay_j^2 + By_j + C + x_j = 0, \tag{4.24}$$

$$Ay_j^2 + By_j + C - x_j - Ay_k^2 + By_k + C + x_k = 0, \tag{4.25}$$

then, these equations can be solved for A and B as follows:

$$A = \frac{x_i - x_j - B(y_i - y_j)}{y_i^2 - y_j^2}, \tag{4.26}$$

$$B = \frac{x_j - x_k - A(y_j^2 - y_k^2)}{y_j - y_k}, \tag{4.27}$$

and substituting Equation 4.27 in Equation 4.26:

$$A = \frac{x_i - x_j - \frac{x_j - x_k - A(y_j^2 - y_k^2)}{y_j - y_k}(y_i - y_j)}{y_i^2 - y_j^2} \tag{4.28}$$

$$A = \frac{(x_i - x_j)(y_j - y_k) - (x_j - x_k)(y_i - y_j)}{(y_i^2 - y_j^2)(y_j - y_k)} + \frac{A(y_j^2 - y_k^2)(y_i - y_j)}{(y_i^2 - y_j^2)(y_j - y_k)}, \tag{4.29}$$

and since $(a^2 - b^2) = (a - b)(a + b)$, Equation 4.29 can be reduced as follows:

$$A\left(1 - \frac{y_j - y_k}{y_i + y_j}\right) = \frac{(x_i - x_j)(y_j - y_k) - (x_j - x_k)(y_j - y_k)}{(y_i - y_j)(y_i + y_j)(y_j - y_k)}, \tag{4.30}$$

next, Equation 4.30 is solved for A as follows:

$$A = \frac{(x_i - x_j)(y_j - y_k) - (x_j - x_k)(y_j - y_k)}{(y_i - y_j)(y_i - y_k)(y_j - y_k)}, \tag{4.31}$$

finally, solving the products in the numerator, the constant A is found as follows:

$$A = \frac{y_k(x_j - x_i) + y_j(x_i - x_k) + y_i(x_k - x_j)}{(y_i - y_j)(y_i - y_k)(y_j - y_k)}. \tag{4.32}$$

For constant B, we will use Equations 4.26 and 4.27. In this constant value we will substitute A in Equation 4.27 as follows:

$$B = \frac{x_j - x_k - \frac{x_i - x_j - B(y_i - y_j)}{y_i^2 - y_j^2}(y_j^2 - y_k^2)}{y_j - y_k}, \tag{4.33}$$

and solving for numerator and denominator:

$$B = \frac{x_j(y_i^2 - y_j^2) - x_k(y_i^2 - y_j^2) - (x_i - x_j)(y_2^2 - y_k^2)}{(y_i^2 - y_j^2)(y_j - y_k)} + \frac{B(y_i - y_j)(y_j^2 - y_k^2)}{(y_i^2 - y_j^2)(y_j - y_k)}, \tag{4.34}$$

then, taking B as common factor:

$$B\left(1 - \frac{(y_i - y_k)(y_j^2 - y_k^2)}{(y_i^2 - y_j^2)(y_j - y_k)}\right) = \frac{x_j(y_i^2 - y_j^2) - x_k(y_i^2 - y_j^2) - (x_i - x_j)(y_2^2 - y_k^2)}{(y_i^2 - y_j^2)(y_j - y_k)} \tag{4.35}$$

$$B = \frac{y_i^2(x_j - x_k) - y_j^2(x_j - x_k) - y_j^2(x_i - x_j) + y_k^2(x_i - x_j)}{(y_i^2 - y_j^2)(y_j - y_k) - (y_i - y_j)(y_j^2 - y_k^2)} \tag{4.36}$$

and, since $(a^2 - b^2) = (a - b)(a + b)$

$$B = \frac{y_i^2(x_j - x_k) - y_j^2(x_j - x_k) - y_j^2(x_i - x_j) + y_k^2(x_i - x_j)}{(y_i - y_j)(y_j - y_k)(y_i + y_j - y_j + y_k)} \tag{4.37}$$

$$B = \frac{y_i^2(x_j - x_k) - y_j^2 x_j + y_j^2 x_k - y_j^2 x_i + y_j^2 x_j + y_k^2(x_i - x_j)}{(y_i - y_j)(y_j - y_k)(y_i + y_j - y_j + y_k)}, \tag{4.38}$$

finally, rearranging Equation 4.38, the constant B can be determined as follows:

$$B = \frac{y_k^2(x_i - x_j) + y_j^2(x_k - x_i) + y_i^2(x_j - x_k)}{(y_i - y_j)(y_i - y_k)(y_j - y_k)}. \tag{4.39}$$

Moreover, the constant value C is obtained by substituting the A and B constants in any of the three Equations 4.21 through 4.23. The resulting equation is given by the following:

$$C = \frac{y_j y_k(y_j - y_k)x_i + y_k y_i(y_k - y_i)x_j + y_i y_j(y_i - y_j)x_k}{(y_i - y_j)(y_i - y_k)(y_j - y_k)}. \tag{4.40}$$

After the constant values have been determined, we can calculate the coordinates of the vertex and the aperture of the parabola if the general formula can be represented in terms of the vertex as follows:

$$(y - k)^2 = 4p(x - h), \tag{4.41}$$

where, the coordinates $(x + h, y + k)$ are the coordinates of the vertex, and $4p$ is known as the aperture. Then, to convert the general equation in this form we will start to represent the general equation in the following form:

$$Ay^2 + By = x - C, \tag{4.42}$$

next, the left part of Equation 4.42 can be taken to a binomial form by complete the perfect square trinomial by, first divide all the equation by A and completing the trinomial as follows:

$$y^2 + \frac{B}{A}y + \left(\frac{B}{2A}\right)^2 = \frac{x - C}{A} + \left(\frac{B}{2A}\right)^2, \tag{4.43}$$

then, Equation 4.42 can be written as follows:

$$\left(y + \frac{B}{2A}\right)^2 = \frac{1}{A}\left(x - \left(C - \frac{B^2}{4A}\right)\right), \tag{4.44}$$

and comparing Equation 4.44 to Equation 4.41, it can be seen that the parameters are determined by the following:

$$y_{vertex} = k = -B/(2A), \quad (4.45)$$

$$x_{vertex} = h = C - B^2/(4A), \quad (4.46)$$

$$4p = 1/A. \quad (4.47)$$

The values A, B, and C of the individual representation are used to calculate the vertex and aperture $4a$ of the parabola represented by Equation 4.20. Moreover, similar to line and circle detection, the Hadamard product between the input and virtual images is used to evaluate the fitness of an individual.

4.5 Computational Experiments

In this section, the method for parabola detection using the UMDA strategy, is applied on synthetic images and medical images of the human retina. In order to assess the performance of the proposed method, it is compared with the HT by applying the algorithm of Clara Sanchez [7] found in the MATLAB central and, by using the HT for parabolic shapes of the software MIPAV® [17], which is available at [18]. The implementations were performed by using the MATLAB version 2013b, on a computer with an Intel Core i5, 4 GB of RAM, and 2.4 GHz processor.

All the computational experiments carried out by the UMDA method were performed using the parameter values presented in Table 4.3.

4.5.1 Application on synthetic images

The first experiment was performed by using the synthetic image of Figure 4.15. This image, was generated drawing randomly located parametric objects such as lines, circles and a parabola. The purpose of this image is to evaluate the algorithm in a controlled input.

Table 4.4 presents the statistical analysis of applying the method to Figure 4.15, using 30 runs. The HT methods [7] obtain the largest time, considering that the result is the execution-time per pixel and this image contains 2246 pixels. Hence, this method depends entirely on the number of pixels of interest. The UMDA method performs a reduction of 94.33% of the execution-time achieved on the MIPAV software; this is because on average only 110 iterations are needed to deliver the best result, as can be seen in Figure 4.15b.

The results of Table 4.3, show a good performance of the proposed algorithm; however, to ensure that the method is robust to different input conditions, the complexity of the image was increased by adding 10% salt & pepper noise, as can be seen in Figure 4.15c.

TABLE 4.3: UMDA parameters for all the computational experiments

Parameter	Value
Number of individuals	15
Selection rate	0.7
Maximum number of generations	20

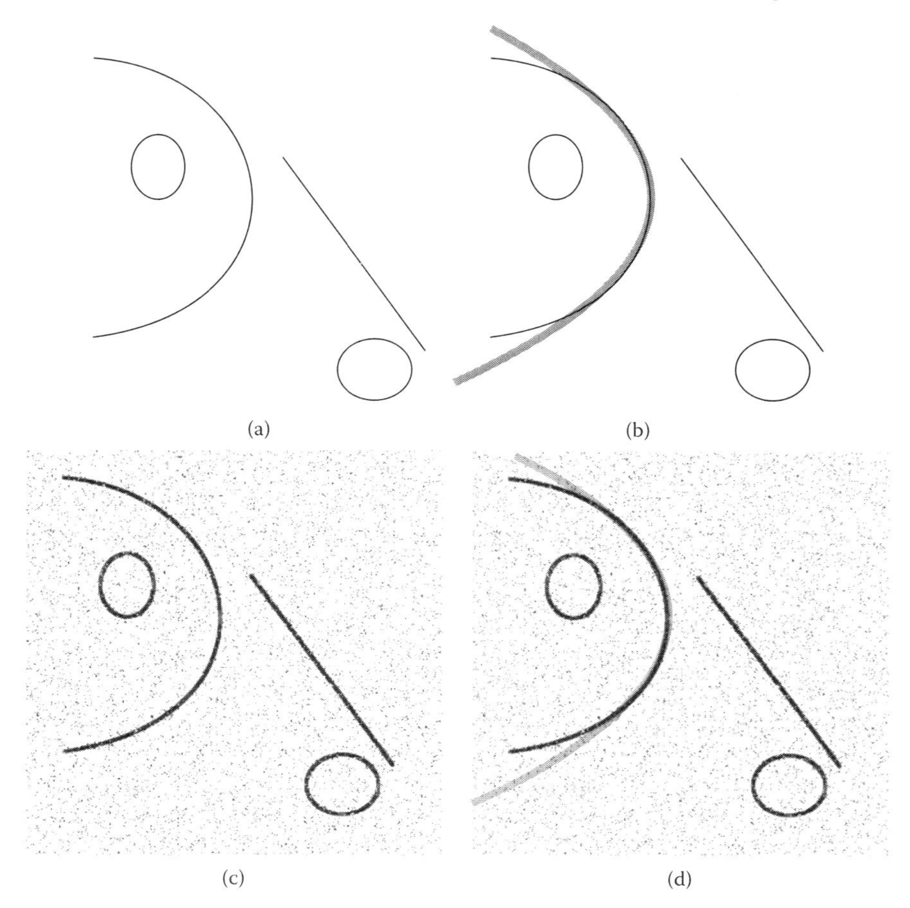

FIGURE 4.15: Parabola detection on synthetic image: (a) original synthetic image; (b) parabola detection result for the image in (a); (c) synthetic image with salt-and-pepper noise; and (d) parabola detection result for the image in (c).

TABLE 4.4: Comparative analysis of execution time using 30 runs on the binary image

Method	Execution time(s)		Number of iterations	
Proposed method	Minimum	0.5854	Minimum	9
	Maximum	5.3987	Maximum	26
	Mean	1.9006	Mean	11
	Median	1.6750	Median	14
Hough transform [7]	9.56 per pixel		-	
MIPAV [17]	33.56		-	

Table 4.5 presents the statistical results of 30 runs of the proposed method applied to Figure 4.15c. To compare the results, the image was tested with the method in [7], since this method works over a single pixel, the results are consistent with Table 4.4. The proposed method achieves a reduction of 96.14% in comparison with the MIPAV software. In this test, the average number of generations to deliver the best result (Figure 4.15d) was 152.

In order to evaluate the performance of the proposed method in terms of computational time with different amount of noise, a test by adding "salt & pepper" noise over the range

TABLE 4.5: Comparative analysis of execution time using 30 runs over the noisy binary image

Method	Execution time(s)		Number of iterations	
Proposed method	Minimum	0.7360	Minimum	11
	Maximum	4.6922	Maximum	25
	Mean	1.9815	Mean	15.27
	Median	1.8793	Median	18
Hough transform [7]	9.56 per pixel		-	
MIPAV [17]	34.92		-	

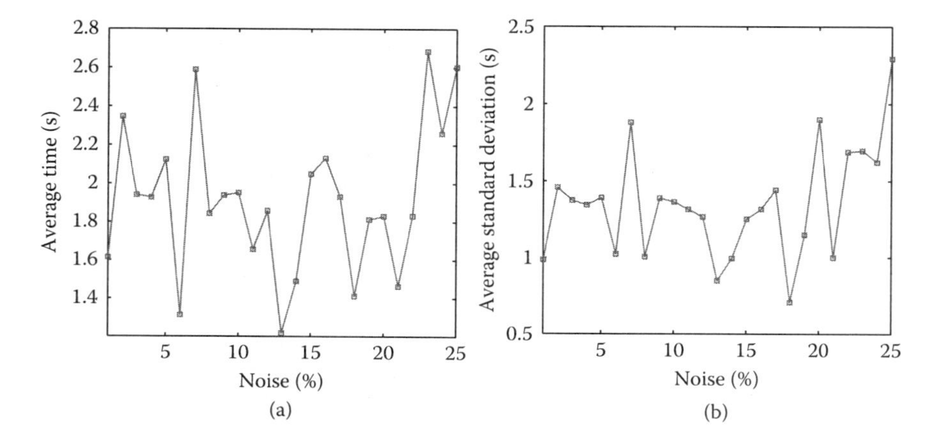

FIGURE 4.16: Performance analysis for the noisy synthetic image using 30 trials: (a) average time in seconds and (b) average standard deviation in seconds.

[1, 25] percent to Figure 4.15 was carried out. Figure 4.16a shows the mean execution-time of the obtained results, it is expected an increasing execution-time with the increasing amount of noise, however, considering that the UMDA method is a stochastic algorithm, there will be cases where the solution will be found in few generations. Figure 4.16b shows the standard deviation of the performed tests, since the graph tends to a constant mean over all the percentage of noise, it can be assumed that the algorithm is stable.

Several tests were performed to validate the algorithm, in most of the cases the algorithm achieves its maximum fitness value before the 30th iteration. Figure 4.17 shows the graph of the fitness value of a random test. It is shown that for this example the graph is stable at the 20th iteration. Since all the performed tests showed a similar behavior, it can be assumed that the chosen fitness function works well for this kind of image.

4.5.2 Retinal fundus images

There are some areas in the medical field that have been used in the detection of parametric objects to analyze some diseases. The detection of parabolas has been applied in retinal funds images with particular importance, since the form of the retinal vessels can be approximated to a parabola and the parameters as the vertex, helps to detect some sickness. In the work of Yu et al. [6], the retinal fundus images are approximated to a parabola to find the vertex of all vessels, this vertex is known as fovea [5]. This study is significant because the position of the vertex can give the grade of diabetic monitor. In the work of Oloumi et al. [19,20], the retinal vessels are approximated to a parabola to monitor the openness of the major temporal arcade (MTA). This study facilitates the quantitative analysis of the

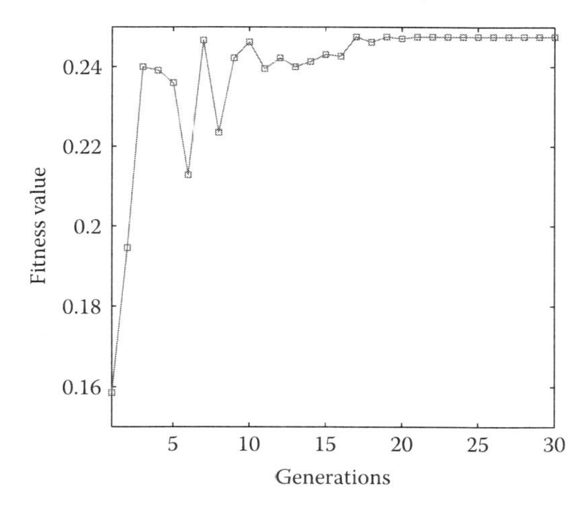

FIGURE 4.17: Fitness value through the generations of a random test for the noisy synthetic image.

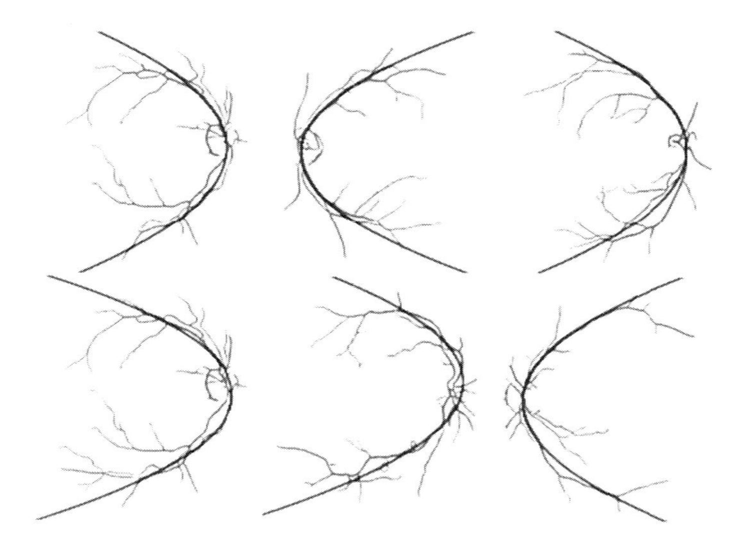

FIGURE 4.18: Approximation of parabolas in retinal fundus images.

MTA and overcomes limitations associated with manual analysis. The mentioned methods are used to search the HT to find a parabola on the images. Then, the method with the UMDA algorithm can be used to find the parametric object in lower computational time than the HT.

The results obtained are shown in Figure 4.18 where the parabola that best fits in the image is shown in black, and the retinal vessels are shown in gray. As in the work of Yu et al. [6], the public database DRIVE [21], which contains a set of 20 retinal fundus images, was used to test the proposed method.

Table 4.6 presents the statistical analysis of applying the method to the DRIVE dataset using 30 runs per image. The HT remains constant with the execution-time per pixel. The proposed method accomplishes the best result with only 174 iterations on average; then, it is possible to reduce the execution time to 90% in comparison with the performance using the MIPAV software.

TABLE 4.6: Comparative analysis of execution time using 30 runs on the binary image of retinal fundus images

Method	Execution time(s)		Number of iterations	
Proposed method	Minimum	1.5743	Minimum	5
	Maximum	7.6356	Maximum	24
	Mean	3.8976	Mean	17.43
	Median	4.7125	Median	18
Hough transform [7]	9.56 per pixel		-	
MIPAV [17]	39.23		-	

TABLE 4.7: Comparative analysis of execution time using 30 runs on the skeleton of the retinal fundus images

Method	Execution time(s)		Number of iterations	
Proposed method	Minimum	1.057	Minimum	4
	Maximum	5.1453	Maximum	23
	Mean	3.7612	Mean	13
	Median	2.7921	Median	15
Hough transform [7]	9.56 per pixel		-	
MIPAV [17]	36.8512		-	

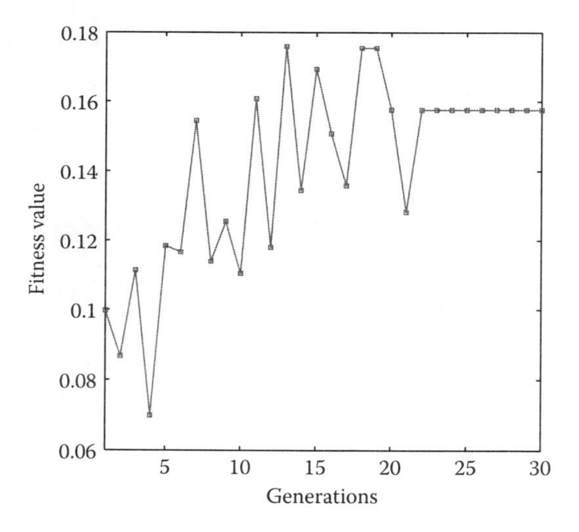

FIGURE 4.19: Fitness through the generations of a random test in a retinal fundus image.

Moreover, since the skeleton of an image passes through the center section of a set of pixels, it is expected that the skeleton contains lower pixels of interest than the binary image. Table 4.7 shows the statistics of the execution-time for the skeleton retinal fundus images. The results are congruent with those of the synthetic image; the proposed method clearly outperforms the execution time of the HT algorithm from Clara Sanchez and the MIPAV software.

The proposed method for parabola detection presents the advantage that the necessary number of iterations is not related to the size of the image, since the algorithm takes three interest pixels to perform the analysis. This assumption is validated with the results shown in Tables 4.6 and 4.7, where the mean iterations for the binary image and skeleton image is less than 30. Figure 4.19 shows a graph of the fitness value for a random test; it can be seen that the number of pixels of interest increases throughout the generations and for this particular example, the maximum value is reached at the 21st iteration. Since all the performed tests have a similar response, the maximization of the fitness function is a valid parameter to be applied to this type of images.

4.6 Conclusion

This chapter presents a novel method for the detection of parametric objects based on estimation of distribution algorithms (EDAs). A comparative analysis is carried out between the proposed method and the Hough transform (HT). The proposed method improves the computational time in the searching of parametric objects in images by selecting three pixels of interest of the input image. The algorithm achieves a significantly lower execution-time (0.008 seconds) than the HT since it is not necessary to calculate the accumulator. In addition, the proposed method is capable of testing several parameters in one run. The HT can vary only two parameters at time, and in the case of circles and parabolas, the third one must be constant during the iterative process. According to the computational experiments, the proposed method outperforms the two comparative methods in terms of execution-time by saving 94% on synthetic images and 90% on retinal fundus images.

Appendix 4.A

MATLAB code for line detection using Hough transform.

```
1   close all;
2   clear all;
3   clc;
4   %%%%%%%%%%%%%%%% Creating the image %%%%%%%%%%%%%%
5   im = zeros(128,128);
6   for i=1:64
7       im(16+i,92-i) = 1;
8       im(16+i,16+i) = 1;
9   end
10  figure;imshow(im);
11  theta = -90:90;
12  max_dist = round(sqrt(size(im,2)^2 + size(im,1)^2));
13  rho = -max_dist:max_dist;
14
15  %%%%%%%%%%%%%%%% Calculating the accumulator %%%%
16  cummulate = zeros(length(rho),length(theta));
17  for x = 1 : size(im,2)
18      for y = 1 : size(im,1)
19          if im(x,y) ~= 0
```

```
20                for z = 1 : length(theta)
21                    r = x*cos(theta(z)*pi/180) + y*sin(theta(z)*pi/180);
22                    [~,cumm_rho] = min(abs(rho - r));
23                    cummulate(cumm_rho,z) = ...
24                        cummulate(cumm_rho,z) + 1;
25                end
26            end
27        end
28    end
29
30    %%%%%%%%%%%%%%%%%%%% Normalizing %%%%%%%%%%%%%%%%%%%
31    cummulate = cummulate/max(max(cummulate));
32
33    %%%%%%%%%%%%%%%%%%%% Displaying %%%%%%%%%%%%%%%%%%%%%
34    figure;
35    mesh(cummulate)
36    xlabel('theta')
37    ylabel('rho')
38    zlabel('Magnitude')
39
40    [~,max_x] = max(max(cummulate));
41    [~,max_y] = max(cummulate(:,max_x));
42
43    k=1;
44    for cumm_x = 1 : size(cummulate,2)
45        for cumm_y = 1 : size(cummulate,1)
46            if cummulate(cumm_y,cumm_x) > cummulate(max_y,max_x)*0.85
47                rho_r(k) = rho(cumm_y);
48                ang_r(k) = theta(cumm_x);
49                k=k+1;
50            end
51        end
52    end
53
54    for i=1:k-1
55        if ang_r(i) == 0
56               x1 = [1 size(im,2)];
57               y1 = [rho_r(i) rho_r(i)];
58               figure(1);
59               line([x1(1) x1(2)],[y1(1) y1(2)])
60        else if ang_r(i) == 90 || ang_r(i) == -90
61                y1 = [1 size(im,2)];
62                x1 = [rho_r(i) rho_r(i)];
63                figure(1);
64                line([x1(1) x1(2)],[y1(1) y1(2)])
65                else
66                y1 = [1 size(im,1)];
67                x1 = (rho_r(i) - y1*sin(ang_r(i)*pi/180))/...
68                cos(ang_r(i)*pi/180);
69                figure(1);
70                line([x1(1) x1(2)],[y1(1) y1(2)])
71                end
72        end
73    end
```

References

1. P. Mukhopadhyay and B. B. Chaudhuri. A survey of Hough transform. *Pattern Recognition*, 48(3):993–1010, 2015.

2. R. O. Duda and P. E. Hart. Use of the Hough transformation to detect lines and curves in pictures. *Communications of the ACM*, 15(1):11–15, 1972.

3. T. J. Atherton and D. J. Kerbyson. Using phase to represent radius in the coherent circle Hough transform. In *IEE Colloquium on Hough Transforms*, pp. 5–11. IET, 1993, London, UK.

4. R. G. Frykberg, L. A. Lavery, H. Pham, C. Harvey, L. Harkless, and A. Veves. Role of neuropathy and high foot pressures in diabetic foot ulceration. *Diabetes Care*, 21(10):1714–1719, 1998.

5. M. E. Gegundez-Arias, D. Marin, J. M. Bravo, and A. Suero. Locating the fovea center position in digital fundus images using thresholding and feature extraction techniques. *Computerized Medical Imaging and Graphics*, 37(5):386–393, 2013.

6. C.-Y. Yu, C.-C. Liu, and S.-S. Yu. A fovea localization scheme using vessel origin-based parabolic model. *Algorithms*, 7(3):456–470, 2014.

7. C. Sanchez. Parabola Detection Using Hough Transform. http://www.mathworks.com /matlabcentral/fileexchange/15841-parabola-detection-using-hough-transform, 2007.

8. P. Larranaga and J. Lozano. *Estimation of Distribution Algorithms: A New Tool for Evolutionary Computation*, vol. 2. New York: Springer Science & Business Media, 2002.

9. P. Larranaga. A review on estimation of distribution algorithms. In *Estimation of Distribution Algorithms*, pp. 57–100. New York: Springer, 2002.

10. M. Hauschild and M. Pelikan. An introduction and survey of estimation of distribution algorithms. *Swarm and Evolutionary Computation*, 1(3):111–128, 2011.

11. M. Pelikan, D. E Goldberg, and F. G Lobo. A survey of optimization by building and using probabilistic models. *Computational Optimization and Applications*, 21(1):5–20, 2002.

12. H. Mühlenbein and G. Paass. From recombination of genes to the estimation of distributions I. Binary parameters. In *PPSN IV Proceedings of the 4th International Conference on Parallel Problem Solving from Nature*, pp. 178–187. Berlin, Germany: Springer, 1996.

13. I. Cruz-Aceves, A. Hernandez-Aguirre, and S. Ivvan Valdez. On the performance of nature inspired algorithms for the automatic segmentation of coronary arteries using Gaussian matched filters. *Applied Soft Computing*, 46(1):665–676, 2016.

14. I. Cruz-Aceves, A. Hernandez-Aguirre, and S. Ivvan Valdez. Automatic coronary artery segmentation based on matched filters and estimation of distribution algorithms. In *Proceedings of the 2015 International Conference on Image Processing, Computer Vision, & Pattern Recognition (IPCV'2015)*, pp. 405–410, 2016.

15. E. Million. The Hadamard product. *Course Notes*, 3, 2007.

16. V. Ayala-Ramirez, C. H. Garcia-Capulin, A. Perez-Garcia, and R. E. Sanchez-Yanez. Circle detection on images using genetic algorithms. *Pattern Recognition Letters*, 27(6):652–657, 2006.

17. M. J. McAuliffe, F. M. Lalonde, D. McGarry, W. Gandler, K. Csaky, and B. L. Trus. Medical image processing, analysis and visualization in clinical research. In *Proceedings of 14th IEEE Symposium on Computer-Based Medical Systems (CBMS 2001)*, pp. 381–386. Bethesda, MD: IEEE, 2001.

18. National Institutes of Health Center for Information Technology. Medical Image Processing, Analysis and Visualization. http://mipav.cit.nih. gov/index.php, 2015.

19. F. Oloumi, R. M. Rangayyan, and A. L. Ells. Parabolic modeling of the major temporal arcade in retinal fundus images. *IEEE Transactions on Instrumentation and Measurement*, 61(7):1825–1838, 2012.

20. F. Oloumi, R. M. Rangayyan, and A. L. Ells. Computer-aided diagnosis of proliferative diabetic retinopathy. In *2012 Annual International Conference of the IEEE Engineering in Medicine and Biology Society*, San Diego, CA, pp. 1438–1441, 2012.

21. J. J. Staal, M. D. Abramoff, M. Niemeijer, M. A. Viergever, and B. van Ginneken. Ridge based vessel segmentation in color images of the retina. *IEEE Transactions on Medical Imaging*, 23(4):501–509, 2004.

Chapter 5

Hybrid Approach towards Malaria Parasites Detection from Thin Blood Smear Image

Sanjay Nag, Nabanita Basu, and Samir Kumar Bandyopadhyay

5.1 Introduction

Malaria is a life-threatening vector-borne parasitic disease, caused by the protozoan parasites of the genus *Plasmodium* and is transmitted through the bite of a female Anopheles mosquito. The disease is most often curable when there is early detection, proper diagnosis of the disease, and application of simple oral drugs. However, the spread of this infective disease is rapid with the abundance of the disease carrying vector, very often leading to pandemic proportions within a very short time. The disease management often fails and gets out of control with an increased number of reported cases within a locality. Due to a lack of health infrastructure to detect, diagnose, and cure suffering patients, it can result in a serious medical issue in underdeveloped nations. The disease is preventable but the inability to eradicate the breeding regions of the mosquito elevates the chances of the disease acquiring epidemic proportions. Apart from a few types of malaria that are life threatening, most types are curable but they cause massive suffering to the infected patient. Complications sometimes lead to increased mortality in regions where high malnutrition is reported especially in rural backward regions of developing nations. Such complications may lead to secondary infections, blood-deficiency-related health issues mostly for the elderly patients, and is a primary reason for child mortality with this disease.

Malaria has been a common life-threatening disease for over 50,000 years, which thereby makes it historically significant. The disease was nicknamed "King of Diseases" [1] due to the high mortality rate that occurred within a short duration. It has been reported as a dreadful disease in warfare zones with more soldiers succumbing to the disease rather than in battle. It was named in 1740 from the Italian word, *malaria*, or 'bad air,' and was also known as the Roman Fever [2] as it originated from marshy regions of Rome. It is considered that the disease actually originated from tropical regions of Africa and later spread to Mediterranean Europe and to tropical regions within Asia, mostly South Asian Countries like India. The incident of the disease spread across the globe, even regions beyond the tropics with travelers porting the disease from endemic zones. Frequent travel to such malaria-infested regions has resulted in the widespread occurrence of the disease, with travelers rather than mosquitoes being the vector for disease transmission.

The origin of the disease was traced back to Africa, which later got propagated to the Mediterranean Europe, Southeast Asia, and India. The identification of the cause and effect of mosquito bites and malaria disease dates back to 800 BC when the Indian sage Dhanvantari wrote that mosquito bites lead to fever and shivering. *Charaka Samhita*, an Indian medical text written in 300 BC, classified fever into five categories: continuous, remittent, quotidian, tertian, and quartan fever. *Susrut Samhita*, written in 100 BC by Susruta, an ancient Indian doctor, stated that fevers can be associated with insect bites. Hippocrates in 400 BC studied malaria in detail and in his text *Hippocratic Corpus* differentiated malaria fever and was the first document to associate changes in human spleen. In 1717, Lancici, a Roman scientist and doctor, reported the spread of malaria by a mosquito and as a preventive measure, he suggested that in the marshy areas the stagnant waters to be drained to eradicate malaria. Malaria was finally identified by French physician Charles Louis Alphonse Laveran in 1880 as some moving organisms in red blood cells from a thin smear slide under a light microscope. He identified the presence of gametocytes and trophozoites in blood cells.

5.1.1 Socioeconomic impact

The disease have profound impact on the livelihood of people in endemic regions of the disease. The extensive economic effect of the disease is not only felt in the household but they tend to affect the financial infrastructure of nations and are definitely responsible for the economic backwardness of the countries. This can further be attributed to the fact that insufficient resources, lack of basic amenities, limited skills, insecurity, and lack of power can be related to poverty that leads to poor access to health care [3]. Poor living conditions like overcrowding and improper housing increase susceptibility towards various infections, including malaria [4]. Malnutrition among women and children results in the poor physical development of children who become prone to a disease like malaria. Illiteracy is profoundly present among the poor which prevents them from being aware of health-related issues. Lack of a sufficient health-care infrastructure available in remote areas without adequate equipment, medicine, or trained staff [5] results in low-quality health service and often becomes insufficient to tackle malaria outbreak. Malaria causes financial loss incurred by the family due to direct costs (cost of treatment and medication), indirect costs (loss of working hand), and opportunity costs (the financial gain by an earning individual). The citation [6] reports that an estimated mean direct cost on malaria treatment by a household can be up to 2%–2.9% of annual income. The disease is responsible for making poor countries even poorer with a decrease in gross domestic product (GDP) value. Studies by [7] suggest that morbidity due to malaria infection reduces the annual per capita growth by 0.25%

points in malaria endemic countries. Economic review by Sachs and Malaney concludes that "where malaria prospers most, human societies have prospered least" [8].

5.1.2 Malaria fact-sheet

Malaria affects 40% of the population across 100 countries who are in danger of getting infected. It is estimated that 500 million get infected annually, resulting in more than one million and less than two million deaths, with infants below the age of 5 years reported as having the highest mortality and greater share in sub-Saharan Africa [9]. Malaria is on top of the list for causing deaths and morbidity in tropical and subtropical regions with 1–2 million deaths per year [10]. The disease is known to infect more than 200 million people each year with the annual mortality rate of approximately three million per annum [11,12]. A malaria report released by WHO in 2011 [13] indicated that almost half the population of the world could have been victims of malaria. The mortality rate of 655,000 people was reported in 2010, among them, 86% were infants below five years [13]. The World Malaria Report of 2015 [14] is a culmination of data taken from 95 countries where malaria persists and another six countries where malaria has recently being eradicated. The report states that approximately 214 million infections (149–303 million) of malaria reported worldwide in 2015. The African continent share among all reported cases is 88%, another 10% reported in Southeast Asia, and only 2% reported case in Eastern Mediterranean Region. In the year 2015, approximately 438,000 mortalities (236,000–635,000) were reported worldwide. Ninety percent mortality was reported from Africa, 7% from Southeast Asia, and 2% from Eastern Mediterranean region.

5.1.3 Malaria parasite

The malaria parasites belong to the phylum Apicomplexa, class Sporozoea, subclass Coccidia, order Eucoccida and suborder Haemosporina. Under this suborder is genus Plasmodium. This genus is characterized by the presence of two hosts in the life cycle with schizogony (asexual cycle) and sporogony (sexual cycle). There are five species that cause a malaria infection in human beings. The disease is most prevalent in the tropical regions of the world between 60°N and 40°S. The parasite resides in two hosts. Within the human host the parasite undergoes an asexual lifecycle and within female Anopheles mosquitoes, it undergoes a sexual life cycle. Malaria protozoa have several forms within their life cycle. The parasite infects the human system as sporozoite. Table 5.1 shows the differences between the different life-forms of the parasite within a human host. Figure 5.1 [15] shows the life-forms in different stages recorded by a digital microcope.

5.1.4 Composition of human blood

In mammals, oxidation of tissues is performed by binding the oxygen molecules to the hemoglobin contained within the circulating discoid shaped and non-nucleated red blood cells (RBCs) or erythrocytes. Malaria infects and lives a parasitic life within the host RBC and destroys the same eventually. For computerized algorithms identification and differentiation of infected and normal RBCs are vital. The possible hinderence in the segmentation process are the presence of white blood cells (WBCs) or leukocytes that are the primary defense system of the body against infections. In normal circulation of blood, there are approximately 7000 WBCs per microliter of blood. They constitute about 1% of the total blood volume in healthy adult individuals. Based on the presence/absence of granules (sacs containing digestive enzymes or other chemical substances) in the cytoplasm, the WBC is

TABLE 5.1: Comparison of morphological characteristics of Plasmodium species during different stages of its life cycle

Plasmodium	*P. vivax*	*P. falciparum*	*P. malariae*	*P. ovale*
Trophozoite	Benign tertian malaria Early trophozoite have blue cytoplasmic ring, red nuclear mass & vacuole. RBC enlarges and irregular with Schüffner dots.	Malignant tertian malaria Early ring forms with fine and uniform cytoplasm ring with nucleus lying outside the ring, often divided into two parts and at opposite poles. RBC remains normal with 6–12 Maurer's cleft.	Quartan malaria They are similar to vivax and assume a band-like shape and coarse brown to black pigment appears in cytoplasm. RBC is not enlarged Ziemann's dot on prolonged staining.	Tertian malaria Early ring forms similar to malariae but without the band shape. Dark brown pigment in cytoplasm. RBC enlarged, irregular in shape with James dots.
Schizont	They almost fill the enlarged RBC; the nucleus is large and lies on the periphery. After nuclear division on average 16 daughter individual form a rossete-like cluster. RBC burst at this stage.	They fill two third of RBC. After nuclear division 8–32 daughter cell produced. RBC remains unenlarged. They burst to release the cells.	The plasmodium fills the RBC. On nuclear division 6–12 daughter cells are arranged around a central mass. RBC remains unenlarged and bursts at maturity.	The plasmodium fills three quarters of RBC. On nuclear division 6–12 daughter cells are arranged irregularly. RBC remains slightly enlarged before bursting.
Merozoite	12–24 cytoplasm containing oval mass.	18–24 cytoplasm containing circular mass.	6–12 cytoplasm containing mass.	6–12 cytoplasm containing crescent mass.
Gametocyte	Spherical in shape and slightly enlarged RBC containing granules.	Host RBC is filled and the gametocyte is crescent shaped.	Same size of the host RBC. They are round in shape.	Round in shape and host RBC is slightly enlarged.

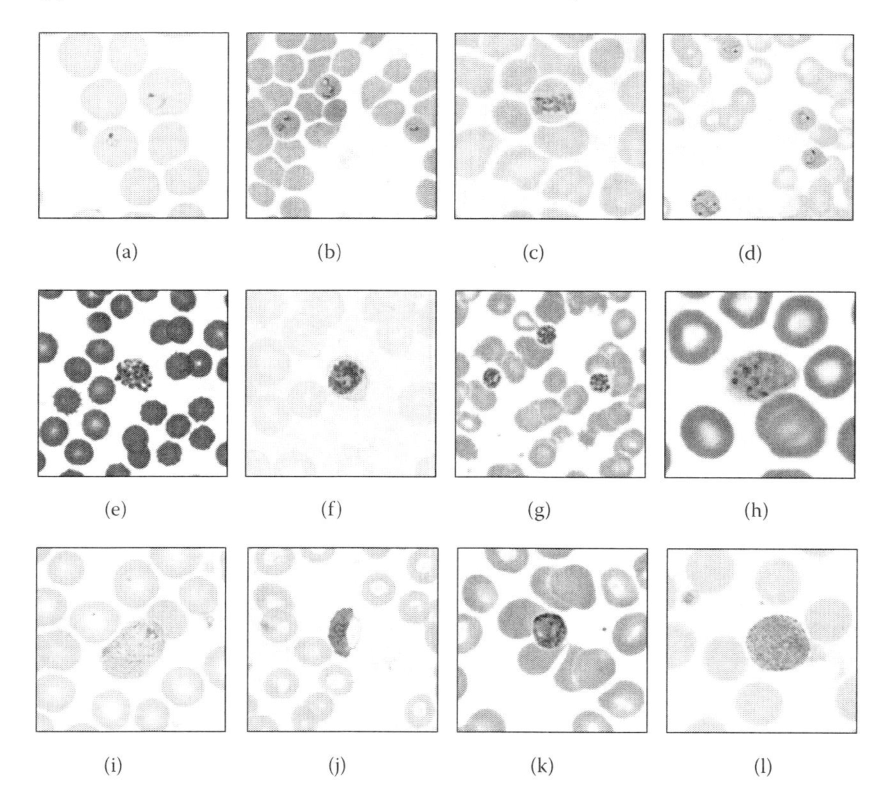

FIGURE 5.1: A Comparison of different asexual life-cycle forms of *Plasmodium* genus. (a)–(d): trophozoite of *P. vivax, P. falciparum, P. malariae* and *P. ovale*, respectively (e)–(h): mature schizont of *P. vivax, P. falciparum, P. malariae* and *P. ovale*, respectively (i)–(l) Gametocyte of *P. vivax, P. falciparum, P. malariae* and *P. ovale*, respectively. (From CDC DPDx. 2016. Available at http://www.dpd.cdc.gov/dpdx/html/imagelibrary/malaria il.htm.)

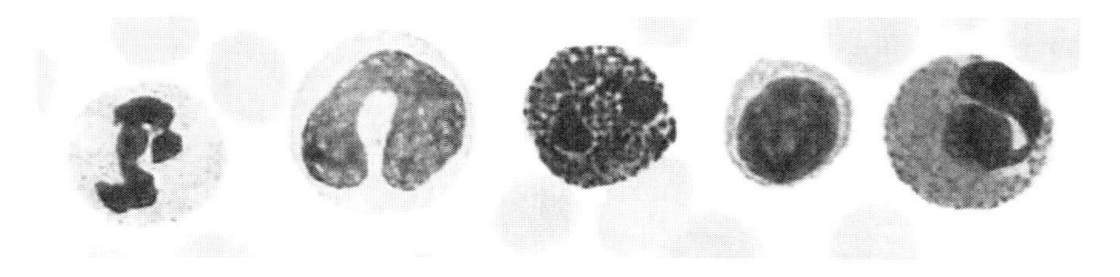

FIGURE 5.2: The figure shows five different types of WBC with neutrophil at the leftmost position, followed by monocyte, basophil, lymphocyte, and eosinophil at the extreme right. This rare image was recorded by a hematologist while scanning through smear slides.

considered to be a granulocyte or agranulocyte. Neutrophils, basophils and eosinophils are granulocytes that differ in the lobular structure of nucleus. Lymphocytes and monocytes have single-lobed nucleus and void of granules and constitutes agranulocytes.

Figure 5.2 shows all the different types of WBC that are found within a smear slide image.

5.1.5 Diagnosis of malaria

The clinical diagnosis of malaria is performed by a medical practitioner. The laboratory diagnosis of malaria involves phlebotomy, and the collected blood sample is analyzed using scientific equipment or chemicals to identify the presence of the malaria parasite. After the discovery of malaria, detection of the parasite was performed through observation with conventional light microscope on blood smear slide stained with Giemsa, Wright's, or Field's stains [16]. Figure 5.3 [17] shows the prepared thick and thin smear slides. This method remains the 'gold standard' for laboratory diagnosis of malaria [18]. Parasitaemia is calculated with both types, thin and thick smear [19]. The WHO practical microscopy guide for malaria lists all the procedures to be followed by laboratory technicians [19]. The system is popular for its simplicity, economical viability, and visual differentiation between normal and parasite-infected RBC. However, the preparation of slides is laborious, time-consuming, and requires skillful laboratory technicians. For the pathologist, detection and species identification at low parasitaemia is challenging.

There are several other methods, like the rapid diagnostic test (RDT) method, to quickly detect malaria. The kit detects the presence of particular species-specific antigens in blood. Other popular methods include the quantitative buffy coat test (QBC) and polymerase chain reaction (PCR) method that can amplify trace amounts of DNA present for the diagnostic purpose. A flow cytometer device is a cell sorter device that is used to detect the presence of malaria pigment haemozoin. A mass spectrometry (MS) device is similarly used to identify haemozoin to detect presence of the malaria parasite.

5.1.6 Digital-microscopy–based CAD methods

The accuracy of detection in the case of conventional microsopic examination is heavily dependent on the heuristic knowledge of the pathologists and shows interoperator/intraoperator variability. Moreover, evaluation by pathologists is time-consuming and prone to error of judgement. Other popular techniques like the use of RDTs provide a rapid diagnosis, but the sensitivity is lower than microscopic evaluation. They are specific for a particular species. Most of the advanced systems discussed require costly setup and equipment. Maintenance of such elaborate systems in remote locations is not possible. Advancement in

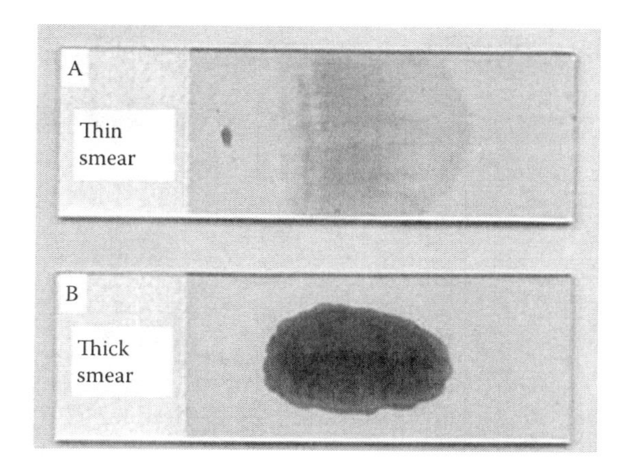

FIGURE 5.3: Showing peripheral smear on glass slide with and without staining taken for the purpose of microscopic studies. (From Web image available at http://textbookhaematology4medical-scientist.blogspot.in/2014/03/thick-and-thin-smears-for-microscopy.html.)

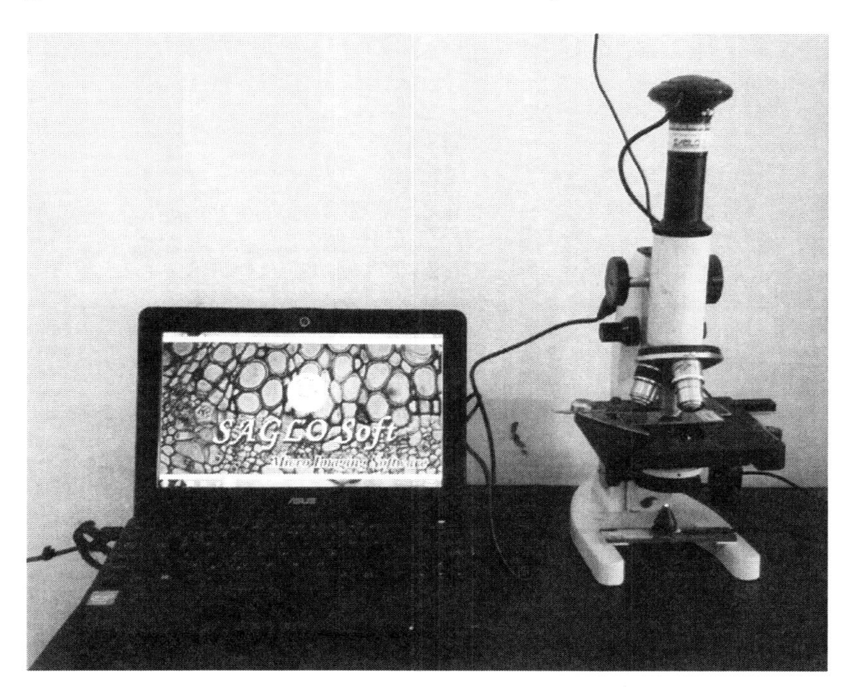

FIGURE 5.4: Digital microscope setup. (From Wikipedia, available at https://en.wikipedia.org/wiki/Digital_microscope#/media/File:Patented_Digital_Micro-imaging_Adaptor_with_SAGLO_Soft_Software_for_Microscopy_developed_by_Inventor_Sachin_G_Lokapure_%28_SAGLO_Research_Equipments%29 3.jpg.)

microscopy, coupled with a digital camera and connecting it to a computer system using firmware has ushered a digital era in a microscopic evaluation. The contemporary digital microscope contains built-in LED lighting systems with an array of light filters and fluorescent light for microscopy as shown in Figure 5.4 [20]. Most of the illumination issues in conventional reflected light microscopy are solved with the use of a built-in illumination system. The modern digital camera has high resolution. Modern microscope systems also contain a motorized stage that can be remotely controlled through software for automated image capturing. The development of CAD software for malaria detection and diagnosis is one of many applications of digital microscopy and contributing towards the era of digital pathology.

This research work intends to propose a method for cellular segmentation and image analysis using machine learning-based hybrid systems. A digital image is used as a source for performing image analysis using image processing techniques to obtain features to be applied to intelligent algorithms. This information will be used to train the system so that the system will derive its own rules to classify the image accordingly.

5.2 Literature Review

The intervention of technology to assist pathologists and medical practitioners plays an important role in the fight against malaria and achieves accurate diagnoses to prevent mortality. The biologists and the chemists were busy discovering new products and means to control the disease. Advancement in microscopy and computer technology has bolstered this effort. Several kinds of literature can be found in the research domain that have contributed to the development of CAD-based systems to detect malaria parasites from a blood smear slide image. Several authors have proposed new algorithms and have compared different methodologies to establish more efficient algorithms to identify and diagnose malaria. Some notable work is described in the following paragraphs in this section.

There are several research works that utilize image processing tools for identifying the malaria parasite from a thin smear digitized slide image. A generalized scheme is shown in Figure 5.5. Other research work utilizes the machine learning technique to segment the image into the cellular component and classifies whether there exists a malaria parasite or they are normal. Different research work has utilized different feature sets to achieve segmentation and classification. Figure 5.6 shows a generalize scheme of a process flow that is applied by different research work. The different feature sets may include a texture-based feature, geometric features, and intensity-based features extracted from different color spaces like RGB, HSV, and Lab color models.

The citation by Selena W.S. Sio et al. [21] proposed the software MalariaCount for automated parasitaemia counts using the image obtained from thin blood smear slide images. The image preprocessing included an adaptive histogram equalization. An edge correlation coefficient was used to extract the edge. The edges are linked to form the

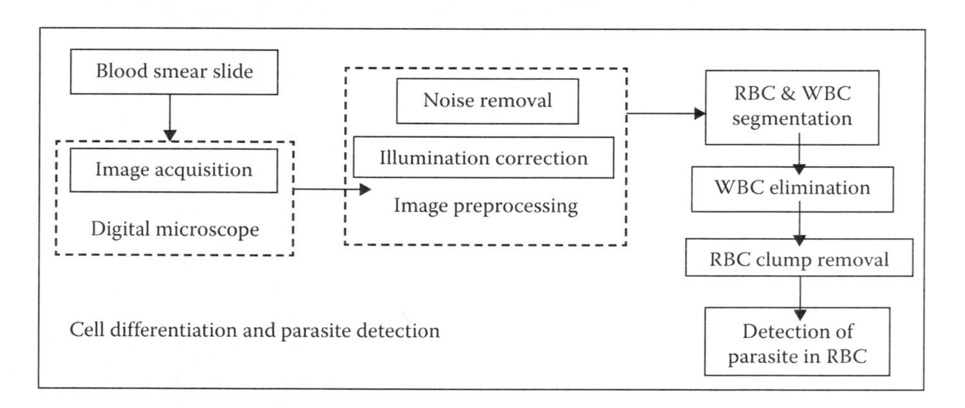

FIGURE 5.5: Generalized schematic diagram of CAD systems that follow simple image processing techniques.

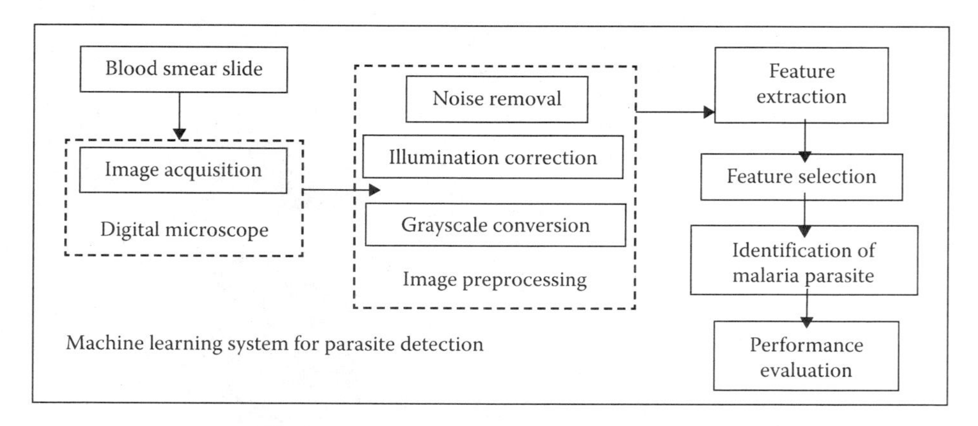

FIGURE 5.6: A generalized schematic diagram of CAD systems that employ machine learning techniques for cell segmentation and parasite detection.

contours for obtaining a closed object. Cell clumping was removed and cells were separated by implementing the clump splitting method proposed in [22]. The image processing was performed using a MATLAB® toolkit. For testing, the authors compared the results of the software with manual counting by experts. The authors used 76 smear slides images. The results, as claimed by the authors, showed a high correlation between manual and MalariaCount software.

The citation by John A. Frean [23] used open-access software, ImageJ, that assists with manual counting for diagnosis of malaria. The objective of the author was to remove the intra- and interobserver variations. Digital images from 20 slides were captured with microscope objective of 50x resolution. A Point Picker plug-in [24] was used to digitally tag the recorded parasites. The Particle Analysis command was used for digitally counting the parasites. The author performed adjustments of brightness and contrast of images for better thresholding results.

The research method proposed by Zheng [25] was used to the detect *falciparum* malaria parasite. The images were first transformed to HSV color space and a mask was generated using the 'V' component. Histogram equalization was done to distribute the grayscale color space. Otsu binarization was performed to create the cell area mask. A canny edge operator was used to obtain the edge image. Further morphological operations of dilations and erosion were performed by the authors. The 'H' and 'S' components were used to identify cell region and parasites. The author then generated six circular ring masks that were slid across the window to identify a circular region of a cell. The percentage of line segments that was greater than 20% we as treated as cell structure. The algorithm was tested with five images with low noise. Results were satisfactory, as claimed by the author.

The authors Somasekar et al., in their research citation [26], have proposed an intensity-based method with the use of morphological operators to differentiate the malaria-infected cells. The image was converted to grayscale image and noise reduction was done using a 5×5 median filter. The intensity-based extraction technique proposed by the authors was to suppress the background and highlight the high-intensity foreground region. Hole filling of cells was done using morphological operators and finally, a morphological erosion operation was performed to isolate the infected cells. The citation reported that the algorithm achieved sensitivity and specificity respective values of 94.87% and 97.3%, respectively, for parasite isolation. According to the authors, the system sometimes failed to distinguish between malaria-infected cells and WBC. This was a major drawback to their proposed algorithm.

The authors Ghosh et al. [27] proposed a method with the objective to enhance the image and filter out unnecessary regions so as to identify the presence of parasites. The image preprocessing was done using the Laplacian filter to sharpen the edges. The image was converted to grayscale. A binary image was obtained by comparing an empirically defined threshold value. The morphological operation of closing was applied in which the first dilation was done followed by erosion to remove small contour regions. A gradient operator was applied to identify rough regions depicted by the presence of parasite over smooth regions. The authors tested the algorithms on 160 images with a detection accuracy of 98.125%.

The authors Cecilia Di Ruberto et al. [28] proposed a composite method to detect and classify malaria parasites. The authors performed cell segmentation by using methods proposed by them in citation [29] where they used the morphological operator and thresholding for segmentation. For classification of parasites, the authors have used two methods involving color histogram similarity and morphological operators. The authors used the HSV color model for segmentation. Morphological area granulometry was implemented with the morphological opening operation to determine the area distribution histogram of RBC. The presence of parasite was determined using thresholding from the histogram. WBC elimination was performed by implementing morphological erosion. Cell clumping was removed with the combination of granulometry and the application of morphological operations. The classification was done with the morphological thinning process called skeletonization. The authors tested the proposed methods on images and compared the results with manual counting by experts providing satisfactory results.

The authors Mehrjou et al. [30] proposed an automatic mechatronics system to detect malaria and to overcome the difficulties of manual methods. The research work focussed on determining parasitaemia by quantifying the number of affected RBC. The author P. T. Suradkar in her research citation [31] proposed a method using color based thresholding. The authors Parkhi et al. [32] proposed an automated parasite detection module based on linear programming. Ghate et al. proposed two different models for automatic parasitaemia estimation algorithms [33]. The first model performs segmentation and dilation and the other model creates a template by using the average of texture features of the dataset. The recognition phase matches the template cell with the test case to find the minimum distance to a label.

The research citation [34] dealt with the problem of detecting the malaria parasite form thick smear images. The authors proposed two separate techniques for segmentation. The first technique involved the use of a morphological method. The slide image was preprocessed to grayscale and was followed by binarization using Zack thresholding. Further, a morphological operation closing with initial dilation followed by erosion was performed. The second technique used by the authors implemented the HSV color model where an empirically derived threshold segmented the background and the target region comprised of malaria-infected cells. Malaria-infected RBCs were segmented based on the porosity value given by the Euler number. Finally, the authors combined the two algorithms. The obtained percentage of detection for the two algorithms and the combined methods were 79%, 90%, and 94.5%, respectively.

The authors J.E. Arco et al. in the research citation [35] proposed a new method based on the image processing technique for enumeration of parasite-infected RBCs. The preprocessing stage included Gaussian filtering for noise reduction followed by adaptive histogram equalization (CLAHE) performed for contrast enhancement. This was followed by image binarization. This was followed by the mathematical morphology operation of closing for hole removal and finally, parasite enumeration. The system yielded an average accuracy value of 96.46%.

Several algorithms have been proposed by authors that implement machine learning methods to identify and classify the malaria parasite. Diaz et al. [36] proposed a semi-automatic method for Parasitaemia quantification from thin blood films. The RGB color space was transformed into chrominance and luminance channels. The luminance values were corrected and added to the chrominance values to obtain the luminance-corrected image. To segregate the erythrocytes, a labeling process was implemented using the k-nearest neighbor (k-NN) classifier that labeled each pixel as erythrocyte or background. The feature was extracted by partitioning the image into the blocks and generating histograms with R and G components, HSV color components, grayscale histogram, and Tamura texture. To reduce the feature space only mean, standard deviation, skewness, kurtosis, and entropy values of the histogram were taken. A total of 25 features were considered. A two-stage classification was implemented to classify abnormal and healthy erythrocyte and stage determination for the abnormal cells. The infected ones were then fed across three learning models for each stage of disease and a fourth learning model for the artifact. The authors have implemented the multilayer perceptron neural network as described in [37] and the support vector machine (SVM) [38] and a nonlinear classifier for evaluation. The SVM with a polynomial kernel of two degrees and complexity of three yielded better results than MLP. This classifier achieved a sensitivity of 94% and a specificity of 99.7% in detection of infected erythrocyte on a dataset of 450 images containing 12,557 erythrocytes and parasitaemia of 5.6%.

The authors F. Boray Tek et al. [39] proposed a new parasite detection algorithm that modified K-Nearest Neighbor (K-NN) classifier and validated the performance using a Bayesian method. The authors further performed three different classifications for detection of parasite, stage of life cycle, and type of infection using a single multiclass classifier. Segmentation was done using the Rao's method [40], using the average cell area value and double thresholding [41] to obtain a foreground mask. The authors implemented area granulometry based cell size estimation studies. For feature extraction, the proposed model extracted a color histogram quantized to 32 colors, local area granulometry and six ratios as shape metrics were used. The system extracted 83 features as proposed by the authors in citation [42]. For a multiclass classification, the authors used K-NN. The authors performed a 20-class (detection, species and stage), 16-class (species and stage), and 4-class species and 4-class stage classification. The identification experiments were performed using the hold-out and leave-one-out cross-validation methods. For the evaluation of proposed model Fisher linear discriminant (FLD) and the back-propagation neural network (BPNN) classifiers were used with sufficient training and test data. The results of the proposed model showed accuracy of 93.3%, sensitivity of 72.4%, and specificity of 97.6%.

The research citation of Savkare et al. [43] calculated parasitaemia from thin slide smear images, using features like color; geometrical features like radius, perimeter, and area; compactness metrics; and statistical features like skewness, kurtosis, energy, and standard deviation, that provided distinct variation between normal and infected cells to be used for training purposes. For classification purposes, a linear SVM was used. All the features were organized in decreasing order and only the most important features were selected for each pair of classes. The authors reported sensitivity of the proposed system to be 93.12%, and specificity is 93.17%. The research citation [44] proposed a similar method for determination of parasitaemia in falciparum malaria infection using the SVM classifier. Morphological features like the geometrical parameters of radius, perimeter, and compactness were used. The intensity histogram of the green plane was used for classification. The mean, variance, skewness, standard deviation, and kurtosis histograms were used as the features for the SVM classifier.

The authors Ahirwar et al. [45] proposed an automated malaria detection and classification method from thin blood smear slide images using artificial neural networks (ANNs). The authors preprocessed the images using the SUSAN approach and then authors employed granulometry. A two-feature set was generated: one was based on the morphological characteristics, color attributes, and texture, and the other feature was the a priori knowledge of different measures of the parasite. The infected RBC was identified using the back-propagation feed forward (BFF) neural network. The color and texture of the parasites were used as training features. The training set included 77 images containing both malaria parasites and normal images.

The authors Nasir et al. in the research citation [46] proposed a method using unsupervised clustering using moving K-means (MKM) using the parameters of the HSI color model The segmentation achieved with the S component of HSI provided the best segmentation results of accuracy 99.49%, sensitivity 92.14%, and specificity 99.79%. The citation of Abdul Nasir et al. [47] performed parasite detection using different color models, namely, RGB, HSI, and C-Y and unsupervised learning using the K-means clustering algorithm. The C-Y model with the S component yielded accuracy and specificity of 99.46% and 99.95%, respectively.

The authors Suryawanshi et al. [48] in the research citation compared the performance of two classifiers for effective identification of the malaria parasite. The authors trained two classifiers, the Euclidean distance classifier and the SVM classifier. Training of the classification model was done with a set of 60 images and testing on a balanced dataset of 30 images of which 15 contained the malaria parasite. For SVM accuracy, sensitivity is 93.33%.

The authors Kurer et al. in the citation [49] proposed a method using the SUSAN filter and the image was trained using the probabilistic neural network (PNN). The authors claimed that the system's overall performance provided a sensitivity of 99% and a positive predictive value of 90%–92%.

The authors Chayadevi et al. [50] in the research citation proposed a method using color parameters and fractal features. The authors extracted 80 features used for training the system. For the purpose of classification, the authors employed four different types of classifiers, namely, the adaptive resonance theory (ART) based neural network, neural network based back-propagation feed forward (NN-BPFF), SVM and K-NN. The authors performed a comparative study based on the performance of the different classifiers. The best results obtained were that of accuracy, 94.45%; precision, 96.41%; specificity, 94.68%; and sensitivity, 94.32%. ROC curve analysis was also performed with the area under the curve of 0.9847.

The author Razzak in his research citation [51] proposed an ANN-based classification model for malaria segmentation and detection. The author obtained GLCM-based 28 texture features. To classify malaria-infected cells, the authors used back propagation ANNs.

The authors Bahendwar et al. [52] proposed a segmentation algorithm for malaria parasite isolation based on ANN. The multilayered neural network model was trained using the RGB feature of erythrocytes to differentiate into erythrocytes region and the image background. Two sets of feature vectors were used with one set of RGB features and the other with RGB and HSI features. The authors observed that the green color component of RGB image as the best feature for segmentation and RGB features excelled over the other set containing HSI features with RGB.

5.3 Methodology

Detection of malaria parasites in thin blood smears can particularly be classified into nine broad phases, namely, data acquisition, image preprocessing, image segmentation, feature selection/extraction, feature standardization and normalization, optimal feature subset selection, shuffling of datapoints, cross-validation, and classification.

Thin blood smear images were acquired from the public MaMic database [53]. Images were acquired at 40x magnification and 25-watt illumination. Once acquired, image noise was corrected. In total, a set of 250 images were used for the study. Of the 250 images acquired, 125 images consisted of blood infected with malaria parasites, while the other images consisted of blood smears taken from individuals not infected with malaria.

Images (of size 1387×932 $pixel^2$) acquired from the MaMic database consisted of salt and pepper noise. To correct salt and pepper noise, 2D median filtering with a 3×3 window was performed. Once noise corrected, the RGB 'JPEG' images were converted to lab color space images. Based on the L, a and b components, unsupervised K-means clustering was performed to segment out RBCs from the Geimsa-stained thin blood smear images. Figure 5.7 illustrates the cluster structures formed based on the pixel values of the image represented in Lab color space.

Based on empirical work, as also experienced, K-means clustering algorithm was designed to produce three clusters (i.e., $K = 3$). While cluster 3 consisted only of the background pixels, cluster 1 consisted of RBCs and cluster 2 consisted of WBCs and platelets.

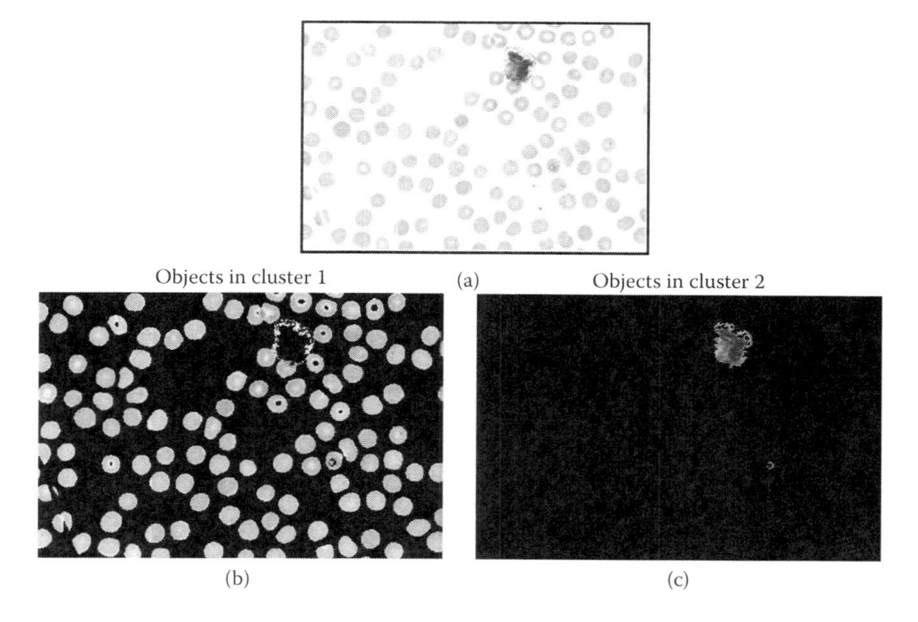

FIGURE 5.7: (a) Thin blood smear image, (b) cluster 1 consisting of red blood cells, (c) cluster 2 consisting white blood cells, platelets, and malaria parasite.

The RBC cluster was binarized. The Sobel operator was used to detect the edges of the RBCs. The image was eroded emulating a disc structure of radius three and neighborhood size four. The number of connected components was enumerated. Disc-based erosion was performed as RBCs have a disc-like structure (see Figure 5.8). Even for thin blood smears, clumping of RBCs is a common occurrence that makes enumeration of RBCs complicated. In coherence with the work of Borah Tek et al. [39], the RBCs were declumped and counted, to obtain a value of the total number of RBCs.

Once WBC outline within the RBC cluster was removed (see Figure 5.8), the RBC within the cluster with suspected presence of malaria parasites were marked out. Algorithmically, the RBCs with holes were marked out based on the Euler number value (see Figure 5.9).

The edges for the distinctly marked out erythrocyte cells with suspected malaria parasites were marked out with the Sobel operator. The suspected presence of malaria parasites within RBC was overlaid on the edge detected image (see Figure 5.10).

Cluster 2, consisting of WBCs, platelets, and malaria parasites, was binarized by dynamic selection of a threshold value (refer Figure 5.11)

The pixels with a value 'one' for binarized cluster 2 and pixels with a value of 'one' in Cluster 1 (i.e., the red regions in Figure 5.10 were marked '1') were compared. If a match is detected or an intersection is detected between the 'hole' cluster for cluster 1 and the pixel marked '1' for cluster 2, malaria parasite is said to be detected and overlaid on the blood smear image. Figure 5.12a and b represent the methodology described and followed for malaria parasite detection present within the RBC.

Figure 5.13 represents the overlay of the malaria parasite present within a red blood cell on the original thin blood smear image.

However, neither of the methods described above can be used to trace malaria at an advanced stage (refer Figure 5.14). Figure 5.14 clearly illustrates that at an advanced stage the malaria parasite engulfs the erythrocyte and emulates the appearance of a WBC.

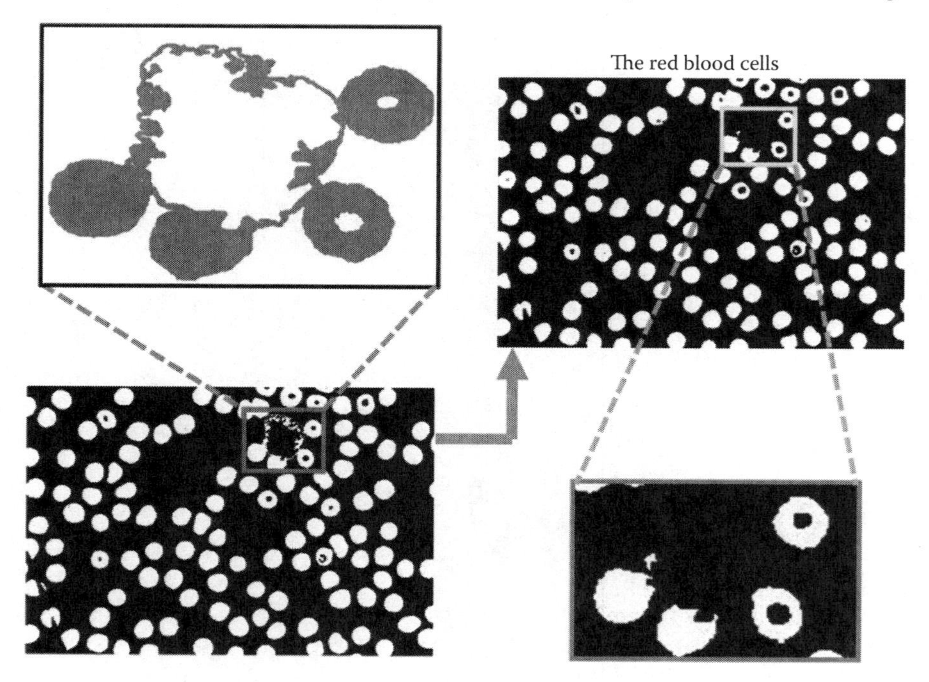

FIGURE 5.8: Disc-shaped erosion was performed to remove trace of WBC outline within the RBC cluster.

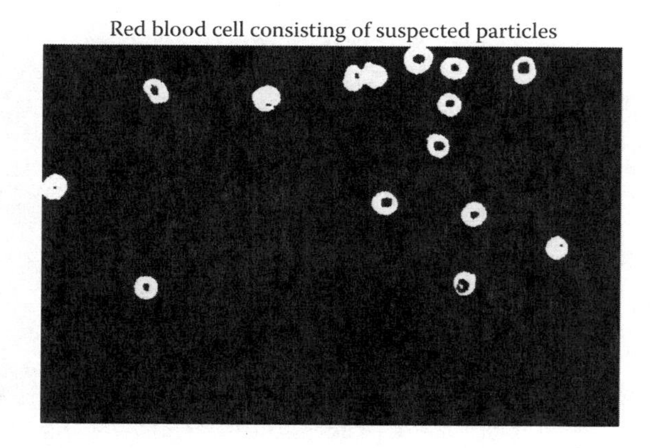

FIGURE 5.9: Red blood cells with suspected malaria parasite were distinctly marked out from the RBC cluster.

Close inspection of Figure 5.14 highlight, that while the WBC has a nucleus, the malaria parasite at an advance stage emulates the granulated texture of the WBC but does not have a nucleus-like structure. This distinction was algorithmically articulated.

As a result of repeated K-means clustering, three clusters were developed. The Cluster consisting of WBC and platelets was further investigated for the presence/absence of the malaria parasite at an advanced stage.

Given that it is difficult to distinguish between platelets, basophil, and malaria parasite at an advanced stage, certain sets of features were developed to distinguish between them. At the very onset, close proximity of erythrocyte to the particles marked out in the WBC, platelet cluster was considered. Morphological features such as dimension of all closed

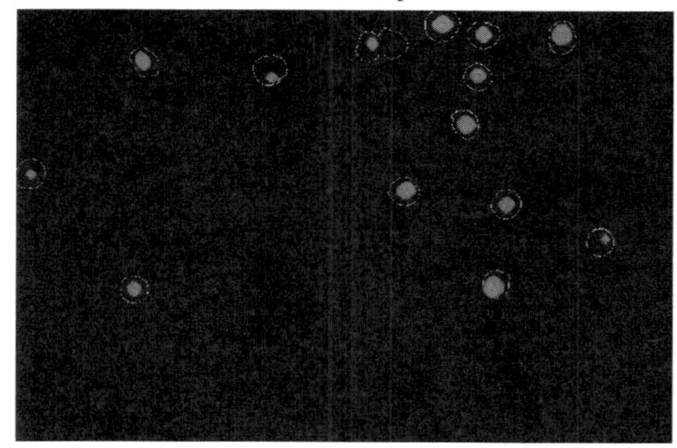

FIGURE 5.10: The suspected malaria parasite present within RBC is marked in gray shading. The edges of the erythrocyte cells that are deemed or suspected to contain suspected malaria parasite are marked out.

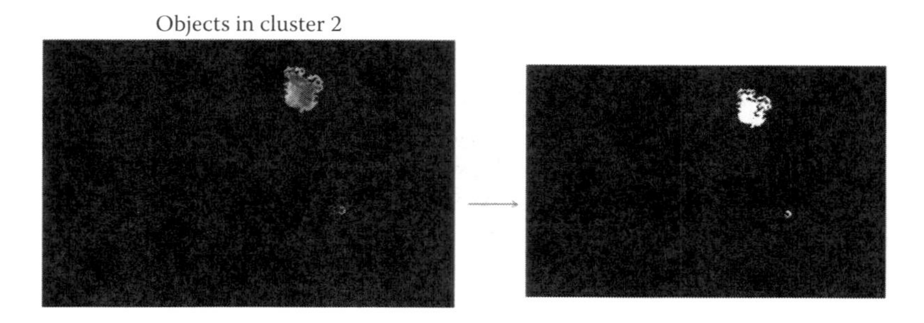

FIGURE 5.11: Cluster 2, consisting of white blood cells, platelets, and malaria parasite(s) (if any), was binarized by dynamic selection of threshold value.

components such as area, perimeter, length, and breadth of each connected component was taken into account.

Again, to entail a particular component as a single connected mass and not a group of small masses, the WBC/platelet cluster (i.e., Cluster 2) was dilated in a disc pattern with radius four and no neighbors. The values were empirically decided and are susceptible to vary for other datasets. By doing so, the connected components selected out by the algorithm resembled the individual connected components marked by an end user based on his/her cognitive ability.

Now for each of the connected components, the closest neighbor detection algorithm was used to detect the presence of closest RBCs. The number of RBCs present within a radius of zero to five pixels for a closed component in cluster 2 was calculated. Also, the Euclidean distance value for the closest RBC was calculated. Again, the length and breadth value for each of the connected components was calculated. Given that the WBC nucleus often deviates from a regular geometrical shape, three length values and three breadth values were taken into account for each of the components. By dividing each closed component into three basic parts in a row major order the median of the breadth value for each of the three parts was calculated. Again, dividing each closed component into three basic parts in column major order, the median of the length value for each of the three parts was calculated. The area and perimeter for each of the connected components was also calculated.

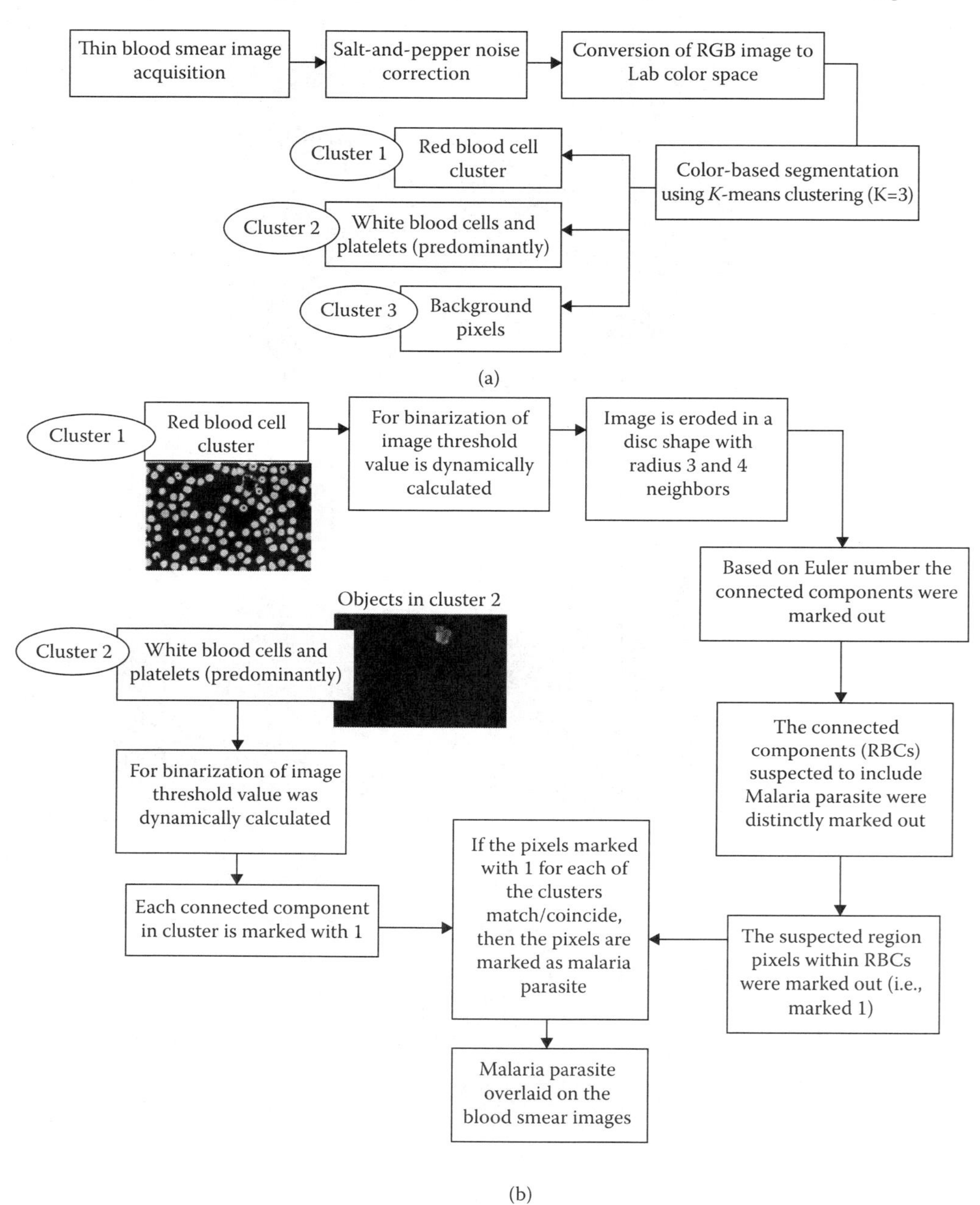

FIGURE 5.12: (a) Graphical representation of the 3-means clustering methodology that was followed for color-based segmentation. (b) Flowchart representation of the algorithm followed for malaria parasite detection within red blood cells.

As has been previously documented (see Figure 5.14), a basic feature that helps distinguish between a basophil and an advanced stage malaria parasite, apart from size of particle, is the presence of a nucleus within a basophil and absence of the same in an advanced stage malaria parasite. Based on the parameter values of the each of the connected components in cluster 2, the segments representing the components were cropped out from the original

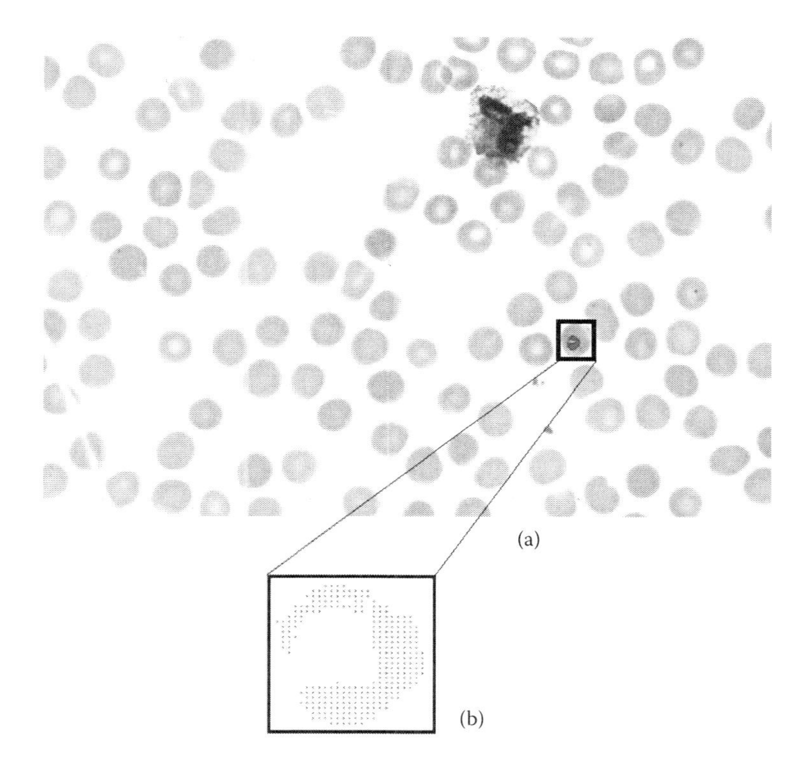

FIGURE 5.13: (a) The malaria parasite was overlaid (highlighted with a rectangular outline) on the thin blood image. (b) The detected malaria parasite was plotted on the X-Y axis for clarity.

blood smear image in true color. Color-based segmentation was performed to confirm absence or presence of nucleus within a connected component (see Figure 5.15).

To sum it all up, the features that were recorded for classification are duly marked out in Table 5.2.

It might be effective to investigate how the aforementioned features individually contribute towards the prediction of whether a closed/connected component in cluster 2 is a malaria parasite or not. The feature vectors are standardized and normalized, so that all feature vectors lie in the range zero to one.

The model developed at the very onset segments the blood smear image into RBCs and non-RBCs. While cluster 1 consists of erythrocyte, cluster 2 was found to consist of WBCs, platelets and malaria parasites (if any).

If a closed component that is part of cluster 2 exists within any of the RBCs in cluster 1, then the malaria parasite is detected. If no such component within RBCs can be traced, then features (as cited in Table 5.3) calculated for all of the closed components in cluster 2. Based on these features a model was developed to predict whether the malaria parasite exists within the thin blood smear image.

Based on the features calculated for each of the thin blood smear images, a supervised system was developed to predict whether cluster 2 consists only of WBCs and platelets (denoted by zero) or whether it consists of malaria parasites (denoted by one) along with WBCs and platelets. Once the features were normalized and standardized, leave-one-out cross-validation was performed to prevent over-fitting of data. As described in Figure 5.16, the data model developed for analyzing whether a thin blood smear image does or does

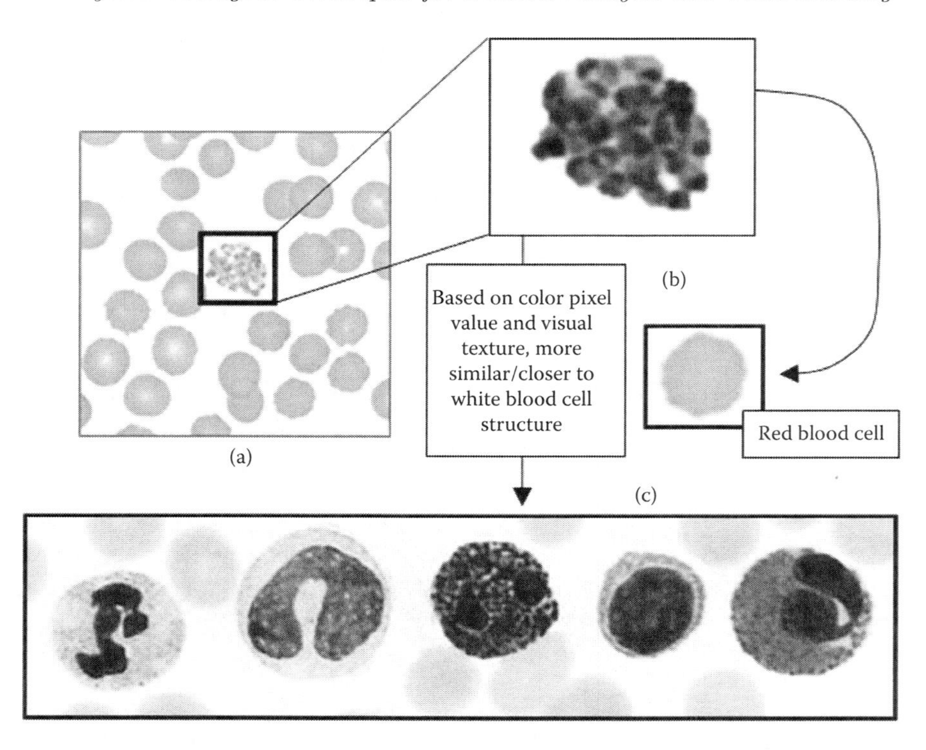

FIGURE 5.14: (a) Thin blood smear image consisting of malaria parasite at an advanced stage. (b) A magnified view of the malaria parasite detected at an advanced stage. (c) In coherence with the image of the different WBCs found in blood, the image as a whole has been used as an illustration to signify that the color pixels of the malaria parasite at an advanced stage are more closer (in terms of Euclidean distance) to the white blood cells as against the red blood cells. Close inspection shows that the malaria parasite at an advanced stage has a granulated texture similar to the texture of the white blood cell.

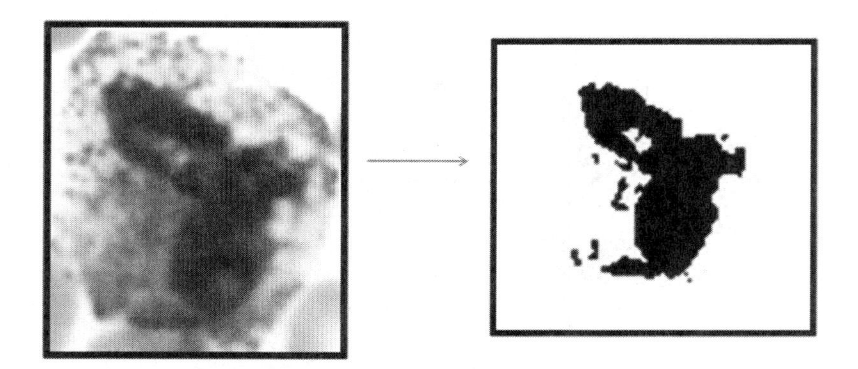

FIGURE 5.15: The segmentation of the nucleus from the white blood cell.

not contain malaria parasites consists of 2 basic parts: a rule-based part and a supervised model. Based on the prediction of the rule-based section, an autonomous decision is made by the system as to whether it wants to explore the supervised learning model or not. If a malaria parasite is detected within RBC cells, the thin blood smear image is predicted

TABLE 5.2: The exhaustive list of the features that were calculated for each of the closed components present within Cluster 2

Features	**Number of features**
Number of red blood cells within a radius of 5 pixels from the center of a closed component in cluster 2	1
The median length value for each of the 3-column major segments	$1x3 = 3$
The median breadth value for each of the 3-row major segments	$1x3 = 3$
Area of each closed component in cluster 2	1
Perimeter of each closed component in cluster 2	1
Eccentricity of each closed component in cluster 2	1
Euclidean distance of the red blood cell that is closest to the closed component in cluster 2	1
Presence/absence of nucleus within a closed component	1
Total	12

TABLE 5.3: The accuracy, sensitivity and specificity values when the neighbors used for classification of a test data-point were varied between 1, 3, and 5

Number of Neighbors	**Accuracy**	**Sensitivity**	**Specificity**
$K = 1$	0.91898	0.92	0.91797
$K = 3$	**0.97826**	**0.98**	**0.97656**
$K = 5$	0.94169	0.95	0.93359

to be obtained from an individual infected with malaria. Again, if no malaria parasite is detected within the RBC cluster, the closed components in cluster 2 shall be investigated using supervised learning techniques before a final conclusion can be arrived at in terms of presence/absence of malaria parasites in a given blood smear image.

All features extracted from cluster 2 were normalized and standardized. Leave-one-out cross-validation was performed for the images. Each image was defined as a set of datapoints where each datapoint particularly represented a closed component for cluster 2. Only the closed components that were not enclosed by an RBC were taken into account. A K-NN classifier was used to predict whether a closed component was a malaria parasite at a matured state (denoted by 1) or not (denoted by 0). So each datapoint from a set of test data points for a particular image was classified based on three of its closest neighbors from the training set, selected on the basis of Euclidean distance.

Each image in the dataset accounted for approximately four closed components in cluster 2, while some of the images recorded six closed components others recorded two and zero. In total the dataset consisted of 1012 datapoints each representing a closed component in cluster 2 that wasn't encapsulated within an RBC or erythrocyte. Of the 1012 datapoints, 500 datapoints represented malaria parasites at an advanced/matured state. The other 512 datapoints represented WBCs and platelets fragments.

If at least one of the datapoints from a set of datapoints representing cluster 2 for a particular image was assigned label 1, then the blood smear image as a whole was considered to be infected by malaria parasites.

The pseudocode for the algorithm developed has been provided in Algorithm 5.1

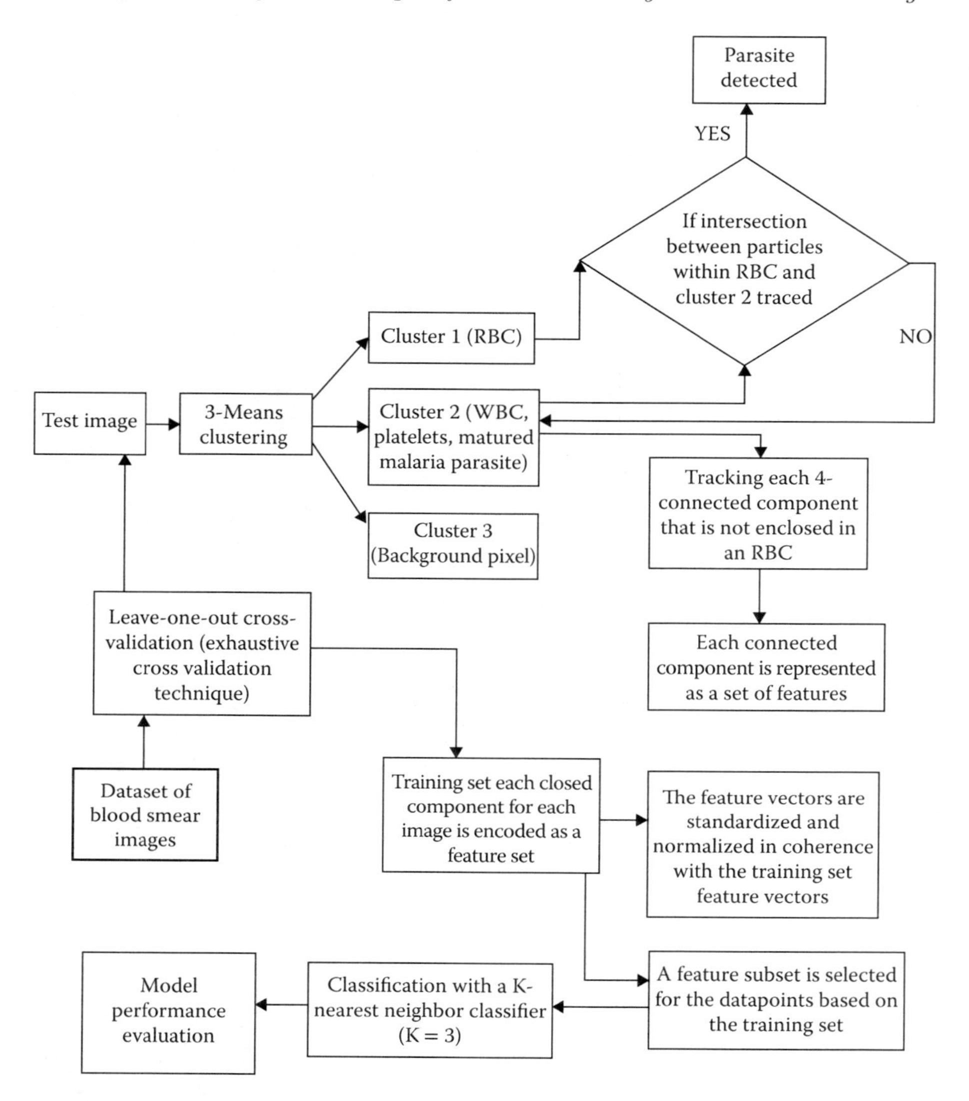

FIGURE 5.16: Data model developed for identification of malaria parasite within a thin blood smear image.

The model that is a hybrid of a rule-based system and a supervised learning system helps overcome the uncertainty that enshrouds the process of malaria parasite detection. If the system used for malaria parasite detection is particularly rule-based then the system fails to generalize to the uncertainty that governs the process of malaria parasite detection. An example case emphasizing the fallibility of the rule-based methodology has been demonstrated in Figure 5.17.

In Figure 5.17, the same cell segment has been displayed at different magnifications ranging from 10X to 100X. In case of a fully rule-based system for each particular magnified image the feature threshold values have to be manually defined based on previous experience of a concerned analyst. Use of a machine learning-based system helps to overcome this constrained hard-coding of rules owing to variation in image magnification.

Algorithm 5.1: Pseudocode

Input: Thin Blood Smear Image/s of size 1387 x 932 pixel^2 each
Output: Prediction whether blood smear image is malaria parasite infected

Step 1: Read image
Step 2: Binarize image using a dynamic threshold value
Step 3: Perform 2D Median filtering with 3x3 overlapping window size
Step 4: If pixel value (Filtered image) = 0
Suppress in original RGB image
Else
Preserve in original RGB image
Step 5: Convert RGB image to Lab Color space.
Step 6: Perform 3-means clustering with selective and repetitive seeding [120 iterations].
Step 7:Store RBC Cluster [Cluster 1], WBC/platelet Cluster [Cluster 2] and background pixel Cluster [Cluster 3].
Step 8: Binarize RBC Cluster [Cluster 1] using dynamic threshold value
Step 9: De-clump by Boray Tek's method
Step 10: Erode binarized Cluster 1 in disc shape with radius 3 px and neighbor 4
Step 11: Consider each red blood cell in Cluster 1 as a separate datapoint
Step 12: Store a Map of the RBC cells and WBC/platelet cells to each blood smear image
Step 13: Calculate Euler number for each RBC datapoint individually
Step 14: If Euler number <= 0
Suspect presence of malaria parasite in RBC
Else
RBC does not contain malaria parasite
Step 15: Store the pixels within RBC that are suspected to contain malaria parasite
Step 16: Binarize Cluster 2 using dynamic thresholding
Step 17: Store 4-connected foreground components in Cluster 2
Step 18: If suspected pixel position (Cluster 1) = 4-connected component pixel position (Cluster 2)
Mark pixel as malaria parasite
Else
Store 4-connected pixel position for Cluster 2
Step 19: Extract features for the stored 4-connected components in Cluster 2
Step 20: Standardize and normalize all the feature vectors
Step 21: Select 10 feature subset from a set of 12 feature vectors
Step 22: Perform Leave-One-Out Cross-Validation on the 1012 Cluster 2 particle dataset
Step 23: Predict a datapoint or closed component in Cluster 2 as malaria parasite or not by using a 3-NN classifier.
Step 24: Based on Step 18, Step 23 and a map created in Step 12, if an image contains a component marked as malaria parasite
Predict the image as a whole is to contain malaria parasite
Else
Predict the image to be taken from an individual not affected by malaria

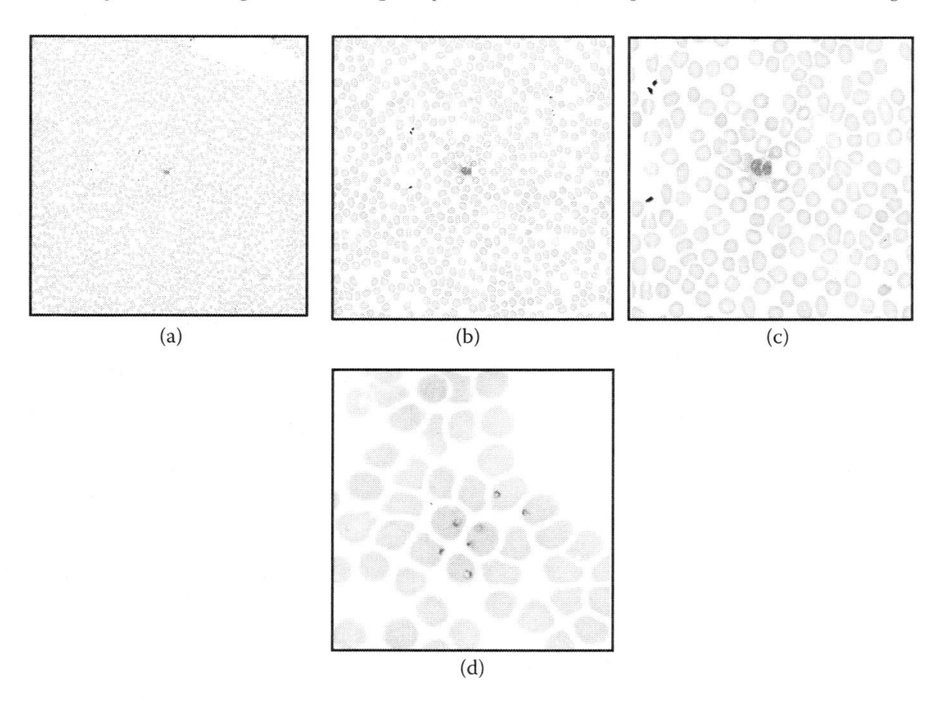

FIGURE 5.17: Thin blood smear image at (a) 10X magnification, (b) 20X magnification, (c) 40X magnification, and (d) 100X magnification. (All the images represent the same blood smear image from the MaMic database at different magnification values.)

5.4 Results and Discussion

It might be interesting to note that the prediction accuracy of the model degrades when all the features (cited in Table 5.2) were taken into account by the classifier for making a prediction. Such degradation in prediction accuracy when all features are taken into account might be due to feature redundancy. That is to say, features that bring in similar information to the classifier often hinder the performance of the classifier. Correlation accounts for the linear relationship between two variables. A strong statistically significant Pearson's rank correlation was predicted between the class label and the number of RBCs near or around a closed component in cluster 2 ($rho = 0.758$, $p = 0.000 < .01$). Pearson's rank correlation was used as the variables under consideration do not follow normal distribution. Again, when the median of the length values for each of the three row major segments for each closed component was considered, a statistically weak, insignificant correlation with the class label ($rho = 0.05$, $p = .534 > .01$) was recorded. The correlation of the area and perimeter values of the closed components with the class labels was found to be statistically significant ($rho = 0.866$, $p = 0.001 < 0.01$). Additionally, the correlation of the eccentricity value of each closed component to the class label was found to be insignificant ($rho = 0.25$, $p = 0.324 > 0.01$) while the distance of the closest RBC to the closed component was found to be significant ($rho = 0.665$, $p = 0.000 < 0.01$). The presence/absence of a nucleus present within a closed component in cluster 2 was found to have a strong correlation with class label and was statistically significant ($rho = 0.925$, $p = 0.000 > 0.01$). Again, correlation of a class label and length, area, perimeter, breadth of nucleus present within a particular closed structure was statistically significant [$rho(length) = 0.730$ ($p = 0.001 < 0.01$),

$rho(area) = 0.643$ ($p = 0.001 < 0.01$), $rho(perimeter) = 0.795$ ($p = 0.000 < 0.01$), $rho(breadth) = 0.712$ ($p = 0.000 < 0.01$)].

The conditional mutual information maximization algorithm was used to find a filter feature set that bring in the maximum information about the concerned class label. This also encompasses or takes into account the nonlinear relationship between a feature vector and the class label.

A minimum error rate was achieved when a subset of 10 features were used for prediction of the class label for a test blood smear image. Again, an exhaustive search over the feature space confirmed that the set of 10 features (i.e. formed by excluding the median length value for the 1st and 3rd half of the WBC) as a wrapper feature set provided the minimum error rate of classification for a $3 - NN$ classifier. Table 5.3 represents the accuracy, sensitivity, and specificity values when the set of 10 features are used for identification of malaria parasites for a thin blood smear image by varying the number of neighbors ($K = 1, 3, 5$) for a test datapoint.

A rule-based methodology was developed for the same set of images. A rule set replacing 3-means clustering was developed and appended to the algorithm developed. A particular set of color intensity values in the HSV color model were assigned to clusters 1, 2, and 3, respectively, by virtue of a set of defined rules. The accuracy, sensitivity, and specificity of assignment of pixels by the rule-based system have been reported against the accuracy, sensitivity, and specificity values for the unsupervised 3-means clustering system (see Table 5.5). The threshold range for the different feature vectors were empirically worked on. It was difficult to narrow in on a suitable threshold range for feature vectors (namely the median of the length values for the three row major segments for the closed components) that did not have a statistically significant correlation with the class labels. Again, it was tedious to tune the model to perform satisfactorily when the magnification of the concerned images was varied as it also lead to re-estimation of the threshold values. For the developed rule-based system, the classification accuracy, sensitivity, and specificity values are represented in Table 5.5.

The method proposed in this work is a hybrid or a generous optimized mixture of a rule-based and machine learning-based approach. While a singularly rule-based system has its drawbacks it might be helpful to analyze the drawbacks of using a machine learning-based

TABLE 5.4: Performance statistics of the proposed hybrid method against other state-of-the-art rule-based and machine-learning algorithms

Algorithm	**Sensitivity (%)**	**Specificity (%)**	**Dataset used**
Boray Tek et al. [39]	74	98	9 images
Makkapati et al. [1]	83	98	55 images
Somasekar et al. [26]	94.87	97.3	76 images
Diaz et al. [36]	94	99.7	100 images
Ghate et al. (Method 1)[33]	81.39	86.49	80 images
Ghate et al. (Method 2)[33]	72.93	75.76	80 images
Suryawanshi et al. [48]	93.33	93.33	30 images
Mashor et al. [46]	92.14	99.79	100 images
Chayadevi et al. [50]	95.2092	96.06758	476 images
Proposed Method	**98.4**	**97.6**	**250 images from MaMic Database**

TABLE 5.5: Comparative representation of the accuracy values of a rule-based and machine learning-based malaria parasite detection algorithm that were tested on 250 image strong MaMic database

RULE-BASED METHOD			
	Accuracy	**Sensitivity**	**Specificity**
Rule-based pixel clustering	0.62	0.63	0.66
Overall classification of malaria parasite	0.741107	0.706	0.77539
MACHINE-LEARNING–BASED METHOD			
3-Means clustering	0.9895	0.9925	0.9832
Overall classification of malaria parasite	0.7260	0.6895	0.7123

methodology singularly. To test the effectiveness of making the algorithm particularly machine learning-based in entirety, certain changes were incorporated in the algorithm developed. At the onset, the rule-based identification of malaria parasites within an by matching the closed component present within RBC to the closed components in cluster 2 was removed. The change that has been introduced into the algorithm has been highlighted in Figure 5.18. All four connected components in cluster 2 were taken into consideration and each connected component was treated as a datapoint that was described by the set of features marked out in Table 5.2. In coherence with the original algorithm described. Based on the feature values and leave-one-out cross-validation, a K-NN classifier was used to classify each of the closed components as a malaria or non-malaria parasite. Similar to the proposed algorithm, the value of K was varied in the range one, three, and five, respectively. The machine learning-based methodology often leads to misidentification of malaria parasites present within red blood corpuscles at the initial stage of infection. The malaria parasite prediction accuracy, Sensitivity and specificity value for the machine learning-based system has been documented in Table 5.5. The aim of the algorithm developed was to identify/predict the presence of malaria parasite in a given thin blood smear image at the initial, advanced, or at a matured stage when the parasite engulfs the RBC. Table 5.4 and Table 5.6 highlight the efficiency of the hybrid algorithm for malaria parasite prediction as opposed to the prediction accuracy of a rule-based or machine learning-based system.

5.5 Comparative Study

The image based performance of the proposed hybrid approach against other rule-based and machine learning-based methodologies proposed by researchers for identification of malaria parasites from blood smear images have been duly documented in Table 5.4.

Table 5.4 represents the proposed method to have better prediction performance over other suggested methods. However, given that all the methods were not tested on the same blood smear dataset, it is difficult to compare the performance of the algorithms under consideration. A serious limitation of comparison of the algorithms proposed lies in the fact that the dataset on which the performance of the algorithm/s have been documented is not publicly available.

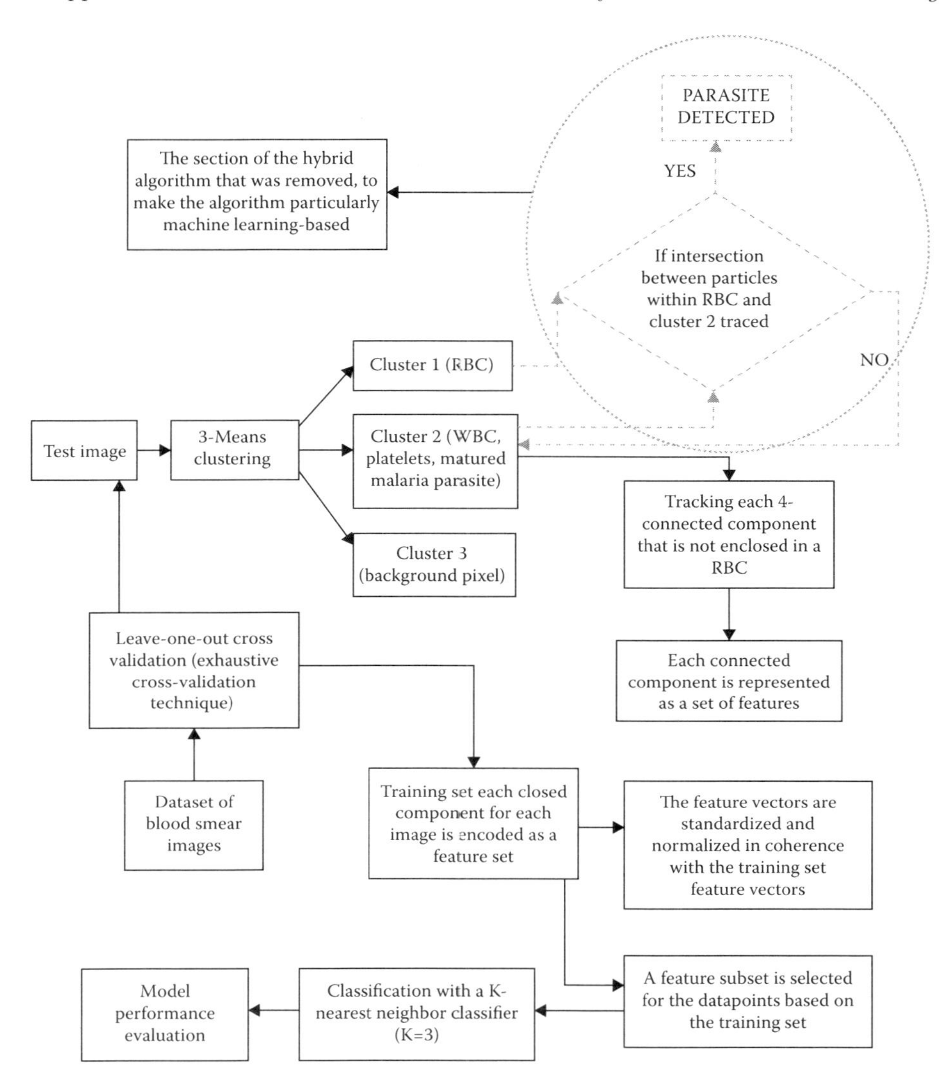

FIGURE 5.18: The circled portion of the flowchart represents the module in hybrid algorithm that was removed in order to make the algorithm particularly machine learning-based.

TABLE 5.6: Comparative study of different methods on 250 images of the MaMic database

Algorithm	**Sensitivity (%)**	**Specificity (%)**	**Dataset used**
Somasekar et al. [26]	90.4	93.6	250 images from MaMic database
Diaz et al. [36]	93.6	88	
Suryawanshi et al. [48]	75.2	92.0	
Mashor et al. [46]	74.4	77.6	
Chayadevi et al. [50]	88.8	85.6	
Proposed Method	**98.4**	**97.6**	

In view of all these constraints, some of the methods put forth by other researchers have been duly implemented and tested on the dataset that was used to test the performance of the proposed hybrid approach. Table 5.6 documents the predictive power of the algorithms for the dataset developed against the performance of the proposed hybrid approach.

The robustness of the proposed algorithm across different blood smear datasets shall warrant further testing.

5.6 Conclusion

The underdeveloped countries still suffer loss of human lives to malaria. The occurrence of malaria has slightly reduced globally with several countries reporting elimination of malaria. The disease remains a major health issue in remote rural areas where it not only effects health but also changes the socioeconomic conditions of poor households. The disease has a huge economic impact on the financial health of countries. Modern approaches to the age-old diagnostic methods have boosted rapid diagnosis of disease resulting in reduction of mortality rates. The advent of the digital microscope and advanced computer technology has ushered in an era of digital pathology. This chapter is particularly aimed at identifying the presence of malaria parasites in thin blood smear digitized images. The images were obtained using digital microscopes and were analyzed using a computer aided system. From a range of 12 features, a feature subset of size 10 was marked out using the CMIM algorithm. This feature subset was used for the classification of a connected component from a test blood smear image. If at least a single connected component from the WBC/platelet cluster for a test image was marked as a malaria parasite (i.e., '1'), The image as a whole was considered to be the blood smear image from an individual infected with malaria. In a nutshell, the system is aimed to predict the presence/absence of malaria parasites within thin blood smear images. The system proposed is still subject to further testing before it can be used as a dependable tool by medical practitioners.

References

1. V. V. Makkapati and R. M. Rao. Segmentation of malaria parasites in peripheral blood smear images. In *Proceedings of IEEE International Conference on Acoustics, Speech and Signal Processing*. IEEE, Taipei, Taiwan, 2009. doi:10.1109/icassp.2009.4959845.

2. Royal Perth Hospital Home, 2016. Available at http://rph.wa.gov.au/ Retrieved on 8 June 2016.

3. S. Coll-Black, A. Bhushan, and K. Fritsch. Integrating poverty and gender into health programs: A sourcebook for health professionals. *Nursing & Health Sciences*, 9(4):246–253, 2007. doi:10.1111/j.1442-2018.2007.00340.x.

4. S. W. Lindsay, M. Jawara, K. Paine, M. Pinder, G. E. Walraven, and P. M. Emerson. Changes in house design reduce exposure to malaria mosquitoes. *Tropical Medicine and International Health*, 8(6):512–517, 2003. doi:10.1046/j.1365-3156.2003.01059.x.

5. Asian Development Bank. Indigenous peoples/ethnic minorities and poverty reduction: Pacific region, Manila, 2002.

6. S. Russell. The economic burden of illness for households in developing countries: A review of studies focusing on malaria, tuberculosis, and human immunodeficiency virus/acquired immunodeficiency syndrome. *American Journal of Tropical Medicine and Hygiene*, 71(2):147–155, 2004.

7. F. D. McCarthy, H. Wolf, and Y. Wu. Malaria and growth (wps2303). Washington, DC: World Bank, Development Research Group, Public Economics (Policy Research Working Paper), 2000.

8. J. Sachs and P. Malaney. The economic and social burden of malaria. *Nature*, 415(6872):680–685, 2002. doi:10.1038/415680a.

9. N. Tangpukdee, C. Duangdee, P. Wilairatana, and S. Krudsood. Malaria diagnosis: A brief review. *Korean Journal of Parasitology*, 47(2):93, 2009. doi:10.3347/kjp.2009.47.2.93.

10. D. L. Hartl. The origin of malaria: Mixed messages from genetic diversity. *Nature Reviews MicroBiology*, 2(1):15–22, 2004. doi:10.1038/nrmicro795.

11. C. Shiff. Integrated approach to malaria control. *Clinical Microbiology Reviews*, 15(2):278–293, 2002. doi:10.1128/cmr.15.2.278-293.2002.

12. R. S. Phillips. Current status of malaria and potential for control. *Clinical Microbiology Reviews*, 14(1):208–226, 2001. doi:10.1128/cmr.14.1.208-226.2001.

13. World Health Organization. Who—World malaria report 2011, 2011. Available at http://www.who.int/malaria/world_malaria_report_2011/en/. Retrieved on 4 February 2016.

14. H. Hisaeda, K. Yasutomo, and K. Himeno. Malaria: Immune evasion by parasites. *The International Journal of Biochemistry & Cell Biology*, 37(4):700–706, 2005. doi:10.1016/j.biocel.2004.10.009.

15. CDC – DPDx. 2016. Available at http://www.dpd.cdc.gov/dpdx/html/imagelibrary /malaria il.htm.

16. D. C. Warhurst and J. E. Williams. Laboratory diagnosis of malaria. *ACP Broadsheet no 148*, 49(7):533–538, 1996.

17. Haematology in a NutShell. Available at http://textbookhaematology4medical-scientist.blogspot.in/2014/03/thick-and-thin-smears-for-microscopy.html, 2016.

18. A. R. Bharti, K. P. Patra, R. Chuquiyauri, M. Kosek, R. H. Gilman, A. Llanos-Cuentas, and J. M. Vinetz. Polymerase chain reaction detection of plasmodium vivax and plasmodium falciparum dna from stored serum samples: Implications for retrospective diagnosis of malaria. *American Journal of Tropical Medicine and Hygiene*, 77(3): 444–446, 2007.

19. M. Guy. Basic malaria microscopy. Part I: Learners guide. *Transactions of the Royal Society of Tropical Medicine and Hygiene*, 86(6):700, 1992. doi:10.1016/0035 -9203(92)90204-p.

20. Wikipedia. Patented Digital Micro-imaging Adaptor with SAGLO Soft Software for Microscopy developed by Inventor Sachin G. Lokapure, Wikimedia Commons, 2016.

21. S. W. Sio, W. Sun, S. Kumar, W. Z. Bin, S. S. Tan, S. H. Ong, and K. S. Tan. Malariacount: An image analysis-based program for the accurate determination of parasitemia. *Journal of Microbiological Methods*, 68(1):11–18, 2007. doi:10.1016/j.mimet.2006.05.017.

22. S. Kumar, S. Ong, S. Ranganath, T. Ong, and F. Chew. A rule-based approach for robust clump splitting. *Pattern Recognition*, 39(6):1088–1098, 2006. doi:10.1016/j.patcog.2005.11.014.

23. J. A. Frean. Reliable enumeration of malaria parasites in thick blood films using digital image analysis. *Malaria Journal*, 8(1):218, 2009. doi:10.1186/1475-2875-8-218.

24. Point Picker. Image J Point Picker plugin. Available at http://bigwww.epfl.ch/thevenaz/pointpicker/, 2013.

25. A. Zheng. *Blood Smear Malaria Parasite Detection*. Department of Electrical Engineering, Stanford University, https://stacks.stanford.edu/file/druid:yj296hj2790/Zheng_Malaria_Red_Blood_Cell_Counter.pdf, 2012.

26. J. Somasekar, B. E. Reddy, E. K. Reddy, and C. H. Lai. An image processing approach for accurate determination of parasitemia in peripheral blood smear images. *IJCA Special Issue on Novel Aspects of Digital Imaging Applications DIA*, pp. 23–28, 2011. doi:10.5120/4153-316.

27. P. Ghosh, D. Bhattacharjee, M. Nasipuri, and D. K. Basu. Medical aid for automatic detection of malaria. *Communications in Computer and Information Science*, 245(1):23–28, 2011. doi:10.1007/978-3-642-27245-5_22.

28. C. Di Ruberto, A. Dempster, S. Khan, and B Jarra. Analysis of infected blood cell images using morphological operators. *Image and Vision Computing*, 20(2):133–146, 2002. doi:10.1016/s0262-8856(01)00092-0.

29. C. D. Ruberto, A. Dempster, S. Khan, and B. Jarra. Segmentation of blood images using morphological operators. In *Proceedings of IPCR, 15th International Conference on Pattern Recognition*, Barcelona, Spain, 2000.

30. A. Mehrjou, T. Abbasian, and M. Izadi. Automatic malaria diagnosis system. In *2013 First RSI/ISM International Conference on Robotics and Mechatronics (ICRoM)*, Tehran, Iran, pp. 205–211, 2013.

31. P. T. Suradkar. Detection of malaria parasite in blood using image processing. *International Journal of Engineering and Innovative Technology (IJEIT)*, 2(10):124–126, 2013.

32. V. Parkhi, P. Pawar, and A. Surve. Computer automation for malaria parasite detection using linear programming. *International Journal of Advanced Research in Electrical, Electronics and Instrumentation Engineering*, 2(5):1984–1988, 2013.

33. D. Ghate, C. Jadhav, and N. U. Rani. Automatic detection of malaria parasite from blood images. *International Journal of Advanced Computer Technology (IJACT)*, 4(1):129–132, 2014.

34. K. Chakraborty, A. Chattopadhyay, A. Chakrabarti, T. Acharya, and A. K. Dasgupta. A combined algorithm for malaria detection from thick smear blood slides. *Journal of Health & Medical Informatics*, 6(1):1–6, 2015. doi:10.4172/2157-7420.1000179.

35. J. Arco, J. Górriz, J. Ramírez, I. Álvarez, and C. Puntonet. Digital image analysis for automatic enumeration of malaria parasites using morphological operations. *Expert Systems with Applications*, 42(6):3041–3047, 2015. doi:10.1016/j.eswa.2014.11.037.

36. G. Díaz, F. A. González, and E. Romero. A semi-automatic method for quantification and classification of erythrocytes infected with malaria parasites in microscopic images. *Journal of Biomedical Informatics*, 42(2):296–307, 2009. doi:10.1016/j.jbi.2008.11.005.

37. A. Carling. *Introducing Neural Networks*. Sigma Press, Wilmslow, UK, 1992.

38. V. N. Vapnik. *The Nature of Statistical Learning Theory*. Springer-Verlag, New York, NY, 1989.

39. F. B. Tek, A. G. Dempster, and İ. Kale. Parasite detection and identification for automated thin blood film malaria diagnosis. *Computer Vision and Image Understanding*, 114(1):21–32, 2010. doi:10.1016/j.cviu.2009.08.003.

40. K. N. R. M. Rao. *Application of Mathematical Morphology to Biomedical Image Processing*, PhD thesis. University of Westminster, London, UK, 2004.

41. P. Soille. *Morphological Image Analysis: Principles and Applications*. Heidelberg, Germany: Springer-Verlag, 2003.

42. F. B. Tek. *Computerised Diagnosis of Malaria*, PhD thesis. University of Westminster, London, UK, 2007.

43. S. S. Savkare and S. P. Narote. Automatic detection of malaria parasites for estimating parasitemia. *International Journal of Computer Science and Security (IJCSS)*, 5(3):310–315, 2011.

44. S. S. Savkare and S. P. Narote. Automatic classification of normal and infected blood cells for parasitemia detection. *International Journal of Computer Science and Network Security (IJCSNS)*, 11(2):94–97, 2011.

45. N. Ahirwar, S. Pattnaik, and B. Acharya. Advance image analysis based system for automatic detection and classification of malaria parasite in blood images. *International Journal of Information Technology and Knowledge Management*, 5(1):59–642, 2012.

46. A. S. Abdul-Nasir, M. Y. Mashor, and Z. Mohamed. Segmentation based approach for detection of malaria parasites using moving K-means clustering. In *IEEE EMBS International Conference on Biomedical Engineering and Sciences*, Langkawi, 2012.

47. A. S. Abdul-Nasir, M. Y. Mashor, and Z. Mohamed. Colour image segmentation approach for detection of malaria parasites using various colour models and K-means clustering. *WSEAS Transactions on Biology and Biomedicine*, 10(1):41–55, 2013.

48. S. Suryawanshi and V. V. Dixit. Comparative study of malaria parasite detection using euclidean distance classifier & SVM *International Journal of Advanced Research in Computer Engineering & Technology (IJARCET)*, 2(11):2994–2997, 2013.

49. D. A. Kurer and V. P. Gejji. Detection of malaria parasites in blood images. *International Journal of Engineering Science and Innovative Technology (IJESIT)*, 3(3):651–656, 2014.

50. M. L. Chayadevi and G. T. Raju. Usage of art for automatic malaria parasite identification based on fractal features. *International Journal of Video & Image Processing and Network Security IJVIPNS-IJENS*, 14(4):7–15, 2014.

51. M. I. Razzak. Automatic detection and classification of malaria parasite. *International journal of Biometrics and Bioinformatics*, 9(1):1–12, 2015.

52. Y. S. Bahendwar and U. K. Chandra. Detection of malaria parasites through medical image segmentation using ann algorithm. *International Journal of Advanced Research in Computer Science and Software Engineering*, 5(7):1063–1067, 2015.

53. WebMicroscope. The MaMic Image Database, Thin smear blood samples for identification of Plasmodium Falciparum, The MaMic 1 Series. Available at http://fimm.webmicroscope.net/Research/Momic/mamic, 2010.

Chapter 6

Noise Removal Techniques in Medical Images

Ranjit Biswas and Sudipta Roy

6.1 Introduction

The success of digital imaging has created a neverending race of applications targeted towards improving the quality of medical services which are composed of digital images such as computed tomography, ultrasonography, mammogram, X-ray, magnetic resonance imaging, positron emission tomography, and single-photon emission computed tomography, etc. The advances in the field of information and communication technology and medical imaging instruments have brought the use of medical images in diagnostics and information technology enabled treatment in medical field to the forefront by producing large quantities of these images [1]. The quest has now changed from acquiring medical images to acquiring high-quality medical images. The quality of an image is judged by the absence of noises, but sadly enough, the instruments producing the images are not free from inducing noises in the image during acquisition process. The presence of noise disorients the decision-making capability of the system due to low visibility of low-contrast objects and also can impact the overall quality of the images. Thus, denoising has become one of the fundamental tasks in the analysis process of medical images [2].

Every type of medical imaging instruments used for acquiring images induces some noises; viz., ultrasonic images are assumed to contain speckle noise and CT images are supposed to be corrupted by Poisson and Gaussian distributed random noise and found in standard X-ray films, etc., as reported by the authors in [3]. The chapter aims to illustrate each and every type of noise and denoising techniques available in recent times to enhance the quality of medical images.

In this chapter, we explore the recent trends in medical image processing: in particular, how different type of noises are removed from the images. The main objective of medical imaging is to acquire a high-resolution image with as many details as possible for the sake of diagnosis, analysis of disease, and for use in other data mining applications. The medical images which are noise free and without artifacts are considered as good-quality images and therefore, denoising becomes an integral part of any medical image processing application [4].

6.2 Background Work

To remove the noise, there are many ways proposed to date. The earlier denoising techniques were based on spatial filtering which were basically blurring techniques. Later, Fourier domain denoising was introduced and this proved to be more efficient most of the time.

The spatial domain contains two types: linear and nonlinear. The linear filter contains mean and Wiener filters which are used most often but there are many other types as well. The most common nonlinear type is the median filter. Generally, the median filter performed well with the low-density noisy image, but they do not usually work well for highly noisy images. A modified technique of dynamic programming for implementing nonlinear smoothing filter was investigated by Ney [5] for better results to remove noise. Later, Kuo et al. [6] developed an efficient median filtering technique, which used additive median filters and then calculated a median value to remove the high-density noise. In 1994, the combined use of the Wiener filter and PSE filter in CT images was investigated to find out the effectiveness of a 3D filter [7] against a 2D filter. An adaptive Wiener filter was implemented by Saluja et al. [8] and a modified noise removal was performed using wavelet transform by a new iterated system. Literature is available in different aspects of image processing works using linear and nonlinear filters.

In the history of the transform domain, preliminary works were starting to be applied to a 1D signal. Later on, for more convenience and clarity, 1D signals were applied in medical images to compress and denoise and in many other emerging field of medical image analysis. Basically, the transform domain filtering contains four types so far: wavelet, ridgelet, curvelet and contourlet transforms. The infrequent nature of wavelet is the foundation of wavelet-based denoising through thresholding [9] and helps in determining threshold between the noisy feature and medical image feature [10]. Earlier, wavelet transform was implemented frequently on unidimensional signal of an image to remove noise [11]. Later, a adaptive wavelet–soft threshold using a data-driven method was introduced [12] which was called the BayesShrink method to compress and remove noise from images for threshold estimation. Therefore, the Bayesian framework was simultaneously used in lossy compression and denoising. Although literature is available on denoising the medical image, the use of the wavelet transform in the diagnosis of fatal diseases in late 1990s. Mojsilovic et al. [13,14] worked on diagnosis and classification of different stages of liver disease using the wavelet transform method. Gorgel et al. [15] worked on breast cancer to enhance the visual quality of the mammographic image based on wavelet transform and homomorphic filtering which provided more visibility for the abnormal tissue regions. VisuShrink, an important thresholding technique, was used for wavelet shrinkage or soft thresholding [16]. Donoho et al. [17] developed another technique called the SureShrink technique, which used the SURE estimator developed by Steins unbiased risk estimator, to estimate the mean of random normal independent variables. Diffusion in the medical image has two types: isotropic and anisotropic diffusion. Zhang [18] implemented isotropic diffusion in the wavelet domain and anisotropic diffusion in the image domain, but the combining method comprised of wavelet domain diffusion and the image domain diffusion proved to be effective to make denoising better. The limitation of wavelet transform is having its suboptimal processing of edge discontinuity [19].

After wavelet transform, ridgelet transform came into existence dealing with line-like phenomena in 2D and plane-like phenomena in 3D [20]. Based on ridgelet, curvelet transform was studied [21,22]. Amoit et al. [23] proposed a curvelet implementation based in contrast enhancement on X-ray images where they used a recursive filter, then spatial fliter, and finally, a line enhancement method on coefficients were used. A method of fusion of the CT image with the MR image [24] and removal of noise from the SAR image [25] was successfully done using the curvelet domain transform. Moore et al. [26] studied and showed 1D Curvelet and Contourlet denoising methods of seismic data. In 2015, Aghazadeh et al. [27] used a two-stage algorithm that showed restoration and segmentation of blurry and noisy images in the curvelet domain where they used the Gaussian scale space method to segment and used Stein's SURE method to threshold. For noise removal from an MRI image, a multiresolution technique based on contourlet transform domain was studied by Anila et al. [28].

6.3 Type of Noise

Noise represents redundant information which degrades the image quality. The type of noise present in the image plays an important role in the denoising process of the image and hence the image quality improves after denoising. Medical images are often corrupted by various types of noise during its acquisition or transmission process. The noises such as Gaussian noise, speckle noise, Poisson noise, and salt-and-pepper noise are the most common noises that corrupt medical images.

6.3.1 Gaussian noise

Gaussian noise is statistical noise having a probability density function (PDF) equal to that of the normal distribution, which is also known as the Gaussian distribution. In this type of noise, random values are added to an image matrix. In other words, the values that the noise can take on are Gaussian-distributed, which has a probability distribution function given by Equation 6.1 [6]:

$$F(g) = \frac{1}{\sqrt{2\pi\alpha^2}} e^{-(g-m)^2/2\alpha^2} \tag{6.1}$$

where g represents the gray level, m is the mean of the function and α is standard deviation in the noise.

6.3.2 Speckle noise

Speckle noise is multiplicative noise: it is in direct proportion to the local gray level in any area. This type of noise occurs mostly in all coherent imaging systems such as acoustics, laser, medical ultrasound, and synthetic aperture radar (SAR) imagery. Speckle noise follows a gamma distribution function given by Equation 6.2 [29]:

$$F(g) = \frac{g^{\alpha-1}}{(a-1)!a^{\alpha}} e^{-\frac{g}{a}} \tag{6.2}$$

where a^{α} is variance and g is gray level.

6.3.3 Poisson noise

The appearance of this noise is seen due to the statistical nature of electromagnetic waves such as X-rays, visible lights, and gamma rays. The X-ray and gamma ray sources emitted the number of photons per unit time. These rays are injected into a patient's body from its source in medical X-rays and gamma rays imaging systems. These sources have a random fluctuation of photons [30].

6.3.4 Salt-and-pepper noise

Salt-and-pepper noise is a form of noise sometimes seen on images. This type of noise consists of random pixels being set to black and white pixels over the image. So this type of noise is termed additive noise. Salt-and-pepper noise is introduced in an image while capturing of images with a camera. Salt-and-pepper noise arises in the medical images due to unexpected disturbances in the image signal and sometimes may be due to an error in data transmission.

6.4 Denoising Techniques

There are two fundamental approaches to image denoising: spatial domain filtering and transform domain filtering methods. Spatial domain filtering works at pixel level and manipulates each pixel. Transform domain converts the image into frequency domain and then manipulates it. There are two spatial domain filtering techniques: the linear filter, viz., mean filter or Wiener filter; and the nonlinear filter, such as median filter. Transform domain filters are based on two newly developed approaches called wavelet transform and curvelet transform.

6.4.1 Spatial domain filtering

In spatial domain, the filtering operation is done on the pixel level, manipulating pixel by pixel. A filter mask is taken and it is shifted from one to another pixel and doing its operations. This smoothes the images and thereby removes the noise in them. The spatial domain contains two types: linear and nonlinear. The linear filter contains the mean filter and Wiener filter which are used the most but other types are also there. The most common nonlinear type is the median filter. In mean filter, the average value of neighbors along with the central pixel is calculated from a window which is of $n \times n$ size, then the central pixel is replaced with that new average value. It is given by Equation 6.3:

$$y[m,n] = mean\{x[i,j],(i,j)\epsilon\omega\} \tag{6.3}$$

where ω are neighborhood pixels.

The Wiener filter is basically used to filter constant power additive noise. It is a pixel wise adaptive level filter. Using the neighborhood of the pixels, the local mean and standard deviation is calculated. The mean is given by Equation 6.4:

$$\mu = \frac{1}{NM} \sum_{(n1,n2\epsilon\eta)} a(n_1.n_2) \tag{6.4}$$

The variance is

$$\alpha^2 = \frac{1}{NM} \sum_{(n1,n2\epsilon\eta)} a^2(n_1.n_2) - \mu^2 \tag{6.5}$$

where η is N-by-M neighborhood of the current pixel. Then using these estimates, the pixel-wise Wiener filter is created, given by Equation 6.6:

$$b(n_1,n_2) = \mu + \frac{\alpha^2 - \nu^2}{\alpha^2}(a(n_1,n_2) - \mu) \tag{6.6}$$

where ν^2 is noise variance.

In the median filter, the pixels in the window are sorted in ascending order, then the central pixel is replaced with that pixel. They are defined by Equation 6.7:

$$y[m,n] = median\{x,[i,j],(i,j)\epsilon\omega\} \tag{6.7}$$

where ω are neighborhood pixels.

6.4.2 Wavelet transform

In wavelet domain denoising, wavelets are used, which is very useful for representing nonlinear signals. Lang et al. [11] used a wavelet analysis of undecimated wavelet transforms on unidimensional signals and showed to remove noise, which was one of the earlier implementations of wavelet in noise removal. A mother wavelet is denoted as ψ. The daughters of the mother wavelets are superpositioned to denote function and then denoted by Equation 6.8:

$$\psi(a,b)(x) = (\frac{a}{\sqrt{a}}) * \psi(\frac{x-b}{a}) \tag{6.8}$$

where a is the dialation parameter and b is the translation parameter. Continuous wavelet transform can be denoted by Equation 6.9:

$$\psi(a,b) = \int_{R^2} f(x)\bar{\psi}_{a,b}(x).d(x) \tag{6.9}$$

Now it is necessary to reconstruct $f(x)$. So the inverse of the transform is done using Equation 6.10:

$$f(x) = \int \int_{-\infty}^{\infty} w(a,b).\psi(a,b)(x).\frac{da.db}{a^2} \quad (6.10)$$

Wavelet is an important tool for the nonlinear representation of signals. It decomposes the noisy image into time and frequency components. The noisy image is decomposed into four subsamples according to their low (L) and high (H) frequency bands called LL, LH, HL, and HH. The LL subsample is further decomposed into four subsamples at level two [8,31] and so on, as per the requirement of the computation.

6.4.3 Curvelet transform

The disadvantage of wavelet denoising is that it does not do a good job while of denoising the curves of an image and results in the loss of details. Stark and Candes [21] solved the problem of the wavelet transform by using the curvelet transform based on ridgelet transform. Ridgelet implementation was done by converting it into radon transform. Figure 6.1 illustrates the overview of discrete curvelet transform. In this process the noisy image is decomposed into sub-bands followed by the spatial partitioning of each sub-band. The ridgelet transform is then applied to each block In the ridgelet transform, the support interval or the scaling is done by using the anisotrophy scaling relationship, denoted by Equation 6.11:

$$width = (length)^2 \quad (6.11)$$

This was done in the first generation of the curvelet transform using the multiscaling ridgelet, where the curve is divided into blocks and the sub-blocks are approximated into a straight line and do ridgelet analysis. The basic curvelet decomposition steps are given as follows (Equation 6.12):

$$f \mapsto (P_0 f, \triangle_1 f, \triangle_2 f, ...) \quad (6.12)$$

where P_0 are sub-band filters, and $\triangle_S$, S >= 0 sub-bands $\triangle_s f$ contain details about 2^{-2s} wide. The smooth windows are $w_Q(x_1, x_2)$ which are localized in diadic squares and which is defined by Equation 6.13:

$$Q = [k_1/2^s, (k_1+1)/2^s] \times [k_2/2^s, (k_2+1)/2^s] \quad (6.13)$$

Then the resulting square is renormalized to unit scale, which is represented by Equation 6.14:

$$g_Q = T_Q^{-1}(w_Q \triangle_s f), Q \epsilon Q_s \quad (6.14)$$

where $(T_Q f)(x_1, x_2) = 2^s f(2^s x_1 - k_1, 2^s x_2 - k_2)$ is a renormalization operator. After the renormalization, the ridgelet transform is done by Equation 6.15:

$$\alpha_\mu =< g_Q, \rho_\lambda > \quad (6.15)$$

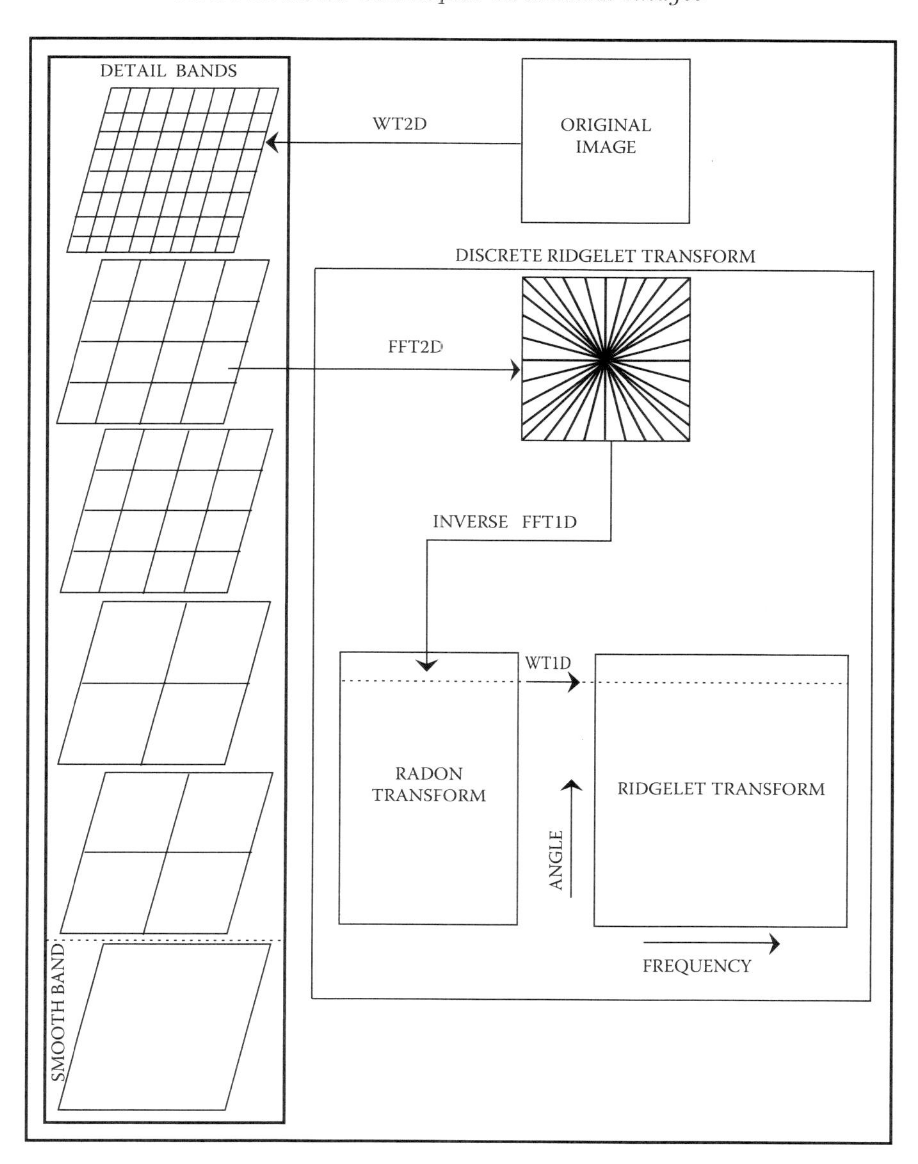

FIGURE 6.1: Block diagram of curvelet transform process. (From J.L. Starck et al., *Transactions on Image Processing*, 11, 670–684, 2002.)

6.5 Thresholding Technique

Thresholding in transform domain is achieved by hard thresholding and soft thresholding to remove unwanted noise signals. Hard thresholding removes all the value after a certain limit and soft thresholding lowers the intensity of noise towards zero values, which is defined by Equations 6.16 and 6.17:

$$y(t)_{Hard} = \begin{cases} x(t), |x(t)| >= T \\ 0, |x(t)| < T \end{cases} \tag{6.16}$$

$$y(t)_{Soft} = \begin{cases} sign(x(t)).(|x(t)| - T), |x(t) >= T| \\ 0, |x(t)| < T \end{cases} \tag{6.17}$$

where T is the threshold value, and x and y are input and output coefficients in the respective transform domain. The threshold value in the wavelet domain is calculated by Donoho et al. [12], as the VisuShrink method, which is based on universal threshold as explained in Equation 6.18:

$$T_{\omega} = \alpha\sqrt{log(N)} \tag{6.18}$$

where T is the threshold value, N is the size of the image, and α is the noise variance. The threshold value in the curvelet transform is calculated by a value of 3*sigma and 4*sigma [21] used for the coarse scale and fine scale elements (Equation 6.19):

$$T_c = 3 * sigma + sigma * (s == length(C)) \tag{6.19}$$

where C is the size of decomposed images, s=2 to length of C (Equation 6.9).

6.6 Proposed Technique

In this chapter, we proposed a new method for medical image denoising which is based on curvelet transform using the concept of thresholding functions combined with the Wiener filter. Initially, a noisy image is acquired with various types of noises with different noise factors. Then, threshold estimation is done followed by the application of the discrete curvelet transform to the acquired noisy image. We apply the wrapping method to implement the curvelet transform to obtain curvelet coefficients of distinct scale and orientations. The curvelet transform decomposes the acquired noisy image into different sub-bands; each sub-band is partitioned into different sub-blocks, the partitions are smoothed, the image is renormalized, and the ridgelet transform is applied. Then we apply the computed threshold value to the curvelet coefficient. After that we apply the inverse transform on the noisy image to get the denoised image. The residual noise, if any, is eliminated by the application of the Wiener filter. A quantitative estimation of the different parameters is done to evaluate the quality of the resulting denoised image. We found that the curvelet transform helps to overcome the problem of the wavelet transform, it also preserves the edge information. The curvelet transform process removes most of the noises present in the acquired image. Also, remaining residual noises are removed using the Wiener filter.

The proposed denoising method consists of the following steps:

Step 1: Read the input medical images.

Step 2: Add the different types of noise.

Step 3: Compute the threshold value.

Step 4: Apply wavelet/curvelet transform to noisy image.

Step 5: Apply thresholding to the wavelet coefficients/curvelet coefficients.

Step 6: Apply inverse wavelet/curvelet transform on step 5.

Step 7: Apply Wiener filter on output of step 6 to obtain denoised image.

Step 8: Calculate and compare PSNR, MSE, and SSIM values of denoised image with noisy image.

6.7 Parameter Estimations

To evaluate the performance of the techniques, we have considered the values of peak signal-to-noise ratio (PSNR), mean square error (MSE), and structural similarity index measure (SSIM), which are defined by Equations 6.20, 6.21, and 6.26,

$$PSNR = 10.log_{10}(\frac{MAX_I^2}{MSE}) \tag{6.20}$$

$$MSE = \frac{1}{mn} \sum_{i=0}^{m-1} \sum_{j=0}^{n-1} [f(i,j) - g(i,j)]^2 \tag{6.21}$$

where mn is size of image, MAXI is maximum probable pixel value of the image, $f(i, j)$ is the noisy image, and $g(i, j)$ is the denoised image. The SSIM accesses luminance, contrast, and structure of an image using a reference image [32].

$$SSIM(x, y) = [l(x, y)]^{\alpha}.[c(x, y)]^{\beta}.[s(x, y)]^{\gamma} \tag{6.22}$$

where,

$$l(x, y) = \frac{2\mu_x\mu_y + C_1}{\mu_x^2 + \mu_y^2 + C_1} \tag{6.23}$$

$$c(x, y) = \frac{2\sigma_x\sigma_y + C_2}{\sigma_x^2 + \sigma_y^2 + C_2} \tag{6.24}$$

$$s(x, y) = \frac{\sigma_{xy} + C_3}{\sigma_x\sigma_y + C_3} \tag{6.25}$$

where are μ_x , μ_y local means, σ_x, σ_y are standard deviations, σ_{xy} is cross covariance for images x, y. If $\alpha = \beta = \gamma = 1$ and $C_2 = C_2/2$, then SSIM can be written as follows:

$$SSIM(x, y) = \frac{(2\mu_x\mu_y + C_1)(2\sigma_{xy} + C_2)}{(\mu_x^2 + \mu_y^2 + C_1)(\sigma_x^2 + \sigma_y^2 + C_2)} \tag{6.26}$$

6.8 Experimentation

In this chapter we considered three sets of medical images: an MRI image of brain image, a mammogram X-ray image, and a thyroid ultrasound image to investigate our proposed method. Various types of noise like white Gaussian, speckle, Poisson, and salt-and-pepper noises are added. The experiments were performed in the MATLAB® environment via the wrapping technique which was implemented based on the curvelet software package. The noise factor of $\sigma = 30$ for Gaussian noise, noise factor $\sigma = 30$ for speckle noise and Poisson noise, and noise density of 0.03 for salt-and-pepper noise are used. Then the various types of denoising techniques were implemented, viz. mean filter, Wiener filter, median filter, wavelet thresholding, curvelet thresholding, and the proposed method curvelet thresholding with Wiener filter.

6.9 Results and Discussion

The performance of the proposed method, with other existing techniques, were evaluated using the quality measures of PSNR, MSE, and SSIM. The experiments were conducted on medical images, viz., MRI brain image, X-ray mammogram image, and ultrasound thyroid image for different types of noises. The tested images were degraded by adding different types of noise, viz., Gaussian noise, speckle noise, Poisson noise, salt-and-pepper noise with different noise factors. Our proposed method showed an improve in performance in context to PSNR, MSE and SSIM values as depicted in Tables 6.1 through 6.3 respectively for the MRI brain image. The PSNR values obtained by proposed method showed an increase in comparison to other techniques (Figure 6.2a). The MSE and SSIM values, respectively, observed in this technique were found to be lower and higher than the other techniques which were

TABLE 6.1: Comparison of PSNR values for brain MRI image

Denoising techniques	Gaussian noise	Speckle noise	Poisson noise	Salt–pepper noise
Mean filter	21.242	28.604	29.155	19.273
Wiener filter	23.676	35.291	35.130	31.886
Median filter	20.725	28.426	28.959	18.751
Wavelet–hard	20.158	25.338	25.679	18.609
Wavelet–soft	19.385	23.089	23.250	17.835
Curvelet–hard	20.941	27.945	28.535	24.427
Curvelet–soft	20.046	24.827	25.077	20.109
Proposed Method	44.571	46.166	46.089	41.146

TABLE 6.2: Comparison of MSE values for brain MRI image

Denoising techniques	Gaussian noise	Speckle noise	Poisson noise	Salt–pepper noise
Mean filter	488.486	89.682	78.987	68.793
Wiener filter	278.900	19.230	19.954	42.118
Median filter	550.280	93.420	82.638	866.990
Wavelet–hard	626.985	190.212	175.861	895.666
Wavelet–soft	749.175	319.252	307.670	1070.597
Curvelet–hard	523.517	104.374	91.113	234.633
Curvelet–soft	643.450	213.968	202.006	634.153
Proposed Method	2.270	1.572	1.600	4.994

TABLE 6.3: Comparison of SSIM values for brain MRI image

Denoising techniques	Gaussian noise	Speckle noise	Poisson noise	Salt-and-pepper noise
Mean filter	0.353	0.908	0.910	0.556
Wiener filter	0.503	0.938	0.930	0.960
Median filter	0.285	0.913	0.914	0.491
Wavelet–hard	0.206	0.699	0.700	0.295
Wavelet–soft	0.173	0.643	0.643	0.234
Curvelet–hard	0.253	0.764	0.766	0.674
Curvelet–soft	0.204	0.684	0.686	0.459
Proposed method	0.990	0.994	0.994	0.970

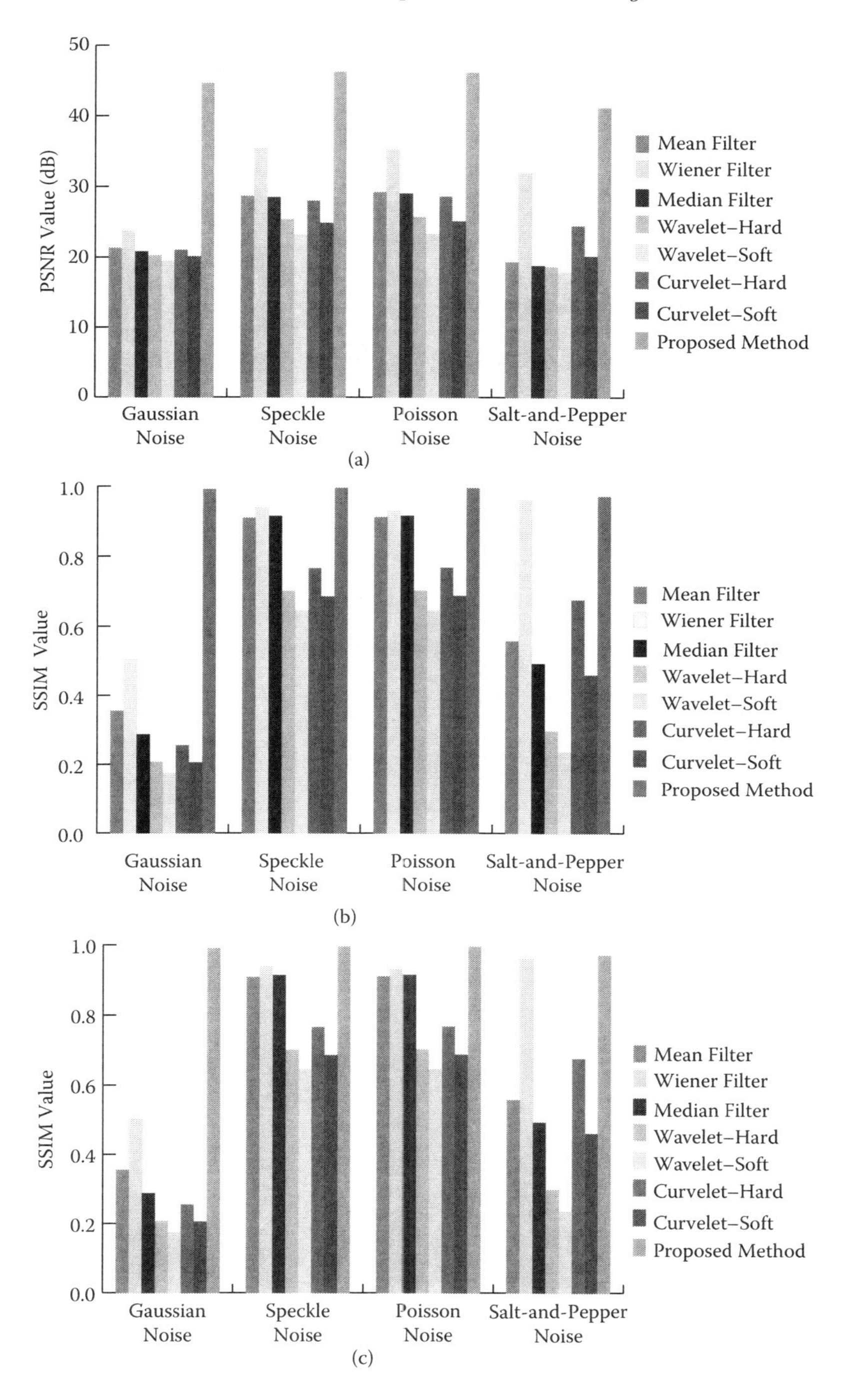

FIGURE 6.2: Graphical representation of different denoising methods obtained for brain MRI image: (a) comparison of PSNR values, (b) comparison of MSE values, and (c) comparison of SSIM values.

experimented in this chapter (Figure 6.2a and c). Figure 6.3 depicts the denoising results on the MRI images in comparison with different methods for Gaussian noise at a standard deviation of $\sigma = 30$. Tables 6.4 through 6.6 respectively report the PSNR, MSE, and SSIM values of mammogram images obtained by the proposed method with other techniques

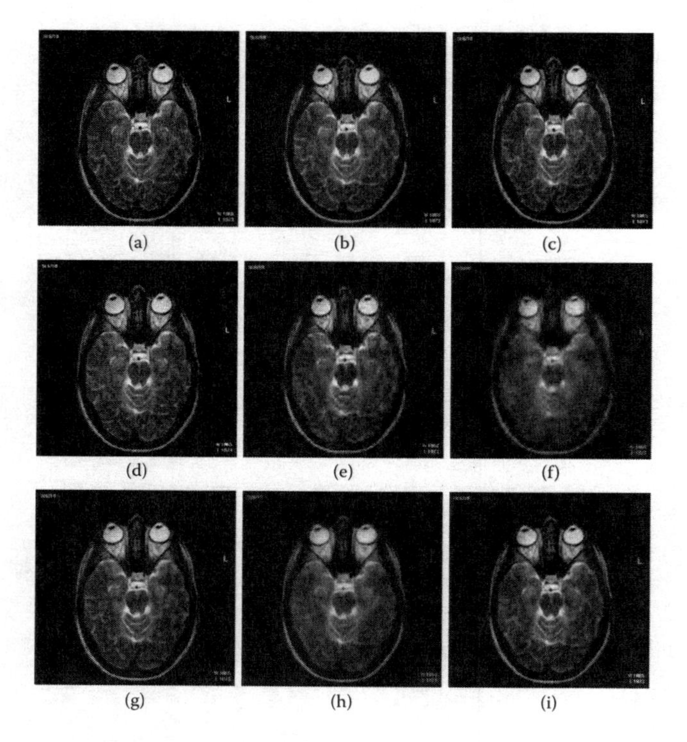

FIGURE 6.3: Visual representation of MRI brain image denoising (Gaussian noise, where $\sigma = 30$): (a) noisy image, (b) mean filter, (c) Wiener filter, (d) median, (e) wavelet–hard, (f) wavelet–soft, (g) curvelet–hard, (h) curvelet–soft, and (i) proposed method.

TABLE 6.4: Comparison of PSNR values for mammogram image

Denoising techniques	Gaussian noise	Speckle noise	Poisson noise	Salt-and-pepper noise
Mean filter	21.175	27.124	29.214	18.925
Wiener filter	23.479	32.413	33.137	31.894
Median filter	20.798	27.593	30.480	19.352
Wavelet–hard	20.565	32.413	28.066	19.216
Wavelet–Soft	20.105	27.595	25.936	18.662
Curvelet–hard	20.779	26.806	28.969	24.566
Curvelet–soft	20.496	25.964	27.658	20.679
Proposed method	46.834	48.599	48.563	36.085

TABLE 6.5: Comparison of MSE values for mammogram image

Denoising techniques	Gaussian noise	Speckle noise	Poisson noise	Salt-and-pepper noise
Mean filter	496.130	126.081	77.922	832.910
Wiener filter	291.880	37.308	31.579	42.040
Median filter	541.170	113.190	58.225	754.800
Wavelet–hard	570.925	37.306	101.514	778.934
Wavelet–soft	634.657	217.595	165.758	884.951
Curvelet–hard	543.489	135.655	82.453	227.247
Curvelet–soft	580.070	164.693	111.503	556.087
Proposed method	1.348	0.898	0.905	16.018

TABLE 6.6: Comparison of SSIM values for mammogram image

Denoising techniques	Gaussian noise	Speckle noise	Poisson noise	Salt-and-pepper noise
Mean filter	0.287	0.771	0.792	0.401
Wiener filter	0.479	0.899	0.856	0.948
Median filter	0.224	0.778	0.792	0.484
Wavelet–hard	0.158	0.899	0.643	0.313
Wavelet–soft	0.147	0.582	0.605	0.279
Curvelet–hard	0.172	0.653	0.677	0.656
Curvelet–soft	0.159	0.604	0.628	0.457
Proposed method	0.991	0.993	0.993	0.953

for different types of noise, which were investigated in this study. Figure 6.4a through c show the graphical plot of PSNR, MSE, and SSIM values for the mammogram image in compression with different techniques for different types of noises. Whereas Figure 6.5 shows the noisy and denoise images of a mammogram image in compression with different methods for speckle noise at noise level $\sigma = 30$; Tables 6.7 through 6.9 depict the PSNR, MSE, and SSIM values for the ultrasound thyroid image of the proposed denoise method and comparison with different techniques. In Figure 6.6a through c shows the plot of PSNR, MSE, and SSIM values for the thyroid image in comparison with different techniques. Figure 6.7 depicts the noisy and denoised images of the ultrasound image obtained by the proposed method in compression with different methods for Poisson noise. Figure 6.8 shows the noisy and denoised MRI images obtained by the proposed method and compression with different techniques for salt-and-pepper noise with noise density 0.03. It was also observed that the proposed methods are not able to completely remove salt-and-pepper noise, but in the case of other noises, viz., Gaussian noise, speckle noise, and Poisson noise it performed well. In Table 6.10 reported the PSNR results obtained by different methods at various noise levels, where $\sigma = 20$, $\sigma = 30$, and $\sigma = 40$, and $\sigma = 50$. Figure 6.9 plots the comparison of various denoising techniques for MRI brain image at different noise levels. From Figure 6.9 it can be observed that the proposed method work better compared to other techniques in case of higher noise level.

6.10 Conclusion

In this chapter, we have studied different types of denoising techniques of spatial domain filtering, viz., mean filter, Wiener filter, median filter, and transform domain filtering, viz., wavelet–hard, wavelet–soft, curvelet–hard, and curvelet–soft thresholding. We have also propose a new denoising method based on the combination of curvelet–hard thresholding and Wiener filter. The overall performance of these techniques are measured in terms of quantitative parameters like PSNR, MSE, and SSIM values. The proposed method has shown a little bit of improvement compared to other techniques which are experimented within the present study. Also, it has been observed that the salt-and-pepper noise is not completely removed by the proposed method, but in the case of other noises like Gaussian noise, Poisson noise, and speckle noise, the proposed method has performed better in denoising the medical image.

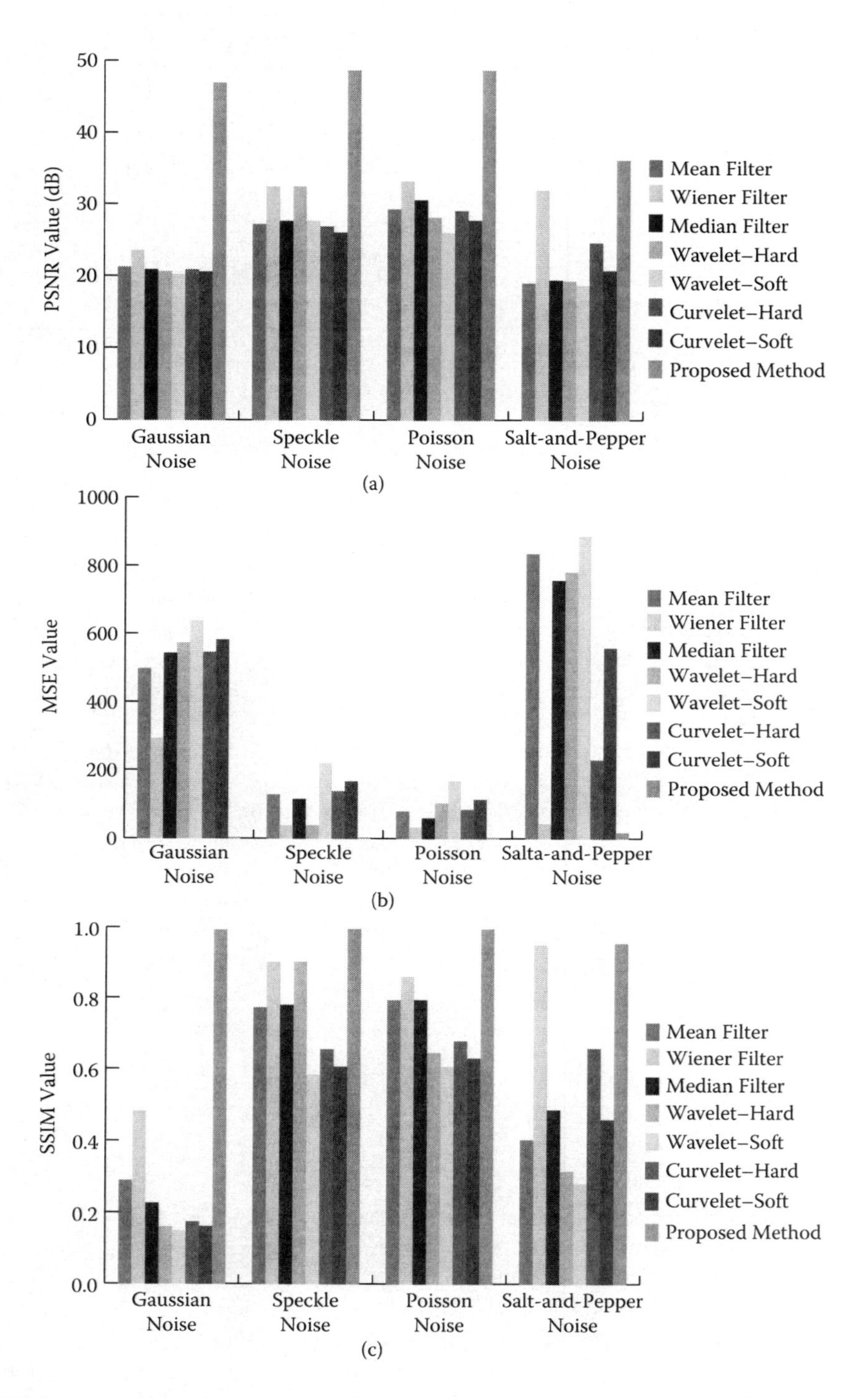

FIGURE 6.4: Graphical representation of different denoising methods obtained for mammogram image: (a) comparison of PSNR values, (b) comparison of MSE values, and (c) comparison of SSIM values.

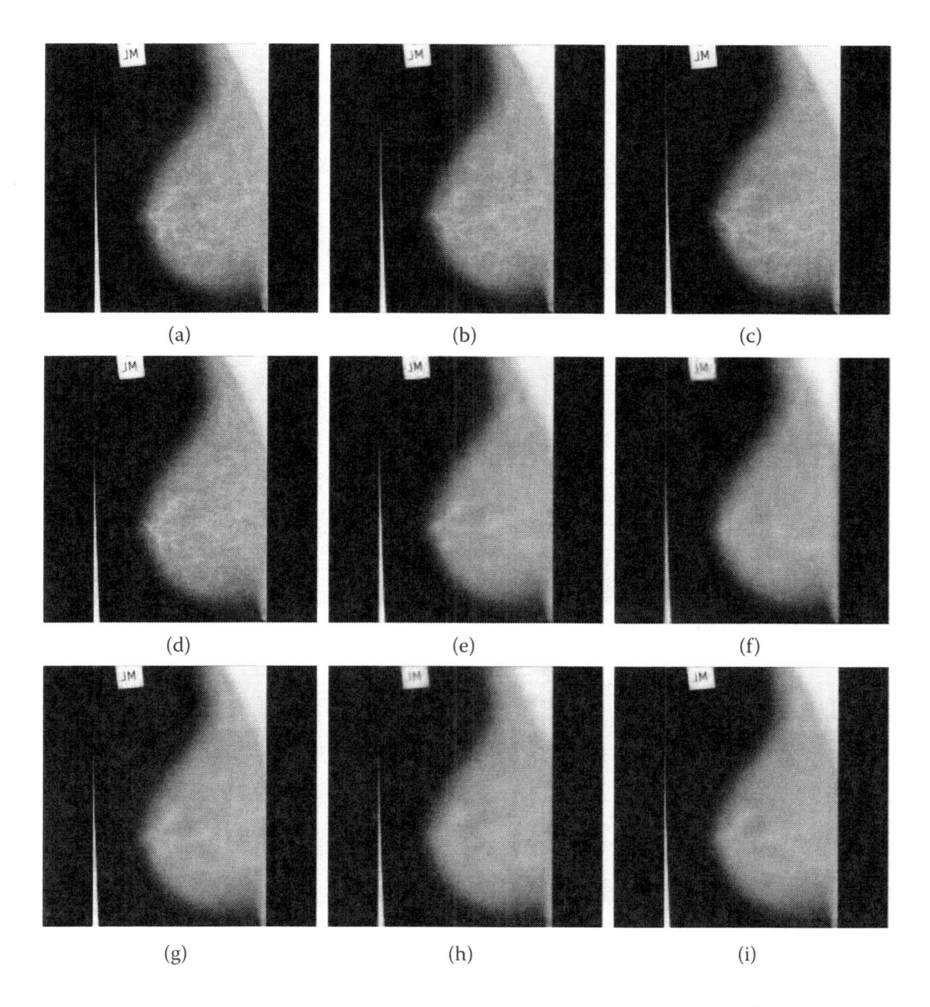

FIGURE 6.5: Visual representation of mammogram image denoising (Poisson noise, where $\sigma = 30$): (a) noisy image, (b) mean filter, (c) Wiener filter, (d) median, (e) wavelet–hard, (f) wavelet–soft, (g) curvelet–hard, (h) curvelet–soft, and (i) proposed method.

TABLE 6.7: Comparison of PSNR values for thyroid ultrasound image

Denoising techniques	Gaussian noise	Speckle noise	Poisson noise	Salt-and-pepper noise
Mean filter	19.951	28.699	29.306	20.023
Wiener filter	21.878	32.119	31.997	30.462
Median filter	19.640	28.871	29.648	19.595
Wavelet–hard	18.823	24.553	24.789	19.15
Wavelet–soft	18.494	23.447	23.626	18.666
Curvelet–hard	20.939	26.230	26.556	23.799
Curvelet–soft	20.049	24.555	24.763	20.312
Proposed method	44.866	47.627	47.703	36.843

TABLE 6.8: Comparison of MSE values for thyroid ultrasound image

Denoising techniques	Gaussian noise	Speckle noise	Poisson noise	Salt-and-pepper noise
Mean filter	657.652	87.727	76.285	646.743
Wiener filter	421.930	39.920	41.054	58.465
Median filter	706.460	84.329	70.519	713.850
Wavelet–hard	852.637	227.920	215.850	790.802
Wavelet–soft	919.864	293.995	282.123	884.000
Curvelet–hard	523.788	154.918	143.698	271.116
Curvelet–soft	642.982	227.797	217.137	605.239
Proposed method	2.121	1.123	1.104	13.452

TABLE 6.9: Comparison of SSIM values for thyroid ultrasound image

Denoising techniques	Gaussian noise	Speckle noise	Poisson noise	Salt-and-pepper noise
Mean filter	0.320	0.796	0.787	0.536
Wiener filter	0.517	0.862	0.840	0.922
Median filter	0.298	0.803	0.795	0.489
Wavelet–hard	0.133	0.493	0.479	0.284
Wavelet–soft	0.110	0.447	0.431	0.237
Curvelet–hard	0.253	0.588	0.579	0.667
Curvelet–soft	0.204	0.494	0.480	0.416
Proposed method	0.987	0.993	0.993	0.967

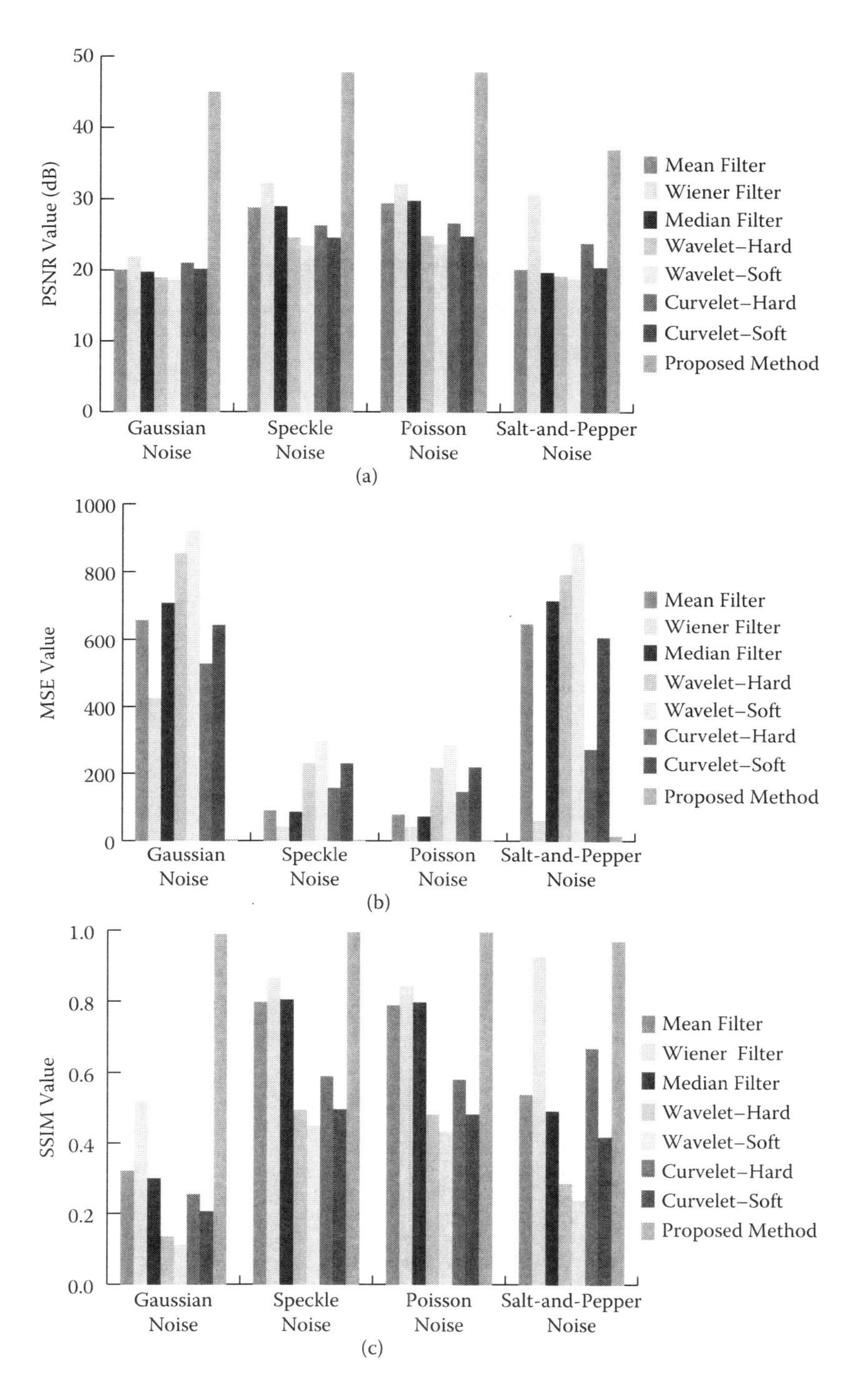

FIGURE 6.6: Graphical representation of different denoising methods obtained for thyroid ultrasound image: (a) comparison of PSNR values, (b) comparison of MSE values, and (c) comparison of SSIM values.

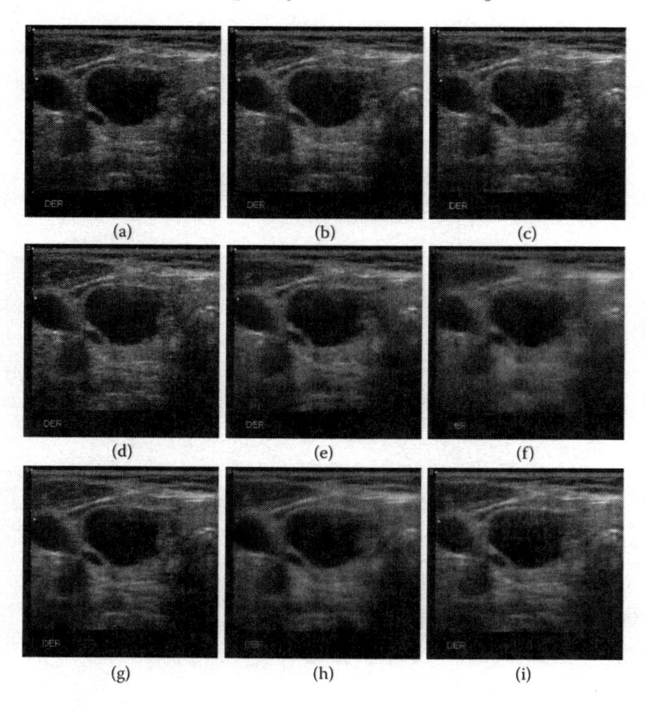

FIGURE 6.7: Visual representation of ultrasound thyroid image denoising (speckle noise): (a) noisy image, (b) mean filter, (c) Wiener filter, (d) median, (e) wavelet–hard, (f) wavelet–soft, (g) curvelet–hard, (h) curvelet–soft, and (i) proposed method.

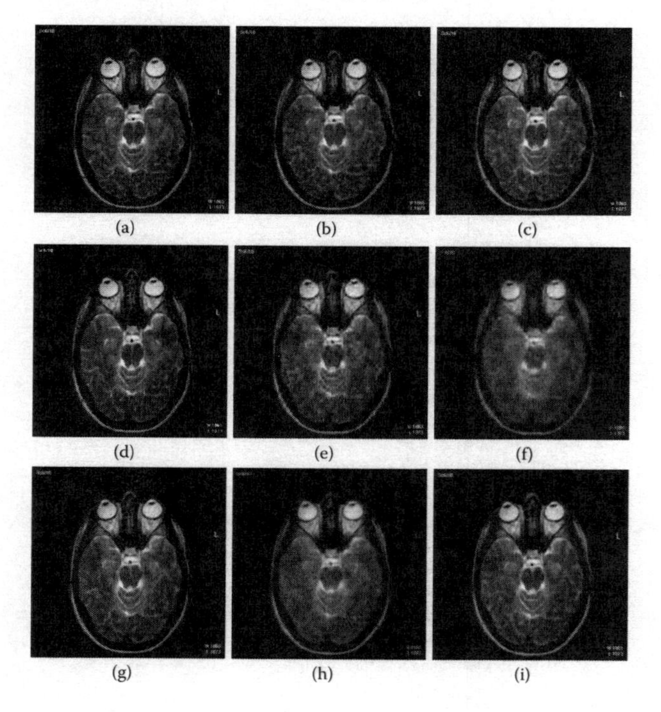

FIGURE 6.8: Visual representation of MRI brain image denoising denoising (salt-and-pepper noise, where noise density = 0.03): (a) noisy image, (b) mean filter, (c) wiener filter, (d) median, (e) wavelet–hard, (f) wavelet–soft, (g) curvelet–hard, (h) curvelet–soft, and (i) proposed method.

TABLE 6.10: Comparative results of the denoise MRI brain image with different techniques at different noise level

Denoising techniques	$\sigma = 20$	$\sigma = 30$	$\sigma = 40$	$\sigma = 40$
Mean filter	24.17	21.27	19.122	17.437
Wiener filter	26.611	23.708	21.552	19.849
Median filter	23.733	20.753	18.56	16.823
Wavelet–hard	23.952	20.875	18.743	17.064
Wavelet–soft	23.529	20.686	18.66	17.037
Curvelet–hard	24.336	20.932	18.608	16.829
Curvelet–soft	23.011	20.05	17.978	16.4
Proposed method	43.364	44.63	45.943	47.543

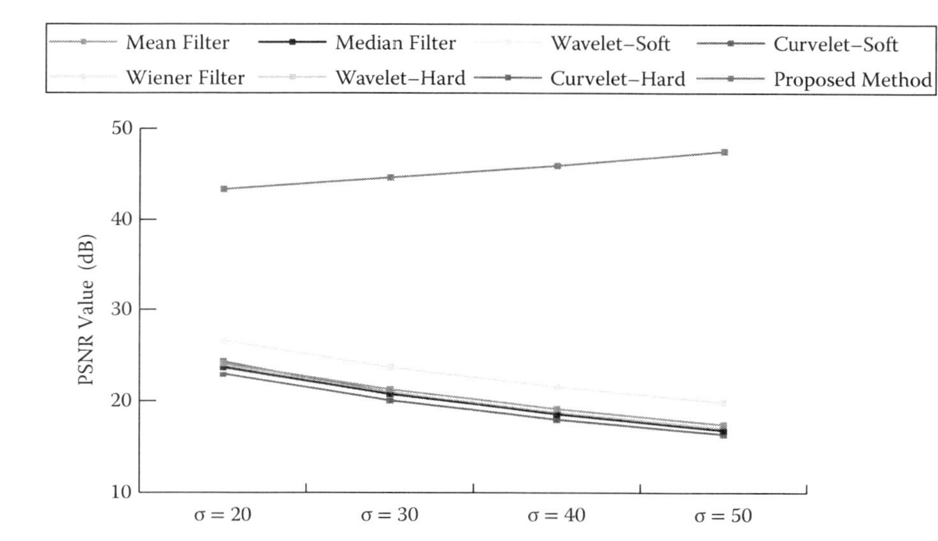

FIGURE 6.9: Comparison of PSNR values for denoising MRI brain image at different noise levels.

References

1. I. Bankman. *Handbook of Medical Imaging.* Academic Press, Orlando, FL, 2000.

2. K. Lu, N. He, and L. Li. Nonlocal means-based denoising for medical images. *Computational and Mathematical Methods in Medicine*, Vol. 2012, Article ID 438617, 7 pages, 2012, doi:10.1155/2012/438617.

3. A. Khare and U.S. Tiwary. A new method for deblurring and denoising of medical images using complex wavelet transform. In *27th Annual Conference of the Engineering in Medicine and Biology, 2005 IEEE*, Shanghai, China, pp. 1897–1900. IEEE, 2006.

4. A. Buades, B. Coll, and J.M. Morel. A review of image denoising algorithms, with a new one. *Multiscale Modeling and Simulation*, 4(2):490–530, 2005.

5. H. Ney. A dynamic programming technique for nonlinear smoothing. In *IEEE International Conference on Acoustics, Speech, and Signal Processing, ICASSP'81*, Atlanta, GA, vol. 6, pp. 62–65. IEEE, 1981.

6. Y.L. Kuo and C.W. Tai. A simple and efficient median filter for removing high-density impulse noise in images. *International Journal of Fuzzy Systems*, 17(1):67–75, 2015.

7. D. Boulfelfel, R.M. Rangayyan, L.J. Hahn, and R. Kloiber. Three-dimensional restoration of single photon emission computed tomography images. *IEEE Transactions on Nuclear Science*, 41(5):1746–1754, 1994.

8. R. Saluja and A. Boyat. Wavelet based image denoising using weighted highpass filtering coefficients and adaptive Wiener filter. In *International Conference on Computer, Communication and Control (IC4), 2015*, Indore, India, pp. 1–6. IEEE, 2015.

9. F. Xiao and Y. Zhang. A comparative study on thresholding methods in waveletbased image denoising. *Elsevier Advanced in Control Engineering and Information Science*, 15:3998–4003, 2011, doi:10.1016/j.proeng.2011.08.749.

10. A. Jaiswal, J. Upadhyay, and A. Somkuwar. Image denoising and quality measurements by using filtering and wavelet based techniques. *International Journal of Electronics and Communications (AEU)*, 68(8):699–705, 2014.

11. M. Lang, H. Guo, J.E. Odegard, C.S. Burrus, and R.O. Wells. Noise reduction using an undecimated discrete wavelet transform. *IEEE Signal Processing Letters*, 3(1):10–12, 1996.

12. S.G. Chang, B. Yu, and M. Vetterli. Adaptive wavelet thresholding for image denoising and compression. *IEEE Transactions on Image Processing*, 9(9):1532–1546, 2000.

13. A. Mojsilovic, M. Popovic, and D. Sevic. Classification of the ultrasound liver images with the 2n × 1-d wavelet transform. In *International Conference on Image Processing, 1996. Proceedings.*, Lausanne, Switzerland, vol. 1, pp. 367–370. IEEE, 1996.

14. H.-J. Zhu. Wavelet based hybrid thresholding method for ultrasonic liver image denoising. *Journal of Shanghai Jiaotong University*, 20(2):135–142, 2015.

15. P. Gorgel, A. Sertbas, and O.N. Ucan. A wavelet based mammographic image denoising and enhancement with homomorphic filtering. *Journal of Medical Systems*, 34(6):993–1002, 2010.

16. D.L. Donoho. De-noising by soft-thresholding. *IEEE Transactions on Information Theory*, 41(3):613–627, 1995.

17. D.L. Donoho and I. M. Johnstone. Adapting to unknown smoothness via wavelet shrinkage. *Journal of the American Statistical Association*, 90(432):1200–1224, 1995.

18. X. Zhang. A denoising approach via wavelet domain diffusion and image domain diffusion. *Multimedia Tools and Applications*, 76(11): 13545–13561, 2016.

19. A.L. Pogam, H. Hanzouli, M. Hatt, C.C.L. Rest, and D. Visvikis. Denoising of PET images by combining wavelets and curvelets for improved preservation of resolution and quantitation. *Medical Image Analysis*, 17(8):877–891, 2013.

20. E.J. Candès and D. L. Donoho. Ridgelets: A key to higher-dimensional intermittency? *Philosophical Transactions of the Royal Society of London A: Mathematical, Physical and Engineering Sciences*, 357(1760):2495–2509, 1999.

21. J.L. Starck, E.J Candès, and D.L. Donoho. The curvelet transform for image denoising. *IEEE Transactions on Image Processing*, 11(6):670–684, 2002.

22. L. Yucheng and L. Yubin. Study on the basic principle and image denoising realization method of curvelet transform. In *2010 International Conference on Multimedia Technology (ICMT),* Ningbo, China, pp. 1–4. IEEE, 2010.

23. C. Amiot, C. Girard, J. Chanussot, J. Pescatore, and M. Desvignes. Curvelet based contrast enhancement in fluoroscopic sequences. *IEEE Transactions on Medical Imaging,* 34(1):137–147, 2015.

24. F.E. Ali, I.M. El-Dokany, A.A. Saad, and F.E. Abd El-Samie. Fusion of MR and CT images using the curvelet transform. In *National Radio Science Conference, 2008 (NRSC 2008),* Tanta, Egypt, pp. 1–8. IEEE, 2008.

25. M.O. Ulfarsson, J.R. Sveinsson, and J.A Benediktsson. Speckle reduction of SAR images in the curvelet domain. In *IEEE International Geoscience and Remote Sensing Symposium, 2002 (IGARSS'02. 2002),* Toronto, Canada, vol. 1, pp. 315–317. IEEE, 2002.

26. R. Moore, S. Ezekiel, and E. Blasch. Denoising one-dimensional signals with curvelets and contourlets. In *IEEE National Aerospace and Electronics Conference (NAECON 2014),* Dayton, OH, pp. 189–194. IEEE, 2014.

27. N. Aghazadeh, F. Akbarifard, and L. S. Cigaroudy. A restoration–segmentation algorithm based on exible arnoldi–tikhonov method and curvelet denoising. *Signal, Image and Video Processing,* 10(5):935–942, 2016.

28. S. Anila, S.S. Sivaraju, and N. Devarajan. A new contourlet based multiresolution approximation for MRI image noise removal. *National Academy Science Letters,* 40(1):39–41, 2017.

29. A.K. Boyat and B.K. Joshi. A review paper: Noise models in digital image processing. *Signal and Image Processing: An International Journal* (SIPIJ), 6(2):63–75, 2015.

30. R. Garg and A. Kumar. Comparison of various noise removals using Bayesian framework. *International Journal of Modern Engineering Research (IJMER),* 2(1):265–270, 2012.

31. A. Joshi, K.A. Boyat, and K.B. Joshi. Impact of wavelet transform and median filtering on removal of salt and pepper noise in digital images. In *IEEE International Conference on Issues and Challenges in Intelligent Computing Techniques (ICICT),* pp. 838–843, Ghaziabad, India, 2014.

32. Z.Wang, A.C. Bovik, H.R. Sheikh, and E.P. Simoncelli. Image quality assessment: From error visibility to structural similarity. *IEEE Transactions on Image Processing,* 13(4):600–612, 2004.

Chapter 7

Face Recognition–Oriented Biometric Security System

Geetika Singh and Indu Chhabra

7.1 Introduction

Face recognition is emerging as one of the most widely used biometric systems. It has several potential applications like authentication, access control, facilitating crime investigations, carrying out secure e-commerce transactions, surveillance, finding missing people, and human computer interaction. Computer-based face recognition methods ascertain the identity of a person on the basis of measurable physical characteristics, referred as facial features, which can be checked automatically. The features extracted, however, are sensitive to different kinds of disparities that a face image undergoes. Faces can be captured in different light conditions, at different poses, with occlusion, and with different facial expressions. Accurate representation of faces under such real-life scenarios is therefore a crucial problem that still needs to be addressed.

The complete face recognition module is comprised of two phases: training phase and the testing phase. Figure 7.1 depicts the phases of a typical face recognition system. The training phase involves training the recognition system for identification of the enrolled users. The testing phase performs identification of the input test face image. Each phase is comprised of three main modules: preprocessing, feature extraction, and classification module. Input into the system is an unknown test image. The preprocessing module focuses on segmenting the individual's face (input test image) and normalizing the segmented face with respect to the geometrical and photometrical properties such as gray-scale conversion, noise removal, and illumination correction. The normalized faces are then fed to the feature extraction module, wherein the important discriminating information is extracted. The classification module of the train phase learns to match the extracted features with the face classes specified as

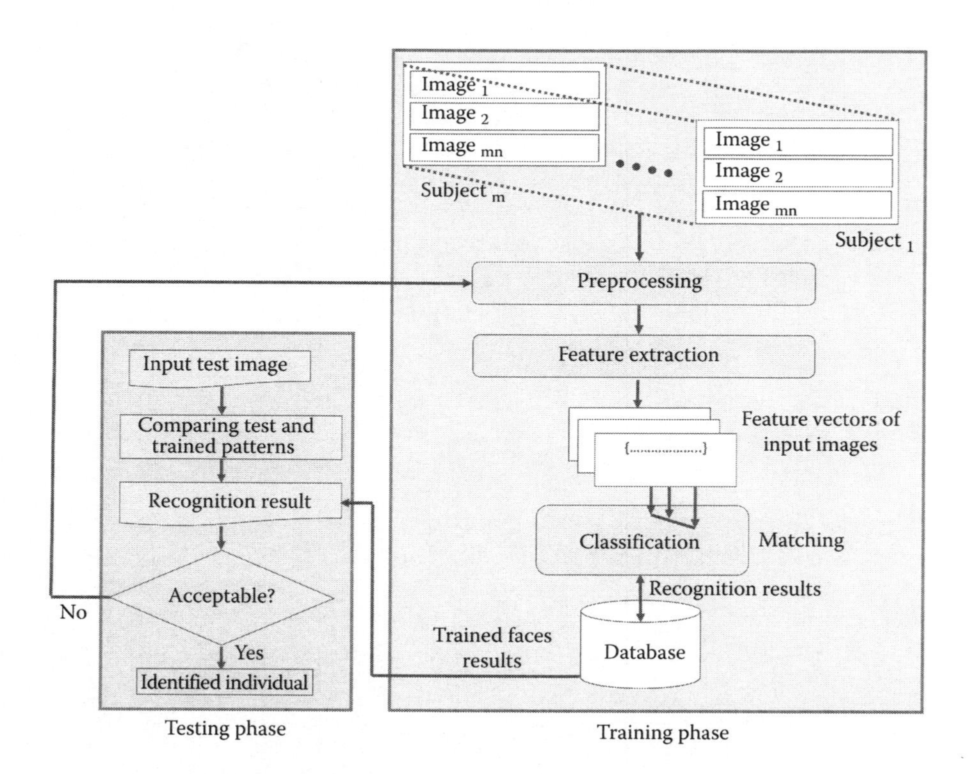

FIGURE 7.1: Schema of the face recognition ensemble.

the correct outputs. In the test phase, the classification module matches the features of the input test image with each face class stored in the database and classifies the face to the desired class based on its deviation from the template classes. The final output from the system is the identification of the recognized individual.

Success of the face recognition system depends entirely on the discriminative and invariant features obtained from the feature extraction techniques to represent the face images. Selection of an appropriate facial feature extraction technique is, therefore, a crucial factor for achieving high recognition accuracy. In this regard, several techniques have been presented in the literature that are categorized into structural and statistical methods. Structural techniques extract the local features of the face through the geometrical relationships of the feature points with the major face components such as width of the eyes and thickness of the eyebrows [1–5]. Statistical methods retrieve the global face information without using any face geometry explicitly. These are classified as subspace-based methods [6–14], histograms-based [15,16], filters [17,18], frequency transforms [19–21], and orthogonal radial invariant moments [22–30]. Subspace-based methods are the dimensionality reduction methods that map the face images to a low dimensional space. These include popular methods of principal component analysis, linear discriminant analysis, and independent component analysis. Histogram-based methods, such as local binary pattern histogram (LBP) and histogram of oriented gradients (HOG), extract the texture properties of an image. Filters, such as Gabor filter–based methods, are band-pass filters which capture the multi-orientational and multiscale information from an image. The spatial-frequency techniques, such as Fourier transform (FT) and discrete cosine transform (DCT), extract facial features at some preferred frequency. The images are first transformed to the frequency domain and thereafter, coefficients of low frequency band are taken as the invariant image features. Moments-based methods are the most widely used face descriptors. These are less susceptible to information redundancy, invariant to the variations of size, pose, scale, illumination, and tilt as well as rotation and are insensitive to noise, which makes them superior over the existing methods. These consist of the well-established techniques like Zernike moments (ZMs) [22–25], pseudo-Zernike moments (PZMs) [26,27], and polar complex exponential transform (PCET) [28–30]. The magnitude of these moments is used as an image descriptor, as it is invariant to rotation and can be made invariant to translation and scale through normalization. Among the moments-based methods, facial shape representation with ZM and PCET has exhibited efficient recognition ability. ZM and PCET coefficients are robust to noise and also possess minimum information redundancy and efficient image reconstruction ability.

It is well known that all the features in the feature vector do not possess same discriminative competence and some are able to differentiate different classes better than others. Careful selection of such a set of features which have high discrimination strength may improve the recognition performance. Therefore, feature selection is crucial in face recognition to generate a compact and highly discriminative feature vector. Significant attempts in this regard are by Dabbaghchian et al. [21] and Shen et al. [31]. In the former one, wavelet moment invariants have been employed for extraction of features and only the within-class information has been used to select the most discriminative features. The method requires large number of training images to perform discrimination of objects. In contrast, the latter approach utilizes discrete cosine transform (DCT) for feature extraction and discrimination power analysis (DPA) for selection of the most discriminative DCT features. The DPA approach has also been applied for measuring the discrimination ability of ZM features wherein only the coefficients with higher discrimination values are utilized for recognition [32]. The resulting discriminative Zernike moment (DZM) technique has been proved to be superior to the DCT-DPA approach, especially in the case of in-plane image rotation and pose variations as DCT features are not rotation invariant by nature.

The present study proposes a discriminative feature selection approach where features are selected based on their calculated discriminative competence values. This discriminative competence is measured based on the extent of variability of the class average for each individual feature. Finally, recognition is performed using only those features that possess the highest discriminative competence. It is projected from the fact that the measure of the average is always the true representative and the balancing point of the estimated feature values for a particular individual. Therefore, the average of different feature values obtained from multiple images of the same face reduces the within-class differences such as those due to different poses, light, and expression conditions. This approach, in contrast to the traditional dimensionality reduction approaches of PCA and LDA, selects features from the original domain without any transformation and thus preserves their originality. It also exhibits lower computational complexity and places no limit on the number of features that can be selected.

The proposed feature selection method is made to select and use only the most discriminative ZM and PCET coefficients for performing face recognition and is referred to as modified discriminative ZM (MDZM) and discriminative PCET (DPCET) respectively. It is also compared with the conventional ZM and PCET techniques and the existing DZM method and is found to provide better results.

The present work makes three main contributions. First, a modified feature selection approach based on statistical analysis to find the most discriminative ZM and PCET coefficients from the traditional feature sets is developed as MDZM and DPCET respectively. Second, efficacy of the DZM, MDZM, and DPCET techniques is exhaustively tested on the benchmark ORL, Yale, and FERET face databases through both the independent as well as hybrid classifiers. These include the nearest neighbor (NN), back-propagation neural network (BPNN), support vector machine (SVM), and adaptive neuro fuzzy inference system (ANFIS). NN and BPNN work as independent classifiers while SVM and ANFIS have been used in a hybrid way. ANFIS combines the properties of both the neural nets and fuzzy techniques and SVM works on the clusters as well as the terminal boundary values. Third, a new self-created face database is also introduced and performance of the implemented techniques is also validated on this database.

7.2 Discriminative Feature Selection

The proposed approach is comprised of three stages: feature extraction using ZM and PCET; feature selection using the proposed discriminative technique; and classification using ED, BPNN, SVM, and ANFIS classifiers. Each of these stages is briefly described in the following subsections.

7.2.1 Feature extraction

Features are extracted from each of the test images using ZM and PCET separately. The extracted feature sets provide the shape characteristics which is essential for the classification of faces. The employed feature extraction methods are discussed in the following two subsections.

7.2.1.1 Zernike moments

ZM features are obtained by transforming an input image on a complex set of orthogonal Zernike polynomials to capture their invariance power [33]. The Zernike radial polynomials of order 'n' and repetition 'm' are computed as:

$$V_{nm}(x,y) = R_{nm}(x,y)e^{im\theta}; \tag{7.1}$$

where

$$i = \sqrt{-1}\ ;\ \theta = tan^{-1}\frac{y}{x}\ and \tag{7.2}$$

$$R_{nm}(x,y) = \sum_{s=0}^{\frac{n-|m|}{2}} \frac{(-1)^s (n-s)!(x^2+y^2)^{\frac{n-2s}{2}}}{s!\left(\frac{n+|m|}{2}-s\right)!\left(\frac{n-|m|}{2}-s\right)!} \tag{7.3}$$

such that $n \geq 0$, $|m| \leq n\ and\ n - |m| = even.$

Orthogonality of this radial kernel allows for maximum separation of data points and thus facilitates differentiation of different image classes. This makes ZM features suitable for object classification applications like character recognition and face recognition. After computing the radial kernel, Zernike moments are calculated for an $N \times N$ digital image through feature calculation of:

$$Z_{nm} = \frac{(n+1)}{\pi}\sum_{i=0}^{N-1}\sum_{j=0}^{N-1} f(x_i,y_j)\, V^*{}_{nm}(x_i,y_j)\, \Delta x_i\, \Delta y_j \tag{7.4}$$

where $V^*{}_{nm}(x,y)$ is the complex conjugate of $V_{nm}(x,y)$;

$$x_i = \frac{2i+1-N}{D}; y_j=\frac{2j+1-N}{D};\ {x_i}^2+{y_j}^2 = 1; \Delta x_i = \Delta y_j = \frac{2}{D}; \tag{7.5}$$

$i = 0, 1, 2....N-1;\ j = 0, 1, ...N-1;$

$$D = \begin{cases} N,\ for\ the\ disc\ inscribed\ within\ the\ image \\ N\sqrt{2},\ for\ the\ outer\ disc\ containing\ the\ complete\ image \end{cases}$$

Magnitude coefficients of these moments are hence used as the features because they are invariant to image rotation. This is a crucial parameter in case the images are aligned. Normalization (Equation 7.5) further makes the Zernike features translation and scale invariant. This whole application of Equations 7.1 through 7.5 makes ZM invariant to different geometrical transformations. Images are reconstructed through the application of inverse transformation as:

$$I_r = \sum_{n=0}^{n_m}\sum_{m=-n}^{n} Z_{nm}V_{nm}(x_i,\, y_j) \tag{7.6}$$

where I_r it the reconstructed image and n_m is the maximum order of moments.

7.2.1.2 Polar complex exponential transform

PCET, a category of 2D polar harmonic transforms, is a wave-based transformation whose kernel definition is harmonic in nature. PCET of order n and repetition l for an $M \times N$ digital image is defined as:

$$M_{nl}=a\sum_{i=0}^{M-1}\sum_{j=0}^{N-1} f(x_i,y_j)\, H^*{}_{nl}(x_i,y_j)\, \Delta x_i \Delta y_j \tag{7.7}$$

such that $|n| + |l| = c$, where $|n| = |l| = 0, 1, 2, \ldots, \infty$ *and* c is the maximum order used for the computation of PCET coefficients;

$$H_{nl}(x, y) = e^{i2\pi nr^2} e^{il\theta}; \tag{7.8}$$

where

$$i = \sqrt{-1}\ ;\ \theta = tan^{-1}\frac{y}{x}\ ;\ \alpha = \frac{1}{\pi} \tag{7.9}$$

$H^*{}_{nl}(x, y)$ is the complex conjugate of $H_{nl}(x, y)$

$$x_i = \frac{2i-1}{M}; y_j = \frac{2j-1}{N};\ {x_i}^2 + {y_j}^2 = 1; \Delta x_i = \frac{2}{M}; \Delta y_j = \frac{2}{N} \tag{7.10}$$

where $i = 0, 1, 2\ldots.M-1$ and $j = 0, 1, \ldots.N-1$.

H_{nl} is the orthogonal PCET kernel which is wave-based and thus accounts for the computational simplicity of this transform. In addition, due to its radial nature, PCET features can be computed up to any higher orders without any numerical instability. Δx_i and Δy_j are the normalization factors and are used to make PCET coefficients translation and scale invariant.

Images are reconstructed through the inverse transformation of:

$$I_r = \sum_{n=-n_m}^{n_m} \sum_{l=-n_m}^{n_m} M_{nl} H_{nl}(x_i, y_j) \tag{7.11}$$

where I_r it the reconstructed image and n_m is the maximum order of moments.

The major advantage of PCET over ZM is that the PCET coefficients can be calculated at even higher orders and thus are free from any numerical instability. Also, PCET focuses on all the face regions and thus, it is capable of capturing even the precise facial details. In contrast, ZM suffers from suppression problem which may cause it to focus only on certain parts of an image and neglect the rest. Thus, PCET generates a comprehensive feature set which greatly facilitates the recognition stage.

7.2.2 Implementation of the developed discriminative feature selection

The discrimination ability of a feature is determined from the extent of its dissimilarity from other face classes. It is computed by estimating the average values of a particular feature with respect to all the face classes separately. Then, the variability of average class values is obtained for that feature. High variability value of a feature indicates that its discrimination strength is higher than the others. This provides two benefits. First, only those features are ultimately selected for which class averages are well a part; and second, the resultant feature set with more variability score indicates its reduced similarity score over other classes. This approach is described as follows:

1. If there are C persons each representing a class and T training images per face class, then there are a total of N training patterns with

$$N = C \times T \tag{7.12}$$

2. Suppose, F features are extracted for each individual image, then this will result in $X = F \times N$ with

$$X = \begin{pmatrix} X_{11} & \dots & X_{N1} \\ \dots & \dots & \dots \\ X_{1F} & \dots & X_{NF} \end{pmatrix} \tag{7.13}$$

where X_{ij} denotes the jth feature of the ith training pattern for $i = 1$ to N and $j = 1$ to F.

3. Then, the discrimination strength of each of these features is evaluated. The mean value for each of the corresponding F extracted features is computed for all the T images of an individual class. As a result, $M_F = F \times C$ matrix (Equation 7.14) consisting of mean feature values (M_{fc}, f = 1,2,...,F and c = 1,2,...C) for each class is generated. This step eliminates any disparities arising due to facial variations within the face images available for an individual, hence it reduces the within-class differences.

$$M_F = \begin{pmatrix} M_{11} & \dots & M_{1C} \\ \dots & \dots & \dots \\ M_{F1} & \dots & M_{FC} \end{pmatrix} \tag{7.14}$$

with,
$M_{fc} = \frac{1}{T} \sum_{t=1}^{T} (X_{tf}),\ f = 1, 2, \dots,\ F;\ c = 1, 2, \dots C$

4. In the next step, variance between the mean feature values is computed. This provides information about the variability between the different face classes. The generated matrix is of size $F \times 1$ and comprises of the between-class mean-feature variance values and is defined as follows:

$$D = Variance(M_F) \qquad \forall\ features\ F \tag{7.15}$$

5. Higher variance value implies high discrimination capability of the corresponding feature. Thus, the variance matrix is sorted in descending order and the top n features with highest discriminative competence are selected. This allows the selection of only those features which have high between-class differences. Figure 7.2 shows the flowchart of the procedure followed.

7.2.3 Classification

For the final step of recognizing faces, classifiers are trained on the extracted feature sets. In the present study, both the traditional as well as intelligent classifiers including the Euclidean distance (ED) measure, multilayer back-propagation neural network (MBPNN), ANFIS, and SVM are utilized.

7.2.3.1 Euclidean distance

ED is the most well-known, the oldest, and the simplest method of classification. It has been used in the present work to capture rotation invariance. As moments-based methods are implemented to capture the tilted as well as rotated faces, the classifier employed should also be capable of representing the rotation invariance of the moments-based features. ED is such a classifier which complements the moments-based features to generate superior results. ED between feature vectors $a = [a_1\ a_2\ \dots\ a_n]$ and

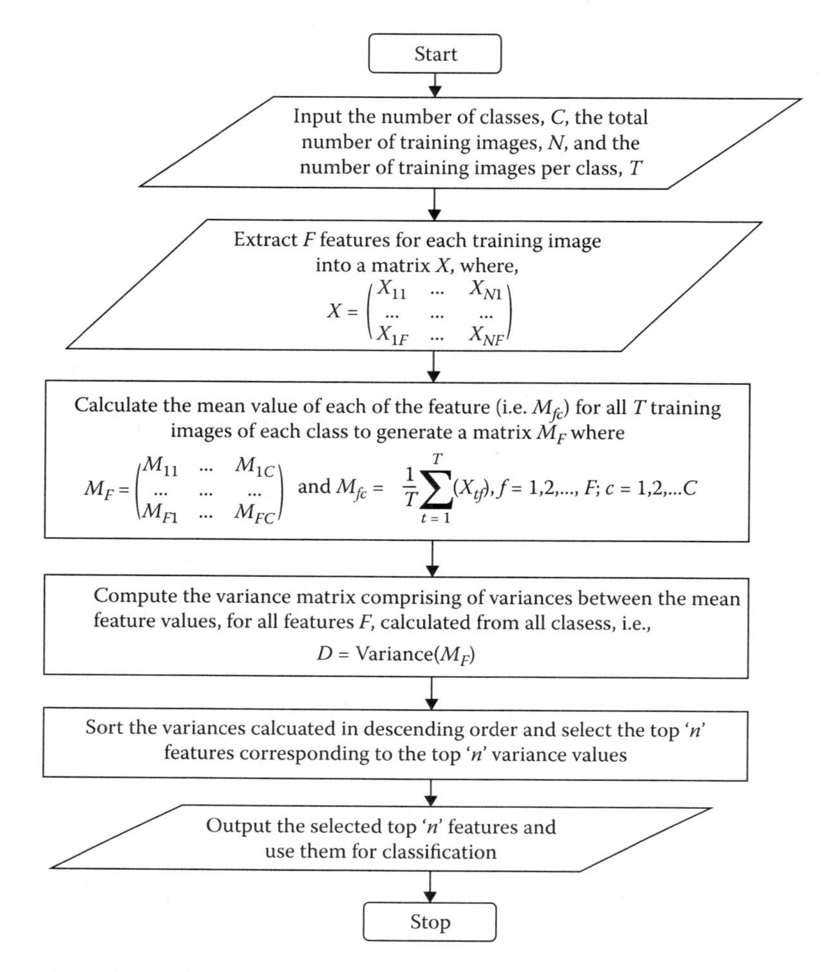

FIGURE 7.2: Flowchart of the proposed process for discriminative competence-based optimal feature selection.

$b = [b_1\ b_2\ \dots\ b_n]$ of database and query images is calculated as:

$$d_{zm}(a, b) = \sqrt{\sum_{i=1}^{n} (a_i - b_i)^2} \tag{7.16}$$

7.2.3.2 Support vector machine

SVM is used to capture the nonlinearity of faces through its kernel capability. As faces are highly nonlinear data, hence, one important property of the classifier must be the clear separation of nonlinear faces into their respective classes. SVM inherently possesses this capability and has proved to be an efficient classifier in this nonlinear context. In addition, it is based on learning algorithms and thus, can also easily separate the unseen data. Algorithmically, SVM is based on the principle of finding the equation of a hyperplane that maximally separates all the data points of various classes. SVM kernels are linear and nonlinear. For the present problem of face recognition, a nonlinear radial basis function (RBF) kernel is found to perform better than the linear kernel due to complex distribution of the nonlinear extracted features. For validation, the libSVM implementation of SVM is used and the parameters of the kernel are identified through the fivefold cross-validation technique [34].

7.2.3.3 Back-propagation neural networks

Neural networks are the mathematical approximators of the human brain operation which synthesize algorithms to depict the human learning process. These are nonlinear models which introduce flexibility in modeling real world complex problems. These are self-adaptive and self-learning maps which learn through experience. This nonlinear mapping system trains a neural network on a given set of inputs with known outputs using the supervised learning mechanism and obtains the good learning and generalization parameters for effective future classification of unrecognized test face images.

7.2.3.4 Adaptive neuro-fuzzy inference system

ANFIS is a hybrid classifier with the capabilities of both the neural nets and fuzzy systems. Thus, it exploits the important properties of self-adaptability and learning capability of the neural networks with the advantage of the fuzzy systems of capturing the prevailing uncertainty and imprecision of the actual real faces. In addition, it also ensures fast convergence to global minima as compared to BPNN, and ensures small convergence errors to introduce better learning ability for a similar network complexity.

7.3 Benchmark Face Databases for Performance Evaluation

ORL, Yale, and FERET represent the benchmark face databases utililized for validation and standardization of the proposed face recognition approaches. Their brief description is given in the following subsections.

7.3.1 AT&T/ORL database

This database consists of 400 face images of 40 individuals (Figure 7.3). There are 10 images in portable gray-map (PGM) format for each person. Images are in grayscale and are 92 × 112 pixels. This database contains pose, expression, lighting, occlusion, scale,

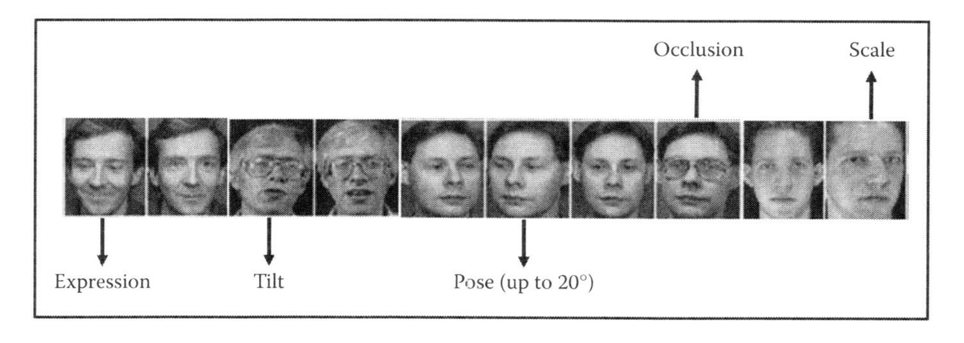

FIGURE 7.3: Images of one subject depicting pose, tilt, position, and scale variations from ORL face database.

and position variations. Poses range from 0° (frontal) to ±20°. Expression variations are presented in a way that either a person is smiling or not smiling and has open/closed eyes. There are also small light variations and occlusion of the eyes with eyeglasses.

7.3.2 Yale database

This database has 165 total face images for 15 persons. Images per person are 11 and are in grayscale form (Figure 7.4). The graphics interchange format (GIF) has been used to store the images. Each image is 243×320 pixels. This database presents major variations of light and facial expressions. Therefore, it is a challenging database to check the robustness of face recognition algorithms especially in case of light changes. To validate the impact of light changes, images have been captured by placing the light source at the left, right, front, and back of each individual. Images have also been captured at different expressions such as sad, happy, and mouth open. There is also occlusion of the eyes with eyeglasses.

7.3.3 FERET

It is a benchmark face database which incorporates significant variations of pose (up to ±90°), illumination, facial expressions, and age. It has 14,051 grayscale images of size 256×384 pixels for 1196 individuals (Figure 7.5). In literature, this database has been widely used by well-known face recognition systems for testing and validation of results.

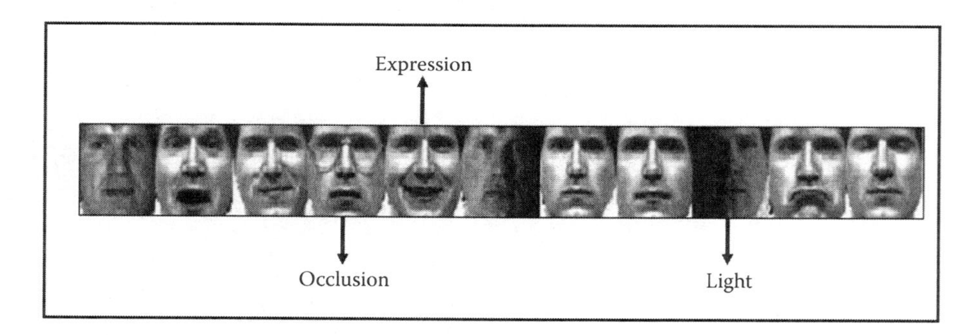

FIGURE 7.4: Cropped face images of one subject with expression, light, and occlusion variations from the Yale face database.

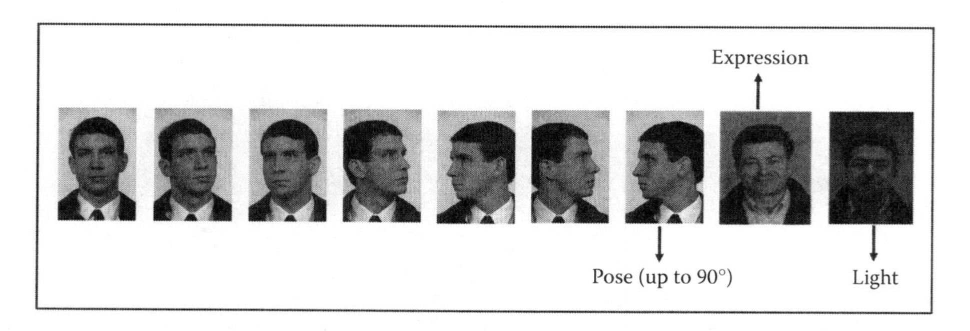

FIGURE 7.5: Face images of one person from the FERET gallery/probe datasets with pose, expression, and light variations.

7.4 Implementation and Evaluation through Standard Datasets

Robustness of the implemented techniques is evaluated against different types of variations present in face images, such as illumination, facial expression, scale, and pose. The performance is analyzed for images captured in a standard environment as well as noisy images. Techniques have been implemented through MATLAB® language source code under Microsoft Windows environment on a Pentium PC with 2.93 GHz CPU and 2 GB RAM.

7.4.1 Experimental setup for analysis

As the efficiency of the face recognition system is greatly affected by the selection of training and test images, the results presented are the means of several possible tests to make the analysis highly representative. In the case of the ORL database, out of 10 images of each person, 5 are used in the training set and the remaining 5 in the test set. For the Yale database, out of 11 images per person, 5 are used in the training set and the remaining 6 in the test set. For the FERET database, experiments are performed on its two subsets. The first subset, referred to as Set 1, in Figure 7.6 has images of randomly selected 100 persons in seven different poses of 0°, ±22.5°, ±67.5°, ±90°. For experimentation on each set, three images per person are used for training and the rest are used for testing. This has resulted in 300 images for training and 400 images for testing.

In the second subset, experiments are performed on its category *'b'* images of 200 subjects, named Set 2 in the study, in Figure 7.7. For each subject, there are 11 images from *ba* to *bk* where *ba* represents the frontal pose with neutral expressions and no light variation. Images *bb* through *be* have pose angles of −15°, −25°, −40° and −60° and are symmetric analogues of images *bf*-*bi* which have pose angles of +15°, +25°, +40° and +60°. The *bj* image has been captured with different expression variations and *bk* under different light conditions.

For simplicity, images in all the three databases have been normalized to 64 × 64 pixels for analysis. The recognition rate (in percentage) is measured as:

$$Recognition\ rate = \frac{N_t - N_r}{N_t} \times 100 \tag{7.17}$$

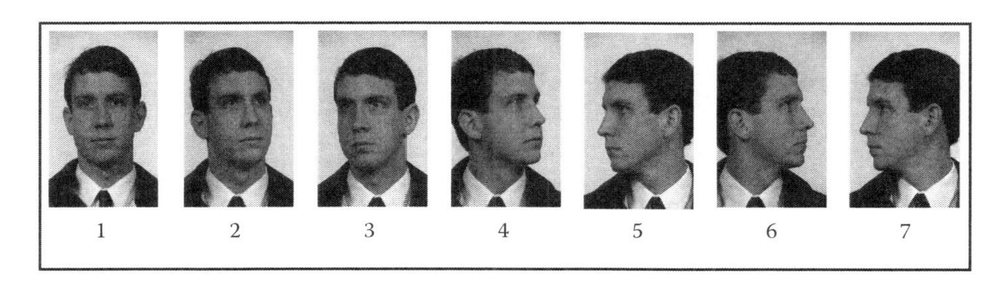

FIGURE 7.6: Face images of one subject from experimental Set 1 of this study.

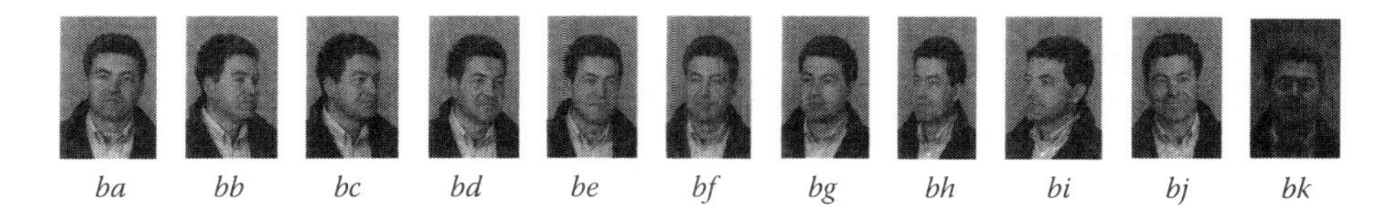

FIGURE 7.7: FERET *b* category images of one subject.

where N_t is the total number of images in the test set and N_r is the number of incorrectly recognized images.

7.4.2 Optimal order selection

It is well known that the order of moments selected to extract the features has a great influence on the recognition accuracy. Further, this selection also depends on the database under consideration [35]. For a database with illumination variations, removing lower-order moments from the feature vector improves the recognition performance while in a database with no such variations, extracting lower-order moments proves to be more effective. Considering this, experiments are performed to analyze the effect of the order of moments to select the discriminative DPCET features for all the three databases. Table 7.1 highlights the different groups of order of moments considered to analyze their effect on recognition performance. The mean recognition results are presented in Figure 7.8.

When experimenting with the Yale database that contains strong illumination variations, discarding the lower-order moments from the feature vector has improved the recognition performance. However, going beyond the maximum moment order of 12, the performance declines gradually. Taking the [lower, higher] moment orders of [3, 12] is found to perform better for the selection of DPCET features on this database. On ORL and FERET databases, lower-order moment features show higher discriminative strength due to less illumination variations. However, going beyond the maximum moment order of 12 for DPCET, lowers the performance; [lower, higher] moment orders of [1, 12] are selected for

TABLE 7.1: Groups of PCET moment orders considered for performance evaluation

Groups $[n_{min}, n_{max}]$	*1*	*2*	*3*	*4*	*5*	*6*	*7*	*8*	*9*
ORL	[2, 12]	[3, 12]	[4, 12]	[5, 12]	[1, 9]	[1, 10]	[1, 11]	**[1, 12]**	[1, 13]
Yale	[1, 13]	[2, 13]	[3, 13]	[4, 13]	[5, 13]	[3, 10]	[3, 11]	**[3, 12]**	[3, 14]
FERET (Set 1)	**[1, 12]**	[2, 12]	[3, 12]	[4, 12]	[5, 12]	[1, 10]	[1, 11]	[1, 13]	[1, 14]

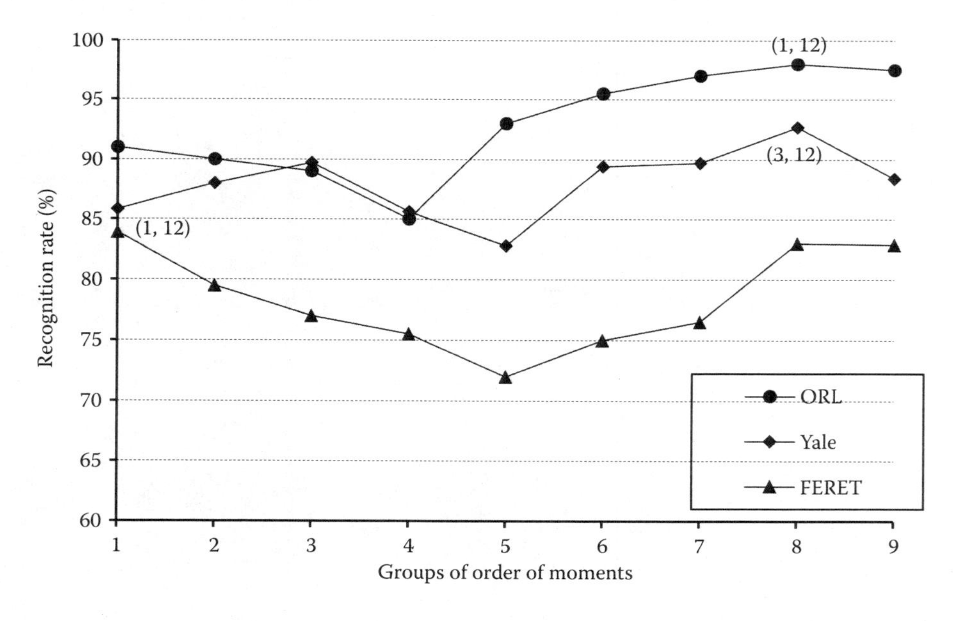

FIGURE 7.8: Performance (mean) of PCET on ORL, Yale, and FERET databases with different groups of order of moments.

extracting the DPCET coefficients on these two databases. For comparison, the established moment orders of [1, 15], [6, 17] and [1, 13] of ZM features are selected for the ORL, Yale, and FERET databases respectively [32].

7.4.3 Proposed versus existing discriminative feature selection

In the existing discriminative feature selection approach, discriminative competence of a feature is estimated through the ratio of its between-class (V_b) to within-class variance values (V_w) [21]. If large within-class variations exist in the image set to be recognized, then the discrimination strength of features may not be calculated correctly with the existing traditional discriminative feature selection approach, as the denominator in the V_b/V_w ratio dominates. To demonstrate its practical implication, the above proposal has been established statistically for the Yale face database which possesses high within-class differences of images due to excessive changes in illumination. Table 7.2 depicts such a scenario. Results have been presented for two ZM feature values for all 15 classes of the database.

Within-class variance V_w of Feature 1 (.025190) is higher than Feature 2 (.013760) which results in the selection of Feature 2 due to the large value of D (V_b/V_w). But in reality, Feature 1 should be selected because of having its large value for between-class variance (V_b). Hence, selection of the wrong feature may deteriorate the recognition rate. In the proposed technique, to maintain the interclass and reduce the intraclass differences, variance of averaged feature values for different images from each individual face class is calculated. The within-class means (M_F) for two ZM features, Feature 1 and Feature 2, are shown in Table 7.2. Variance between within-class means for Feature 1 (0.0089087) is higher than that of Feature 2 (0.000297). Hence, this has resulted in the selection of Feature 1 over Feature 2, which is expected also, as the between-class variance of Feature 1, is higher in comparison to Feature 2.

7.4.4 Evaluation with different number of features selected

The performance of the designed MDZM and DPCET approaches in terms of the number of features used in the recognition process has also been compared. As feature selection is database dependent, discriminant coefficients have been estimated separately for each

TABLE 7.2: Effect of using both existing and proposed feature selection approaches for estimating discriminative competence of two ZM features for varying illumination conditions

	Feature 1	Feature 2	Result
Absolute value of a sample image	0.207567	0.280607	Feature 2 is selected as its D value is higher.
Within class variance (V_w)	0.025190	0.013760	
Between-class variance (V_b)	1.110270	0.679560	
Discriminative competence i.e., $D = V_b/V_w$ ***(Existing)*** [21]	44.08	**49.39**	
Within-class mean of a sample class	0.227720	0.201270	**Feature 1 is selected (as desired).**
Discriminative competence = Variance between within-class means (i.e., D = Variance (M_F)) ***(Proposed)***	**0.008908**	0.000297	

database. The mean recognition rates for DZM, MDZM, and DPCET on different databases are shown in Table 7.3.

From the results, it is clear that MDZM and DPCET provide 0.5% and 1% improvement on ORL, 1.2% and 3.1% on Yale, 1.5% and 0.6% on FERET (Set 1), and 1.6% and 1.5% on FERET (Set 2) databases respectively with reduced features. This improved accuracy with a smaller number of features has been achieved owing to the identification and utilization of only those discriminative coefficients that exhibit minimum within-class variations while maintaining the between-class differences. The conventional PCET approach in general provides a large set of features for image representation, some of which might not be discriminative and are unstable. DPCET overcomes this limitation by utilizing only those features which possess the highest discriminative competence.

Table 7.4 depicts the results obtained with different classifiers. It is observed that both the traditional ED metric and the intelligent BPNN classifiers achieve similar recognition

TABLE 7.3: Mean recognition rate of DZM, MDZM, and DPCET with different number of features (Nof) selected on ORL, Yale, and FERET databases

		ORL				*Yale*	
Nof	*DZM*	*MDZM*	*DPCET*	*Nof*	*DZM*	*MDZM*	*DPCET*
14	84.0	88.0	87.5	23	77.5	78.0	81.9
18	90.0	91.0	90.5	28	77.5	78.6	84.8
23	93.0	93.5	92	34	82.4	83.7	87.0
28	93.5	95.5	93	40	84.8	86.9	90.6
34	94.0	***97.0***	95	42	86.5	89.4	***91.7***
40	***96.5***	97.0	97	52	87.3	***89.8***	91.5
48	96.0	97.0	***97.5***	65	88.2	89.8	91.0
52	96.0	97.0	97.5	78	***88.6***	89.8	91.0
		FERET (Set 1)				**FERET (Set 2)**	
Nof	*DZM*	*MDZM*	*DPCET*	*Nof*	*DZM*	*MDZM*	*DPCET*
23	60.0	75.4	58.7	23	57.5	61.1	60.7
28	67.5	80.2	63.5	28	59.0	63.6	61.2
33	70.0	80.3	66.8	37	61.3	65.4	63.8
40	75.0	***80.6***	70.0	44	66.2	70.7	67.3
42	77.5	79.7	75.7	49	68.9	***74.3***	71.5
47	***79.1***	79.7	79.4	51	71.1	73.8	***74.2***
58	78.3	79.3	***79.7***	53	***72.7***	73.0	74.0
62	78.7	78.9	79.4	57	72.0	73.0	74.0

TABLE 7.4: Mean recognition rate of DZM, MDZM, and DPCET with different classifiers

		ORL			*Yale*	
Method	*SVM*	*ANFIS*	*BPNN*	*SVM*	*ANFIS*	*BPNN*
DZM	97.5	95.5	97.0	91.1	88.0	89.1
MDZM	98.0	96.0	97.0	93.2	89.2	90.0
DPCET	***99.0***	96.5	98.0	***94.6***	91.0	91.0
		FERET (Set 1)			**FERET (Set 2)**	
Method	*SVM*	*ANFIS*	*BPNN*	*SVM*	*ANFIS*	*BPNN*
DZM	82.3	78.8	79.1	75.6	72.0	73.0
MDZM	***84.4***	80.2	81.0	76.7	74.0	74.5
DPCET	84.1	79.4	79.8	***76.8***	74.0	74.4

accuracy. This is because the moment coefficients are rotation invariant features and ED is inherently rotation invariant, making it difficult to achieve results comparable to that of BPNN. SVM classification results have shown significant improvement in performance by up to 5.6% over both ED and BPNN classifiers. However, improved results have not been obtained using the ANFIS classifier because the outputs of this classifier are not whole numbers and have to be rounded off to determine the class labels, which introduces some numerical instability. In a nutshell, MDZM exhibits an improvement of up to 1.8% and 3.1% at lower dimensionality over DZM and ZM respectively. DPCET further supplements the recognition rate by about 0.2% to 1.9%, respectively, at reduced computational complexity. Both MDZM and DPCET achieve high classification accuracy for lighting, rotation, scale, and pose variations.

7.5 Recognition Performance on Self-Generated Database

After establishing the efficacy of the developed techniques on benchmark databases available for face recognition research, a self-designed database has been employed to validate their performance in-house. This database not only evaluates the developed techniques under diverse pose and light conditions but also verifies their performance for a real-life scenario. The classification process is realized through the SVM algorithm. The collected database is comprised of 15 images per person for a total of 10 individuals. Each image has a size of 128×128 pixels and is stored in the grayscale format as jpeg (joint photographic experts group) file. The individuals have been captured under varying light and pose conditions. The poses range from $\pm 10^\circ$ to $\pm 60^\circ$ ($\pm 10^\circ$, $\pm 20^\circ$, $\pm 30^\circ$, $\pm 45^\circ$ and $\pm 60^\circ$).

Thus, there are a total of 10 posed images for each person. These posed images have been captured by placing the camera at the respective directions with respect to the individual. Posed images from this database are shown in Figure 7.9.

Light variations have been captured in four images, with the light source placed to the left, right, front, and back of the person being photographed. A neutral frontal face image at 0° (no pose variation) with no light variation has also been taken for each person. The images with light intensity variations have been captured by placing the light source in the desired position. These are shown in Figure 7.10.

The entire facial set of each individual has been captured in one sitting. Care has been taken to ensure uniform facial expressions of each person throughout the sessions. The illumination of the photography room has also been kept constant.

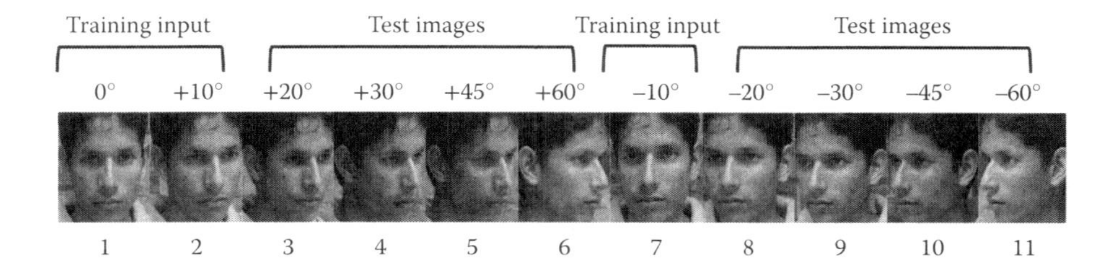

FIGURE 7.9: Self-generated database: face images captured under different pose variations.

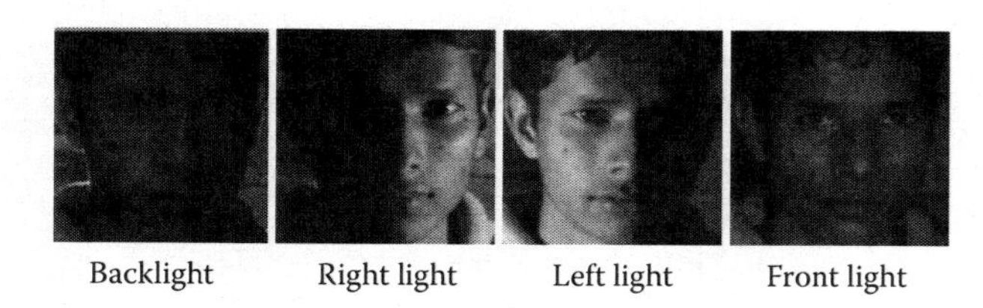

FIGURE 7.10: Self-generated database: face images captured under different light variations.

TABLE 7.5: Mean recognition rate of DZM, MDZM, and DPCET in case of pose variations on self-generated database

Experimental Datasets		Method		
Training	**Test Set**	***DZM***	***MDZM***	***DPCET***
1,2,7 (±10°)	1,2,3,7,8 (±20°)	91.0	93.0	93.1
1,2,7 (±10°)	1,2,3,4,7,8,9 (±30°)	84.8	85.6	87.0
1,2,7 (±10°)	1,2,3,4,5,7,8,9,10 (±45°)	78.2	79.1	79.5
1,2,7 (±10°)	1-11 (±60°)	69.8	71.6	71.7
1 (0°)	1,2,7 (±20°)	87.9	88.0	88.0
1 (0°)	1,2,3,7,8 (±30°)	75.0	76.0	76.2
1 (0°)	1,2,3,4,7,8,9 (±45°)	68.1	70.0	70.5

7.5.1 Pose invariance analysis

To test the performance in case pose variations, experiments have been performed with different training and test sets. The optimal derived moment orders of [1, 13], [1, 13] and [1, 12] are used for selecting DZM, MDZM, and DPCET features in this case. First, the classifier is trained to recognize face images with pose changes of a maximum up to ±10°. Testing has then been performed to identify face images with poses ±20°, ±30°, ±45° and ±60°. This results in three images in the training set and five, seven, nine and eleven images respectively in the test sets for the experimentation. In the next experiments, the classifier has been trained for features of only frontal face images while testing has been done for face images with poses ±20°, ±30° and ±45°. Thus, there is only one image in the training set and three, five, and seven images in the test sets respectively. Table 7.5 depicts the configuration of both training and sets and presents the recognition results of the executed experiments. It is observed that both MDZM and DPCET provide a performance improvement of about 2% over the existing technique.

7.5.2 Light invariance analysis

Experiments have also been performed for testing the performance of the developed techniques for different light variations. Obtained optimal moment orders of [6, 17], [6, 17] and [3, 12] are used for selecting DZM, MDZM, and DPCET features. Lower-order moments are not considered in this case because these are more prone to light changes whereas higher-order moments remain more stable. For evaluation, first, training has been done on one image randomly selected from the images with light changes for each subject. However, testing has been carried out for five images including the frontal image with no light effects and the four images with light variations. In the next experiment, training has been done on two randomly selected images and testing on all the five images as considered in the first experiment. Results of this experimentation are shown in Table 7.6. From the results it is

TABLE 7.6: Mean recognition rate of DZM, MDZM, and DPCET in case of light variations on self-generated database

Method	Training Set	
	Random 1 image	*Random 2 images*
DZM	71.0	86.2
MDZM	74.0	88.0
DPCET	75.1	88.2

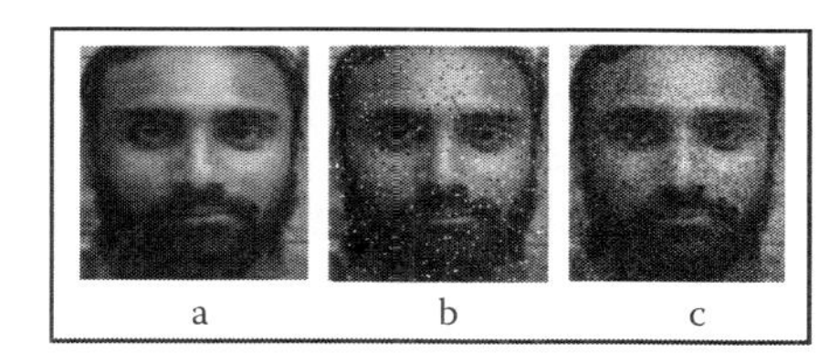

FIGURE 7.11: (a) Original image, (b) image with added salt-pepper noise, and (c) Image with added Gaussian noise.

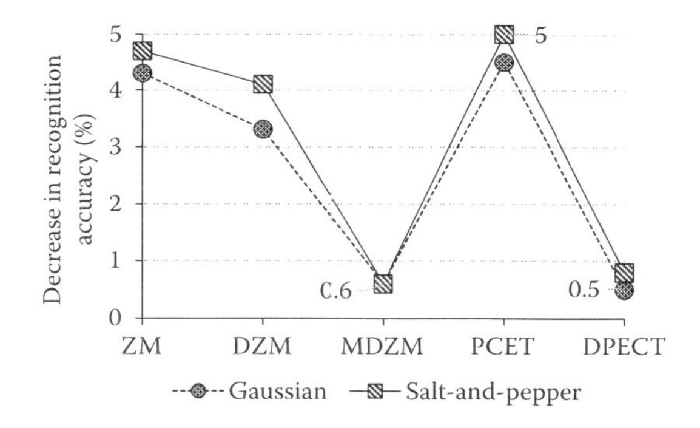

FIGURE 7.12: Percentage decrease in recognition accuracy of implemented techniques in case of noise variations.

clear that MDZM supplements the recognition rate by up to 3% and DPCET by up to 4% over the traditional technique.

7.6 Robustness to Noise Variations

Efficacy of the implemented approaches against noise variations is tested for two types of noise, namely, Gaussian and salt-and-pepper noise. To estimate the effect of noise during image acquisition due to illumination and temperature conditions, Gaussian noise of 0 mean and 0.01 variance is added in the first set of experimental test images whereas training is done on images with no added noise. To handle the modalities of data transmission, salt-and-pepper noise of density 0.05 is added to the test images in the second set of experiments. Figure 7.11 shows the face images with added Gaussian and salt-and-pepper noise.

Figure 7.12 shows the decrease in recognition accuracy (in %) of different techniques on self-created database. Proposed techniques perform better and exhibit maximum degradation of up to 0.6% and 0.5% by MDZM and DPCET respectively as compared to other approaches which show degradation of up to 5% in recognition accuracy. This is due to the fact that the proposed technique takes care of the data values that are closest to the center of gravity. It also withstands the instabilities in the measured feature values for an individual and acts as an unbiased estimator.

7.7 Time Complexity Analysis

The recognition time required by the conventional and the proposed methods for face recognition has been compared. Table 7.7 depicts the average time required for feature extraction, selection, and classification of the developed methods. The time taken by DZM and MDZM is slightly more than ZM due to the selection technique employed for reducing the optimal feature set of the conventional approach. Thus, the DZM techniques improve the accurate recognition rate but at the cost of a slight increase in computation time. However, MDZM performs better than DZM as it possesses lower dimensionality and therefore requires less matching time during recognition stage. The computation time of PCET is low due to its harmonic radial function. In light of this fact, DPCET requires lower time in contrast to the ZM-based methods. Further, the space complexity of the DPCET technique is decreased from that of PCET by reducing the size of the feature vector. This indicates that the proposed methods achieve a good balance between speed and dimensionality and are suitable in real-time environments as well as on devices with low computational power.

7.8 Performance Comparison with Other Methods

The performance of the proposed MDZM and DPCET techniques has been compared to other recent and well-known approaches on both the ORL and Yale databases. Table 7.8 shows the recognition rates of the implemented approaches as well those of recent methods. In case of ORL database, five images of each person are used for training and the other five images are used for testing. For the Yale database, out of 11 images available per person, 5 are used for training and the other 6 are used for testing. Recognition rates are averaged for different possible combinations of training and test sets. The results show that the recognition rate of MDZM is higher as compared to moments-based and discriminative approaches including CZM, CWM, discriminative DCT, DZM, and DZM + FLD. DPCET provides the highest recognition rate among all methods on both datasets.

TABLE 7.7: Recognition time (in seconds) exhausted by the implemented approaches

Database		Time (in seconds)				
		ZM	*DZM*	*MDZM*	*PCET*	*DPCET*
ORL		59.63	61.20	61.03	18.10	***17.91***
Yale		20.51	21.98	20.79	8.58	***8.59***
FERET	**Set1**	66.02	68.14	67.34	28.34	***28.20***
	Set2	133.95	134.68	134.20	56.38	***56.50***

TABLE 7.8: Performance comparison of the proposed approach with recent methods on ORL and Yale databases

Method	*Yale*	*ORL*
2DPCA with feature fusion [32]	87.4	95.1
D-LDA [11]	93.2	92.5
2D-DWLDA [11]	89.3	94.0
CZM [25]	-	96.5
Intrinsicfaces [36]	74.0	97.0
Discriminative DCT [21]	89.0	96.0
DZM [32]	91.1	95.9
DZM with FLD [32]	89.6	96.3
2D2PCA [12]	-	90.5
Wavelet with LDA [37]	-	97.1
LDP [38]	-	90.5
2DPCA with LDA [14]	-	91.2
Gabor with LBP and LPQ [39]	90.7	-
GELM [40]	82.3	-
Gabor with DSNPE [41]	93.5	-
CWM [42]	-	96.0
Proposed MDZM	***93.2***	***98.0***
Proposed DPCET	***94.6***	***99.0***

7.9 Conclusion

In this study, the performance of conventional ZM, PCET, and DZM approaches are compared to that of the proposed MDZM and DPCET approaches. This comparison is achieved using different classifiers including Euclidean distance (ED), back-propagation neural network (BPNN), adaptive neuro-fuzzy inference system (ANFIS), and support vector machine (SVM). The recognition performance of these approaches is analyzed on different types of variations present in face images such as illumination, expression, scale, and rotated poses. All these experiments are performed on three benchmark face databases, such as the Yale database with expression and illumination variation, the ORL database with scale and small pose variation, the FERET database with pose (yaw) variation, and also on a self-generated database with light and pose changes. The selection of discriminative features in the feature vector has significantly improved the recognition performance at reduced dimensions. The proposed approaches focus on minimizing the within-class variations while maintaining the between-class differences by selecting the coefficients which have well-separated class means. ED and BPNN have provided with similar results. High classification accuracy is achieved with the SVM algorithm. MDZM shows an improvement of 3.1% over the existing ZM and DZM techniques. DPCET has overcome the limitation of PCET to generate a plenty of features, some of which might not even be effecting the recognition performance, by performing recognition only on the selected most discriminative coefficients. It has maintained a high classification of accuracy for lighting, rotation, and pose variations and has further enhanced the recognition rate by up to 1.9%. These techniques have also operated well in case of noise variations as the proposed feature selection approach takes care of the instabilities in the estimated feature coefficients. Results also imply that careful selection of both PCET- and ZM-based approaches has reduced their dimensionality and discrimination analysis has increased their recognition performance with a slight increase in computation time. However, as DPCET is computationally very efficient, it maintains a good balance between dimensionality as well as speed which makes it suitable for real-time environments.

Acknowledgment

The authors are grateful to the National Institute of Standards and Technology, AT&T Laboratories and Computer Vision Laboratory, Computer Science and Engineering Department, U.C. San Diego for providing the FERET, ORL, and Yale face databases respectively. During the course of this study, the first author was funded by INSPIRE–Junior Research Fellowship (IF120810) from the Department of Science and Technology, Govt. of India.

References

1. T Kanade. Picture processing system by computer complex recognition. PhD thesis, Department of Information Science, Kyoto University, Japan, 1973.

2. BS Manjunath, R Chellappa, and C von der Malsburg. A feature based approach to face recognition. In *Proceedings of the IEEE Computer Society Conference on Computer Vision and Pattern Recognition, 1992 (CVPR'92, 1992)*, Champaign, IL, pp. 373–378. IEEE, 1992.

3. R Brunelli and T Poggio. Face recognition: Features versus templates. *IEEE Transactions on Pattern Analysis and Machine Intelligence*, 15(10):1042–1052, 1993.

4. M Lades, JC Vorbruggen, J Buhmann, J Lange, C von der Malsburg, RP Wurtz, and W Konen. Distortion invariant object recognition in the dynamic link architecture. *IEEE Transactions on Computers*, 42(3):300–311, 1993.

5. IJ Cox, J Ghosn, and PN Yianilos. Feature-based face recognition using mixture-distance. In *Proceedings of the IEEE Computer Society Conference on Computer Vision and Pattern Recognition, 1996 (CVPR'96, 1996)*, San Francisco, CA, pp. 209–216. IEEE, 1996.

6. M Turk and A Pentland. Eigenfaces for recognition. *Journal of Cognitive Neuroscience*, 3(1):71–86, 1991.

7. PN Belhumeur, JP Hespanha, and DJ Kriegman. Eigenfaces vs. Fisherfaces: Recognition using class specific linear projection. *IEEE Transactions on Pattern Analysis and Machine Intelligence*, 19(7):711–720, 1997.

8. Q Liu, R Huang, H Lu, and S Ma. Face recognition using kernel-based Fisher discriminant analysis. In *Proceedings of the Fifth IEEE International Conference on Automatic Face and Gesture Recognition, 2002*, Washington, DC, pp. 197–201. IEEE, 2002.

9. MS Bartlett, JR Movellan, and TJ Sejnowski. Face recognition by independent component analysis. *IEEE Transactions on Neural Networks*, 13(6):1450–1464, 2002.

10. S Martin. An approximate version of kernel PCA. In *2006 5th International Conference on Machine Learning and Applications (ICMLA'06)*, Orlando, FL, pp. 239–244. IEEE, 2006.

11. R Zhi and Q Ruan. Two-dimensional direct and weighted linear discriminant analysis for face recognition. *Neurocomputing*, 71(16):3607–3611, 2008.

12. G Huang. Fusion (2D) 2 PCA LDA: A new method for face recognition. *Applied Mathematics and Computation*, 216(11):3195–3199, 2010.

13. GDC Cavalcanti, TI Ren, and JF Pereira. Weighted modular image principal component analysis for face recognition. *Expert Systems with Applications*, 40(12):4971–4977, 2013.

14. H Ren and H Ji. Nonparametric subspace analysis fused to 2DPCA for face recognition. *Optik-International Journal for Light and Electron Optics*, 125(8):1922–1925, 2014.

15. T Ahonen, A Hadid, and M Pietikäinen. Face recognition with local binary patterns. In *European Conference on Computer Vision*, Prague, Czech Republic, pp. 469–481. Springer, 2004.

16. O Déniz, G Bueno, J Salido, and F De la Torre. Face recognition using histograms of oriented gradients. *Pattern Recognition Letters*, 32(12):1598–1603, 2011.

17. A-A Bhuiyan and CH Liu. On face recognition using Gabor filters. *World Academy of Science, Engineering and Technology*, 28:51–56, 2007.

18. V Štruc, R Gajšek, and N Pavešic. Principal Gabor filters for face recognition. In *Proceedings of the 3rd IEEE International Conference on Biometrics: Theory, Applications and Systems*, Washington, DC, pp. 113–118. IEEE Press, 2009.

19. H Spies and I Ricketts. Face recognition in Fourier space. In *Vision Interface*, vol. 2000, pp. 38–44, 2000.

20. ZM Hafed and MD Levine. Face recognition using the discrete cosine transform. *International Journal of Computer Vision*, 43(3):167–188, 2001.

21. S Dabbaghchian, MP Ghaemmaghami, and A Aghagolzadeh. Feature extraction using discrete cosine transform and discrimination power analysis with a face recognition technology. *Pattern Recognition*, 43(4):1431–1440, 2010.

22. NH Foon, Y-H Pang, ATB Jin, and DNC Ling. An efficient method for human face recognition using wavelet transform and Zernike moments. In *Proceedings of the International Conference on Computer Graphics, Imaging and Visualization, 2004 (CGIV 2004),* Penang, Malaysia pp. 65–69. IEEE, 2004.

23. A Wiliem, VK Madasu, WW Boles, and PKDV Yarlagadda. A feature based face recognition technique using Zernike moments. In P Mendis, J Lai, E Dawson, and H Abbas, editors, *RNSA Security Technology Conference 2007*, pp. 341–355, Melbourne, Australia, 2007. Australian Homeland Research Centre.

24. C Singh, N Mittal, and E Walia. Face recognition using Zernike and complex Zernike moment features. *Pattern Recognition and Image Analysis*, 21(1):71–81, 2011.

25. C Singh, E Walia, and N Mittal. Rotation invariant complex Zernike moments features and their applications to human face and character recognition. *IET Computer Vision*, 5(5):255–265, 2011.

26. J Haddadnia, K Faez, and M Ahmadi. An efficient human face recognition system using pseudo Zernike moment invariant and radial basis function neural network. *International Journal of Pattern Recognition and Artificial Intelligence*, 17(1):41–62, 2003.

27. Y-H Pang, ABJ Teoh, and DCL Ngo. A discriminant pseudo Zernike moments in face recognition. *Journal of Research and Practice in Information Technology*, 38(2):197, 2006.

28. P-T Yap, X Jiang, and AC Kot. Two-dimensional polar harmonic transforms for invariant image representation. *IEEE Transactions on Pattern Analysis and Machine Intelligence*, 32(7):1259–1270, 2010.

29. M Liu, X Jiang, AC Kot, and P-T Yap. Application of polar harmonic transforms to fingerprint classification. *Emerging Topics in Computer Vision and its Applications*, 24(1):1–9, 2011.

30. G Singh and I Chhabra. Human face recognition through moment descriptors. In *2014 Recent Advances in Engineering and Computational Sciences (RAECS)*, Chandigarh, India, pp. 1–6. IEEE, 2014.

31. D Shen and HHS Ip. Discriminative wavelet shape descriptors for recognition of 2-D patterns. *Pattern Recognition*, 32(2):151–165, 1999.

32. C Singh, E Walia, and N Mittal. Discriminative Zernike and pseudo Zernike moments for face recognition. *International Journal of Computer Vision and Image Processing (IJCVIP)*, 2(2):12–35, 2012.

33. C-Y Wee and R Paramesran. On the computational aspects of Zernike moments. *Image and Vision Computing*, 25(6):967–980, 2007.

34. C-C Chang and C-J Lin. Libsvm: A library for support vector machines. *ACM Transactions on Intelligent Systems and Technology (TIST)*, 2(3):27, 2011.

35. C Singh, E Walia, and N Mittal. Robust two-stage face recognition approach using global and local features. *The Visual Computer*, 28(11):1085–1098, 2012.

36. Y Wang and Y Wu. Face recognition using intrinsicfaces. *Pattern Recognition*, 43(10):3580–3590, 2010.

37. Z-H Huang, W-J Li, J Wang, and T Zhang. Face recognition based on pixel-level and feature-level fusion of the top-levels wavelet sub-bands. *Information Fusion*, 22:95–104, 2015.

38. P Huang, C Chen, Z Tang, and Z Yang. Feature extraction using local structure preserving discriminant analysis. *Neurocomputing*, 140:104–113, 2014.

39. S-R Zhou, J-P Yin, and J-M Zhang. Local binary pattern (LBP) and local phase quantization (LBQ) based On Gabor filter for face representation. *Neurocomputing*, 116:260–264, 2013.

40. Y Peng, S Wang, X Long, and B-L Lu. Discriminative graph regularized extreme learning machine and its application to face recognition. *Neurocomputing*, 149:340–353, 2015.

41. G-F Lu, Z Jin, and J Zou. Face recognition using discriminant sparsity neighborhood preserving embedding. *Knowledge-Based Systems*, 31:119–127, 2012.

42. C Singh and AM Sahan. Face recognition using complex wavelet moments. *Optics & Laser Technology*, 47:256–267, 2013.

Chapter 8

Optimization of Hand Motion Recognition System Based on 2D HMM Approach Using ABC Algorithm

K. Martin Sagayam, D. Jude Hemanth, Yadav Nitesh Ramprasad, and Raghav Menon

8.1 Introduction

Gestures are bodily actions which are used for communication without the using words. Hand expressions can be used for creating gestures for human and computer interactions in a virtual environment to make the system more reliable without using any mechanical devices. The idea of the proposed work is to optimize hand gesture recognition using computer vision, machine learning, pattern recognition, and an image processing algorithm. It creates a better bridge between human and machine with better recognition in less time. Rautaray and Agrawal (2011) have demonstrated the virtual games through hand motion recognition by incorporating the virtual and real content. The system has adopted with an input device to interact with the system. Zyda (2005) has proposed advance gaming in a virtual environment with greater effort in the development of video games through the complex background with different lighting conditions. The public has an interest in playing games during leisure time to relax and become more comfortable with a virtual environment without being connected to the machine.

8.1.1 Literature survey

Human computer interaction (HCI) has a wide range of applications for virtual gaming, medical surgery, constructing buildings, etc. using hand gestures without any physical contact with machines, as proposed by Alon et al. (2009). Hand gesture recognition is a group of patterns generated in a spatio-temporal domain. Ganapathi et al. (2010) have explained that the hand gesture can be implemented in different ways and can reduce the complex usage of input devices in virtual reality applications. It can be implemented in various steps, depending upon the problem, to provide a better solution, Ganapathi et al. (2010); Elmezain et al. (2008). It is used in the development of digital houses and virtual surgeries and treatments, machine learning, and robotic language, as presented by Elmezain et al. (2008). Hand gesture recognition involves several modules like detection, tracking, and recognition, Ren et al. (2013). Followed by these three modules, segmentation and classification have to be performed for all image processing applications. There are some techniques used to classify the data such as support vector machine (SVM), hidden Markov model (HMM), dynamic time warping (DTW), artificial neural network (ANN), etc., proposed by Raheja et al. (2015). Among these techniques, HMM is considered to be accurate and produces a better recognition rate than any other algorithms suggested by Raheja et al. (2015). It is more realistic and immerses input for hand gesture recognition and reduces complexity. The recent development in hand gesture recognition involves the usage of data gloves. The bottleneck of data gloves has bends in the finger while folding; this leads to short circuitry criticized by Ren et al. (2013). This led to the development of the kinect sensor and leap motion sensor for analysis of 3-D data using property algorithms for feature selection, scene analysis, motion tracking, and structural tracking suggested by Raheja et al. (2015). The latest improvements in hand gesture are left sector tracking and right sector tracking for virtual reality applications, Ren et al. (2013); Raheja et al. (2015). The back end of the kinect sensor and leap motion sensor are built in by the stochastic mathematical model to get a better recognition rate.

Mantecon et al. (2015a) have proposed an idea for the depth image acquired by sensors that provide a better quality of hand image. The feature descriptor based on depth spatiograms of the required pattern is discriminative, but reduces the dimensionality without loss of content. Mantecon et al. (2015b) also extend his work by using a sparse representation of hand features, which has been trained and tested by the SVM classifier. El-Zaart and Ghosn (2013) have proposed the a model for detecting the affected portion from the MRI image by perfect thresholding value. The random projection technique has provided a better solution using L1-normalization for segmenting an object under different occlusion and illumination conditions suggested by Boyali and Kavakli (2012).

Song et al. (2013) have investigated a gesture recognition algorithm with a less optimized state in HMM while training and testing by using an ant colony algorithm. This produces a better recognition rate and accuracy even in complex gestures and background. Li et al. (2016) have proposed an entropy-based K-means clustering method used to reduce the state variable in HMM by using a data-driven technique. The optimum path in the HMM model is to determine by the expectation maximization (EM) algorithm or the Baum–Welch algorithm. Further, to improve the performance measures of HMM, use the artificial bee colony (ABC) algorithm for obtaining the best cost value of the state sequence. Sheenu et al. (2015) have presented a novel technique for hand gestures in complex background, illumination, translation, and rotations. In order to solve these conditions, apply sequential minimal optimization (SMO) to improve the robustness of the system. Liu et al. (2015) have investigated the shape based descriptor of hand gestures with limited feature points. The feature points provide the details of tips and valley points from the hand in different angular positions by using the curve fitting method. To optimize the weight and the cut-off coefficient value of hand gestures, use the particle swarm optimization (PSO) algorithm.

8.1.2 Related works

A detailed survey of hand motion recognition for virtual reality application is discussed here. This section describes about the various techniques used in feature extraction, classification, and optimization respectively.

8.1.2.1 Feature extraction

8.1.2.1.1 Scale invariant feature transform (SIFT)

The SIFT approach is used to generate image feature vector points based on rotation, scaling, and translation of hand gestures, with dynamic background and illumination conditions suggested by Pandita and Narote (2013). The objective of an approach is to extract the minimal useful content from the entire portion of an object and match the template with the image in the database. The following steps are used to determine the SIFT features.

1. ***Detection of scale-space local extrema***: This step is used to find the scaling and locations of the hand motion in different angular positions. This can be found by the scale-space function S given in Equation 8.1,

$$S(x, y, \sigma) = g(x, y, \sigma) * I(x, y) \tag{8.1}$$

where $g(x, y, \sigma)$ represents a Gaussian function, $I(x, y)$ is the input data, and $*$ represents the convolution operator. Taking the difference between two inputs data at S_{n-1} and S_n or S_n and S_{n+1} respectively shown in Equation 8.1. It helps to locate the neighborhood feature point in the coordinate system using the difference of the Gaussian (DoG) function given in Equation 8.2:

$$D(x, y, \sigma) = S(x, y, k\sigma) - S(x, y, \sigma) \tag{8.2}$$

In order to get the desired feature vector by comparing maxima and minima points in the scale-space extrema, takes a lot of time to compute with many feature points from large scale space as shown in Figure 8.1.

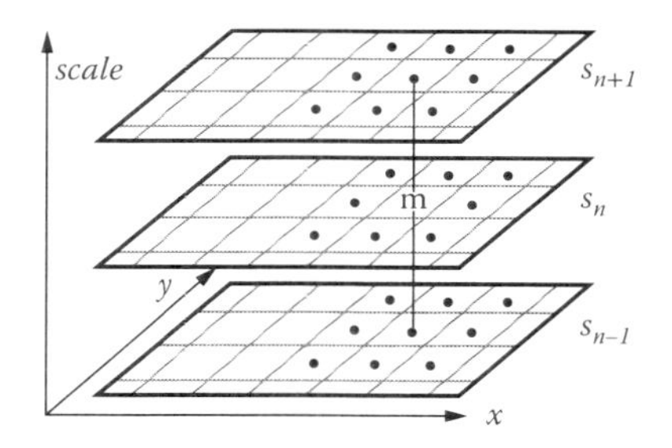

FIGURE 8.1: DoG in neighborhood of scale-space.

2. ***Localization***: After detecting local extrema, the 2D quadratic function is used to be located by the fitting function at the sample point. A second-order Taylor series is used with the quadratic function at the same sample point. Then the low frequency components of the feature points are discarded.

3. ***Orientation***: After locating the SIFT features of the hand image, find the magnitude and orientation of the image gradient by using Equations 8.3 and 8.4:

$$|m(x,y)| = \sqrt{\sigma(S(x+1,y) - S(x-1,y))^2 + (S(x,y+1) - S(x,y-1))^2} \quad (8.3)$$

$$\theta(x,y) = tan^{-1}((S(x,y+1) - S(x,y-1)/(S(x+1,y) - S(x-1,y)) \quad (8.4)$$

The image gradient weighted by the Gaussian function depends on the feature vector points. The condition for local extrema has been decided by the maxima and minima of the difference of the Gaussian (DoG) function of each point.

4. ***Descriptor***: The key descriptor point is transformed from the image gradient shown in Figure 8.2. It shows the significant shape of the object with respect to the rotation and illumination condition.

8.1.2.1.2 Speeded up robust feature (SURF)

SURF is similar to the SIFT approach yet its way of algorithm processing is different, Sheta et al. (2012). It is generally a detector and a descriptor used to identify the major points of interest, thereby converting the target image into coordinates. The algorithm is broadly classified into three major parts:

1. ***Interest point detection***: SURF employs a blob to identify the point of interest. These points of interest are arranged in terms of a Hessian matrix. The feature vector point is derived by the Hessian matrix of the second order Gaussian function derivation of hand image at point $C(x, y)$ in scaling of γ represent in 2D quantity.

2. ***Local neighborhood descriptor***: This descriptor is used to produce an accurate description of an image by describing the intensity of the pixel distribution around the reference point or point of interest. Moreover, a higher version of SURF called U-SURF is used in cases where the camera remains horizontal for faster computational speed.

3. ***Matching***: By carefully comparing the results from the descriptors applied on various images, the pairs that are matching can be found in Sheta et al. (2012) as shown in Figure 8.3.

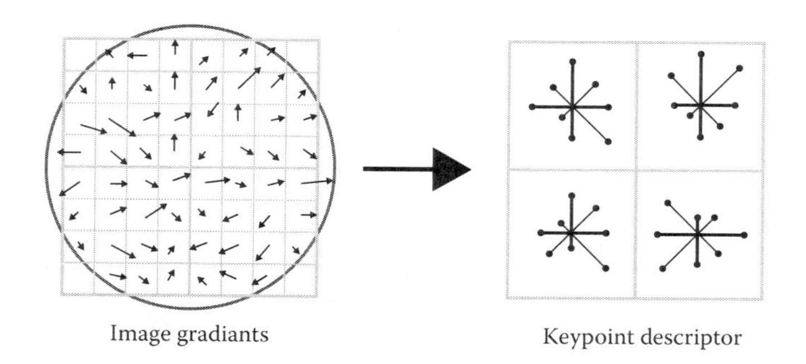

FIGURE 8.2: Assigning orientation.

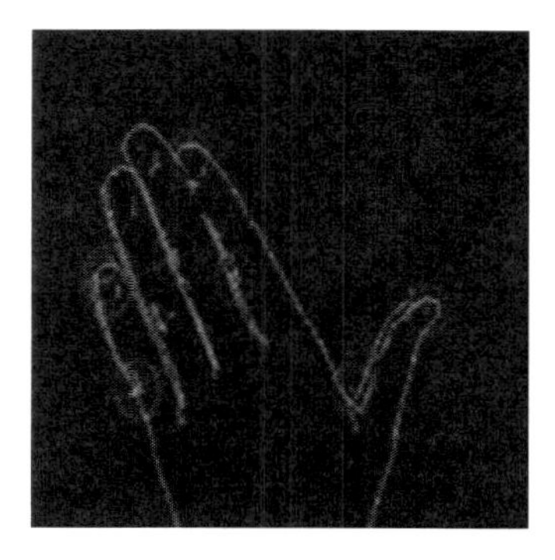

FIGURE 8.3: Valley and peak points in the image were found by SURF.

8.1.2.1.3 Zernike moments

From the compressed data, feature vector points has been extracted by using the Zernike coefficient $V(\rho,\theta)$ of n-order with m-recursive steps are allowed, which has been defined by Equations 8.5 and 8.6, demonstrated by Gu and Su (2008):

$$f(x) = \sum_{x,y=1}^{\infty} p(x,y)V(\rho,\theta), \rho \leq 1 \tag{8.5}$$

$$A_{n,m} = \frac{n+1}{\pi} f(x) \tag{8.6}$$

where, $n \pm |m|$ = even or odd, $\rho = \sqrt{x^2+y^2}$ and $\theta = arctan(y/x)$. Hand images rotate invariantly with respect to their center point of an object. This technique has more comfort in determining the complex shape and angular position, even if in the randomize projection of a 3-D plane.

8.1.2.1.4 Otsu's thresholding method

Otsu (1979) have proposed an idea of 1D Savitzky–Golay filter can perfectly stretched put to a 2D polynomial fitting. The function $f(x,y)$ represents 2D polynomial fitting used for detecting meaningful feature vector point:

$$f(x,y) = C_{nm} + C_{(n+1)m}x + C_{n(m+1)}y + C_{n(m+1)(n+1)}xy + C_{(n+2)m}x^2 + C_{mn}y^2 \tag{8.7}$$

where x and y represents the 2D data, C represents the column vector coefficient and n and m denotes rows and columns in column vector matrix. Vala and Baxi (2013) have discussed the detection of static hand gesture in a fixed background and illumination condition. The Otsu's thresholding method is used to identify the position of the hand boundary in complex back ground. The color image is convert into a grayscale image and then into a binary value by applying Otsu's thresholding. By seeing the binary values of the input data, easily identified the contour of the hand figure shows the binary value of a hand gesture containing '0' and '1' to represent the black and white of the hand image respectively (Figure 8.4).

8.1.2.2 Classifiers

8.1.2.2.1 Support vector machine (SVM)

A support vector machine is used in various applications, preferably for pattern recognition, machine learning, and virtual reality by Ben-Hur et al. (2001); Gaonkar and Davatzikos

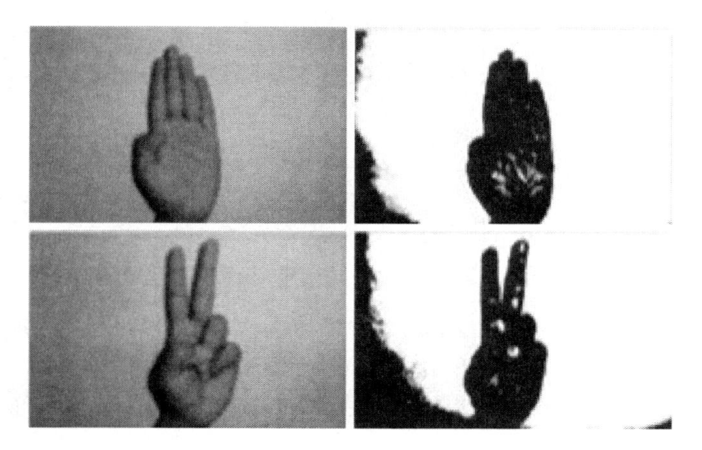

FIGURE 8.4: Detection of hand portion by using Otsu's threshold method.

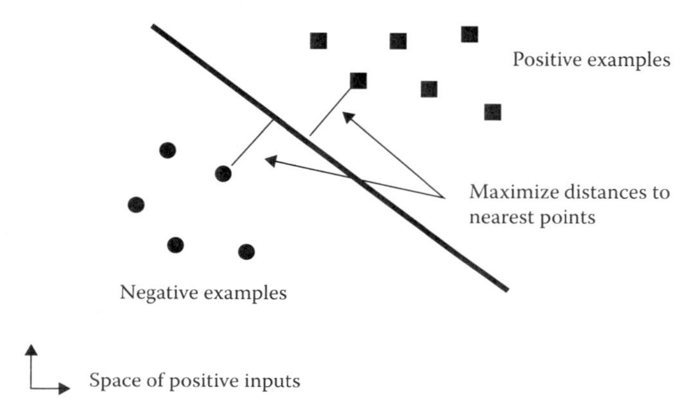

FIGURE 8.5: Linear classification using support vector machine (SVM).

(2013); Chih-wei (2002). It can be a nonlinear classification in addition to a linear classification. This uses a technique called the kernel trick to perform efficient classification. This trick enables the mapping of input into high dimension feature spaces by Chih-wei (2002); Shalev-Shwartz et al. (2011). SVMs have a wide range of applications in robotics, medical diagnosis through feature extraction, virtual gaming, and machine learning by Chih-wei (2002). It can be used to analyze the data for classification and regression by Gaonkar and Davatzikos (2013) as shown in (Figure 8.5). These can be done by constructing a hyper plane or infinite dimensional space. The mappings are done in such a way that the products can be given in terms of variables using kernel function. According to Dominic et al. (1991) and Forouzanfar et al. (2010), the product of hyper planes with a vectors in the free space coordinated is constant. Let vectors be a_i and x_i with hyper plane A, as given by Equation 8.8:

$$\sum_{i}^{n} a_i k(x_i, A) = const \tag{8.8}$$

This equation supports both linear and nonlinear SVMs for classification of pattern recognition Lee et al. (2004).

8.1.2.2.2 Dynamic time warping (DTW)

DTW resembles other classifiers and gives more accuracy in pattern recognition applications. It wraps or combines the time sequence of joint positions to reference time sequences and produce a similar value, Keogh and Ratanamahatana (2004). It reduces the complexity of the pattern of the given input sequence. The wrapping of time axis is must to achieve a better alignment for gesture recognition Keogh and Ratanamahatana (2004); Lemire (2009). Let us consider Q and C of length m and n respectively, where $Q = q_1, q_2, \ldots, q_n$; $C = c_1, c_2, \ldots, c_n$. These two sequences are modeled by using DTW by constructing an $n \times m$ matrix by Gupta et al. (1996). The wrapping between the sequences Q and C is given by a set of matrix elements 'w', Wang et al. (2013); Gupta et al. (1996) the wrap output is represented by $W = W_1, W_2, \ldots, W_n, \ldots, W_k$. The wrapping path is subjected to several constraints by Gupta et al. (1996). They support boundary conditions, continuity, and monotonicity property. Even though these constraints are considered for many wrapping paths, the cost-efficient wrap is considered of k factor for different lengths given in Equation 8.9:

$$DTW(Q, C) = min[\sqrt{\sum_{k=1}^{k} W_k/k}] \tag{8.9}$$

8.1.2.2.3 Artificial Neural Network (ANN)

An artificial neural network (ANN) is used as part of a classifier method in pattern recognition applications. It is constructed by many neurons in that network, are interconnected with each other in which data's are transferred only from the input layer through the hidden layers to the output layer. Elkan (2003); Drake et al. (2012). The network formed is represented by the pattern of the gesture followed by the feature extraction points. Another way of constructing the network is by using its transition states in the ANN model suggested by Amorim and Mirkin (2012). This shows the architecture is well-equipped by 'n' number of states connected like a neuron connection in the brain. Coates (2011) has demonstrated the maximum possible case of occurrence to get the output transition state probability. This probability makes the system produce better decision making. Graves and Schmidhuber (2009) have applied it to offline handwriting recognition with mutlidimensional recurrent neural network.

8.1.2.2.4 Random tree decision method

In computer vision, the random decision forest method is used in multiclass classifiers for pattern recognition applications. Sharp (2008) has proposed an idea for classification of real-time patterns by using random tree methods. It follows the basic three steps to achieve maximum speed:

1. It classifies the multiclass feature points but not to train several binary feature points for this problem.
2. It usually works faster in both the training and testing phase.
3. It works for multiclass feature point simultaneously.

Goussies et al. (2014) have investigated a novel algorithm with the decision forest method that works faster than the conventional method and acquires less error rate.

8.1.2.3 Optimization techniques

8.1.2.3.1 Ant colony algorithm

Song et al. (2013) have suggested that using the gesture recognition system with the ant learning algorithm (ALA) eliminates some limits with the recent algorithms, such as HMM. It requires only minimized learning instances and reduces the computational complexity of the system. Blum (2005) has described the origin of ant colony optimization (ACO). Consider that an ant searches for food and stores it in the nest. It decides the path from the source point to the target food location and again from target food location to the source point. It identifies the shortest path between the source point and target food location. This leads to time reduction for the entire process. Let the path followed by the ant in graph model $G = (N, M)$ where N denotes the node pair for two entity points. $N = (N_n, N_f)$ where N_n denotes the nest of ants, and N_f denotes food. The link between two entities is represented by $M = (m_1, m_2)$ where m_1 has a sequence length of l_1; m_2 has a sequence length of l_2, $l_2 > l_1$. Let S_1 denote the shortest path between N_n and N_f and S_1 denote the longest path.

8.1.2.3.2 Particle swarm optimization (PSO)

Nyirarugira et al. (2016) have investigated that hand gestures in 3-D spatio-temporal representation are very difficult to implement in the vision-based detection method. This affects the performance metrics of the system. The traditional algorithm decided the cost function through the travelling sales man problem (TSP). The cost function has decided the system performance. Nyirarugira et al. (2016) suggested the new optimization technique called particle swarm optimization (PSO). This decides the best cost function between two entity points so that the system performance tends to be high. The following steps show how to spot through the hand gestures with PSO.

1. **Gesture spotting** is defined as detecting meaningful information in a time series. To detect hand gesture, first check whether a particle has already reached to the referred template.

$$a_o \leftarrow P_b^a est == |T^a|, \forall N \tag{8.10}$$

2. To detect the non-gesture a_o denoted by the normal gestures $a_o \neq \Phi_{<sub>}$.

3. Subgesture spotting is detected at this condition by Equation 8.11.

$$\Delta F = \begin{bmatrix} a_s^o up \\ P_{best} \end{bmatrix} == 0 \tag{8.11}$$

8.1.2.3.3 Sequential minimal optimization (SMO)

Platt (1998) has proposed a newer version of the SVM for training, testing, and optimization called as sequential minimal optimization (SMO). Especially in the training and testing phase, the SVM requires a very large set of quadratic programming (QP) problems. SMO divides QP into the smallest QP problem as possible. These values have a pair of Lagrange multipliers, because it satisfies linear equality condition. At each step, SMO finds the pair of Lagrange multipliers together to get the optimum value as shown in Figure 8.6. This avoids time consumption in numerical optimization by QP in the branching loop. This leads to reducing the memory size but SMO takes a large set of training data for computing with large scaling rather than SVM.

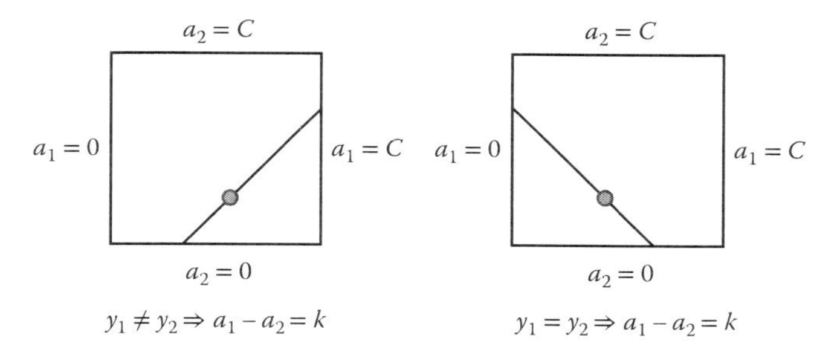

FIGURE 8.6: Linear equality conditions for SMO.

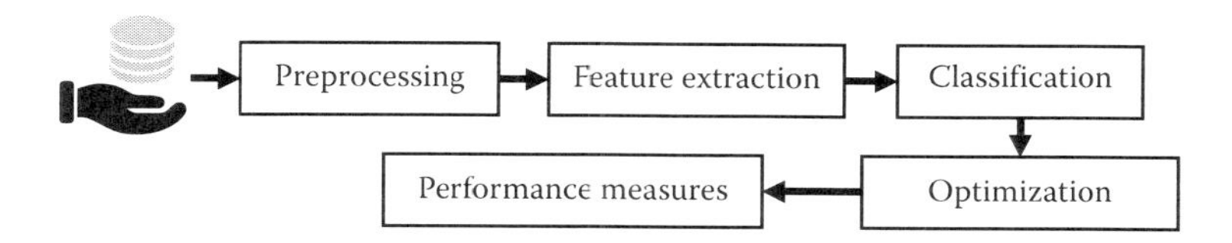

FIGURE 8.7: Outline of hand motion recognition system.

8.1.3 Methodology used

Figure 8.7 describes about the outline of the hand motion recognition system (HMRS) for virtual reality application. There are four main stages: pre-processing, feature extraction, classification, and optimization. Initially, the image database has to create with the system which consists of 360 images of five different classes with 60 orientations. This hand image data is taken from the Imperial College, London website. Here, edge detection technique is used to locate the hand portion from the entire image with complex background by using four different types of operators known as Sobel, Prewitt, Canny, and Robert. After detection of hand portion, extract the feature vector points from the hand gesture with different orientation, shape, and position, etc. as discussed by Gu and Su (2008). The next step is to classify the hand gesture by using the stochastic mathematical model called 2D hidden Markov model (2D HMM). It produces the delayed classified output due to the more state sequence in each transition state probability. To reduce the state variable in HMM, the optimization technique called artificial bee colony (ABC) algorithm is preferred.

8.2 Database

The dataset consists of 360 image sequences of six different classes with 60 arbitrary shapes, positions, angles, velocity, etc. The variation of temporal patterns in each class has created the dataset. The typical, six actions used in the gaming in virtual environment by Ma et al. (2013) are (1) flat leftward, (2) flat rightward, (3) flat contract, (4) V-shape leftward, (5) V-shape rightward, and (6) V-shape contract, as shown in Figure 8.8.

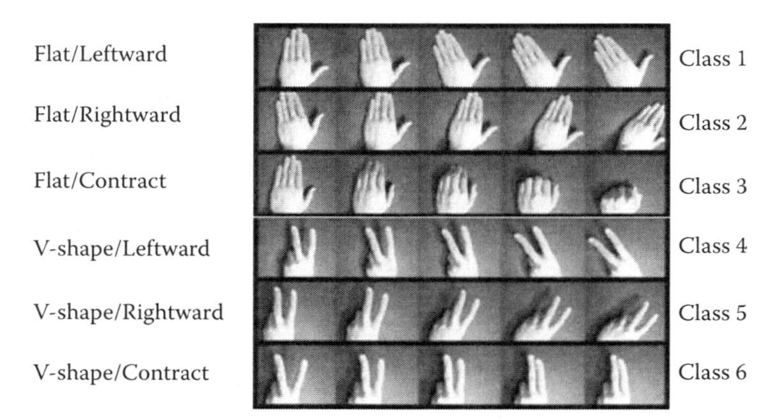

FIGURE 8.8: Hand dataset: six different gesture classes with three different motions and shapes.

8.3 Preprocessing

In any pattern recognition application, before finding the feature vector points, the most essential step is preprocessing of the input data. Preprocessing is used to transform data into other form that reduces noise without loss of informations, suggested by Hanegan (2010). There are various methods used for preprocessing: edge detection, filtering, histogram equalization, HSI color space and desaturation. Canny has proposed how to segment the object from the complex background, which also identifies and locate the sharp discontinuities in hand gesture in both fixed and dynamic background conditions. In pattern recognition or classification, edges are very important to identify and it is important to recognize the boundary between adjacent points suggested by Panda and Patnaik (2010). Gonzales and Wood (2008) have investigated in his book about the various types of operators in edge detection, in which the Sobel operator provides a better quality of output image. Figure 8.9 shows kernel values of preprocessed data by using a Sobel operator. Even a Sobel operator does not provide a good quality of output image. Gonzales and Wood (2008) have suggested low-pass filters or smoothing filters to suppress the high-frequency component of an image. Figure 8.10 shows the kernel value of a low-pass filter for the improvement of the pixel value for colored images (red, green, and blue, or RGB). Yoon et al. (2009) have suggested another way of preprocessing by using the histogram equalization. It is used to represent the pixel distribution of the image which is directly linked with the vision of a hand image. The histogram of an image equalizes the image contrast to enhance the image quality. Gonzales and Wood (2008) have proposed an idea about the concept of image thresholding in hue, saturation, and intensity (HSI) color in a 3-D coordinate system. This balances the lighting condition in the dynamic background. This adjusts the hue, saturation, and intensity of a hand image from the space coordinate of the system. Burdick (1997) has discussed the occurrence of the illumination condition while converting from the original RGB to HIS after filtering or histogram and before edge detection. Thresholding is performed by setting the hue value to < 43. This produces a binary image where '1' represents the area of hand image in white background and '0' represents the other area of image in black background. Lionnie et al. (2012) have described desaturation in image preprocessing technique. Desaturation is a process of conversion of an input image into a grayscale image by removing

1 –	0	1 +
2 –	0	2 +
1 –	0	1 +

–1	–2	–1
0	0	0
+1	+2	+1

FIGURE 8.9: Gradient matrix for Sobel operator.

$$\frac{1}{9} \times \begin{bmatrix} 1 & 1 & 1 \\ 1 & 1 & 1 \\ 1 & 1 & 1 \end{bmatrix}$$

FIGURE 8.10: Kernel value of filtering.

color content. This can be done by HIS color space, so it easily takes the intensity layer (I) from the image. Thus, it produces the high-quality image for the next step, such as feature extraction.

8.4 Feature Extraction

In the image processing application, feature extraction is used as a dimensionality reduction technique. This tends to reduce the large set of feature vector points into the desired feature point to represent the input data. This leads the reduction of memory size and power. The subsection has explained histogram features used in the proposed system.

8.4.1 Histogram of oriented gradient (HOG)

In image processing, feature points are very important for identifying the shape of an image under various conditions. Using a histogram of oriented gradients (HOG), it is possible to obtain the feature set with greater precision. It works on the uniformly spaced dense grids lapping over contrast normalizations. The descriptor is an advanced technique which resembles existing methods of edge orientation histograms which work on the overlapping local contrast normalizations on a uniformly spaced grid. Dalal and Triggs (2005) discovered various methods for block normalization. They derived an equation for reducing dimensionality of an image. Let u be the non-normalized vector of histogram features of a block $||u||k$ is k_{norm} for $k = 1, 2$ and e is a constant for negligible values. Thus, the normalization factor is expressed in Equation 8.12:

$$L_{1-sqrt} = \sqrt{\frac{u}{||u||_1 + e}} \tag{8.12}$$

Dalal and Triggs (2005) also discovered L_{1-norm} and L_{2-norm} provide identical performances, whereas the L_{1-norm} delivers less dependable performance. Therefore, a remarkable improvement over non-normalized data is obtained by these methods. The feature point represents the segmented region of a hand portion having differences in brightness and orientation as shown in Figure 8.11.

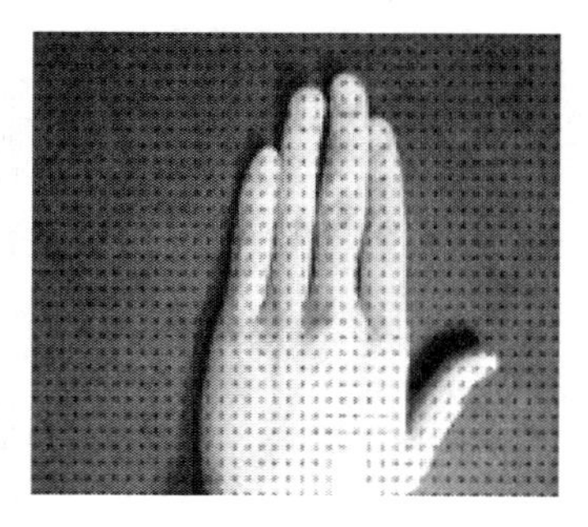

FIGURE 8.11: HOG feature with a difference in brightness.

8.5 Classifier

In statistical analysis, classification is the process of identifying to which set the following observation belongs. As far as the pattern recognition is considered, the classifiers can be used in both supervised and unsupervised learning. Supervised learning labels the data in the group whereas unsupervised learning is not labeled for data in the group. The following subsection explains the 2D HMM approach which is applied to pattern recognition for trained data.

8.5.1 2D hidden Markov model (2D HMM)

HMM is the best output for different patterns compared with the other classifiers suggested by Miya (2008). In traditional HMM, the major drawback in the system is scaling, as suggested by Bobulski (2015). It could not take the image in two dimensions. It first converts from 2D form into 1D form, leading to a loss of information in the given data suggested by Miya (2008). In order to avoid the scalability issue, 2D HMM has been used by Bobulski (2015). The 2D HMM reduces the complexity for input hand gestures in 2D qualities. It gives a more realistic interaction compared to traditional devices. The image or scene captured is taken as patterns are assigned by different classes based on the number of states in HMM by Sathis and Gururaj (2003); Lanchantin and Wojciech (2005). The 2D HMM consists of a set of hidden state, initial state probabilities, and output probabilities for an each individual state. Boudaren (2012) considered the hidden state as 'x' and input as 'i'. Let 'i' vary from $1, \ldots, N$. It is defined as $X = x_i$, $i = 1, \ldots, N$. Thus, the transition probabilities are given in Equation 8.13:

$$Y = [a_{ij} = P(x_j@t + 1/x_i@t)] \tag{8.13}$$

The probability of the transition state depends on the initial probabilities $P_i = P(x_i@t = 1)$ of the state sequence given by Rabiner (1989). The 2D Viterbi algorithm decides the optimized path from the initial state to the end state by Coates (2011); Rumelhart (1986) as shown in Figure 8.12.

8.6 Optimization Technique

In pattern recognition applications, optimization plays a vital role to reduce the number of state sequences in the classifier part. During the training and testing phase, the state

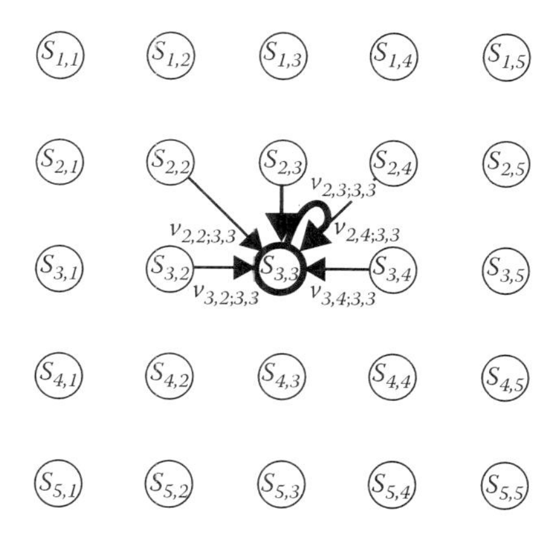

FIGURE 8.12: Structure of 2D HMM.

variable requires more path to travel from one state to another state. This affects the system performance. In the following subsection, we explain the ABC optimization technique used in the proposed system.

8.6.1 Artificial bee colony (ABC) algorithm

Kumar and Kumar (2010) have discussed the heuristic-based approach for optimization problems that solves the high dimensionality of state sequences in the building block of the 2D HMM. The decoding problem of the HMM classifier can be implemented by using 2D Viterbi decoding algorithm. Liwicki et al. (2009) have explained how to find the shortest path in the state sequence in HMM by using the expectation maximization (EM) approach. Goussies et al. (2014) have also extended the work by using the Baum–Welch procedure or Viterbi algorithm. They have explained how the system decides gesture recognition with the optimum state sequence in HMM. First, select the model parameters of HMM and then apply the optimization technique. There are some steps to follow to optimize the state variable sequence in HMM, by Kumar and Kumar (2010):

- **Initialization phase**: Assume the size of the state sequence generated by bees represented by x_i has N variables, where its randomized value ranges from 0 to 1, which has lower bound 'l' and upper bound 'u' of the optimum solution in each search:

$$x_i = 1 + rand(0,1) * (u - 1) \tag{8.14}$$

- **Employed bee phase**: Identify the best state probability from one state to an adjacent state by using Equation 8.15:

$$\nu_{me} = x_{ie} + \Psi_{me}(x_{me} - x_{ke}) \tag{8.15}$$

  From the randomized state sequence, determine the best fitness function.
- **Onlooker bee phase**: This step uses the same Equation 8.15 but gives more priority to the randomized state sequence which has a high fitness function.
- **Scout phase**: This step repeats until get the new optimum solution x_i, by using Equation 8.14.

8.7 Experimental Results and Discussions

A detailed analysis has been made for the entire process of the hand motion recognition system. The system consists of four major steps: preprocessing, feature extraction, classification, and optimization.

8.7.1 Results of image preprocessing

The hand image has to be preprocessed using the edge detection technique. It consists of four different kernel functions, including Sobel, Prewitt, Canny, and Robert. It subtracts the background of a static hand gesture with the constant illumination condition. It detects the edge portion of the hand and filters the unwanted frequency component of image data. In each frame the hand image orients in a different angular position by these kernel functions. Figure 8.13 shows the GUI of the preprocessed hand image data. In the preprocessing stage, filter the unwanted frequency in the hand image and locate the object in a free space coordinate system. The contour of an object has to detect various kernel functions with respect to the gradient operator such as Sobel, Prewitt, Canny, and Robert. Among these operators, Sobel performs better than the others in this task. The sample preprocessed data from the available dataset is shown in Figure 8.14.

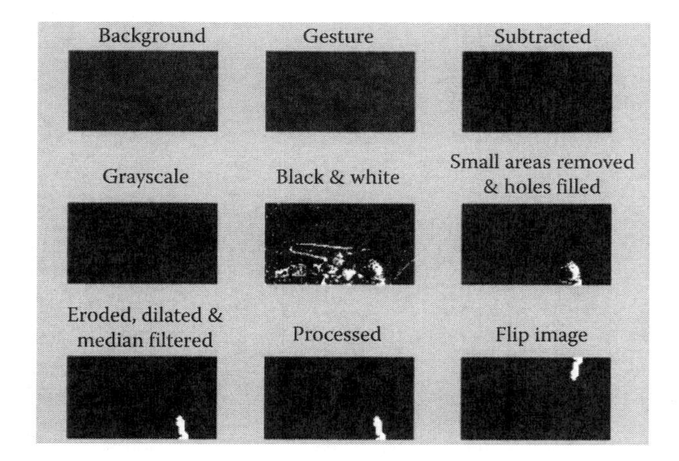

FIGURE 8.13: Output of image acquisition, edge detection, and preprocessing.

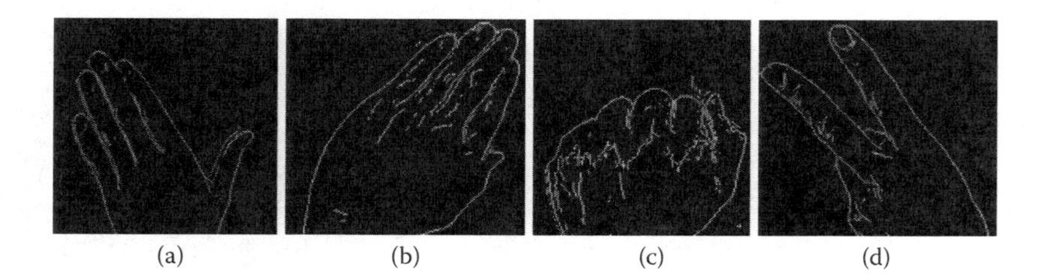

FIGURE 8.14: Preprocessed data by using Sobel operator: (a) flat-to-left (b) flat-to-right (c) flat-to-contract (d) V-shape left.

8.7.2 Result of feature extraction

The next step is to extract the feature point from the preprocessed or compressed data. Here, five techniques used for feature extraction of hand gestures: SIFT, SURF, HOG feature, Zernike moment, and Otsu's thresholding method. Detailed analysis has been done with these techniques by doing an experiment on each method. Among these techniques, this work concentrates on the Zernike moment and Otsu's thresholding method. Figure 8.4 shows the binary value of the hand image represented by '1' and '0' that is a white and a black background respectively. This method easily segments the hand portion from the background of the image with minimum noise factor.

Hand gestures are rotated over different angular positions with velocity. Locate the hand portion in the 3-D coordinate system, which provides meaningful information to the system. Determine the Zernike moment (n,m) with respect to the centroid of the hand gesture. By using Equations 8.12 and 8.13 calculate the amplitude and phase of the moment of the hand. The elapsed time per image to rotate is 0.43103 seconds. If the hand gesture perfectly matches with the image in the database, then $A = 0$ and $\Phi = 0$ as shown in Figure 8.15. Shows that orientation of various hand gestures using Zernike moment, if the hand gesture pattern is not perfectly matched then A and Φ will produce some different values.

8.7.3 Results of classifier

In this proposed work, 2D HMM results have been discussed in terms of performance measures. Initially, we have to analyze for single states how many possible transition states and observation states are available. The architecture of HMM is more complicated with hidden states for single transition of signal. Figures 8.16 through 8.18 show the graph between number of states versus time for $N = 10$, $N = 20$, and $N = 30$. The forward and backward scheme is followed in HMM. Alpha and beta value decide the forward and backward path respectively. The final updated value is decided by gamma function which provides all possible paths from starting point to ending point. Figure 8.19 shows the graph for the updated value of the forward-backward scheme with respect to time. The transition state probability varies exponentially with respect to time. It allows the trained data of the system to HMM model. The path traveled by state variable from one state to another state has an observation state and a transition state. The system has to make a decision about the optimum path trajectory by using the forward-backward procedure or the Baum–Welch

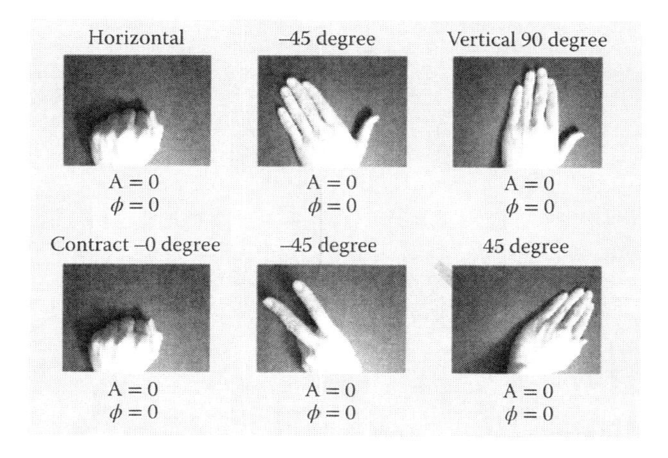

FIGURE 8.15: Orientation of various hand gestures using Zernike moment.

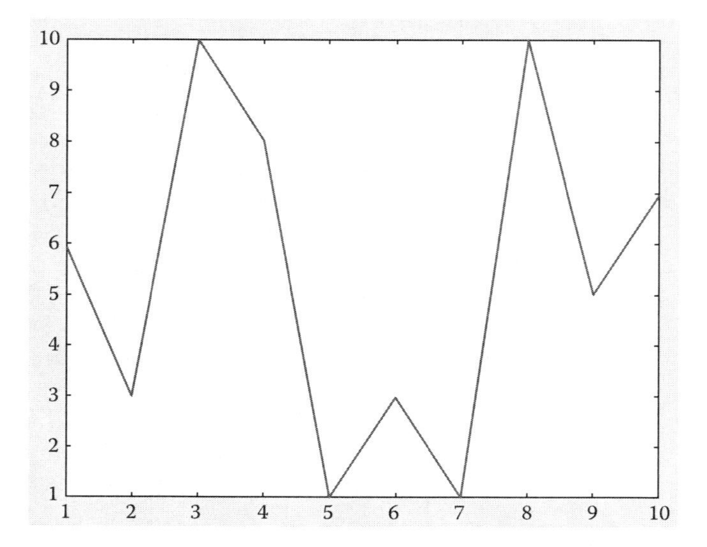

FIGURE 8.16: Number of states versus time in HMM for N = 10.

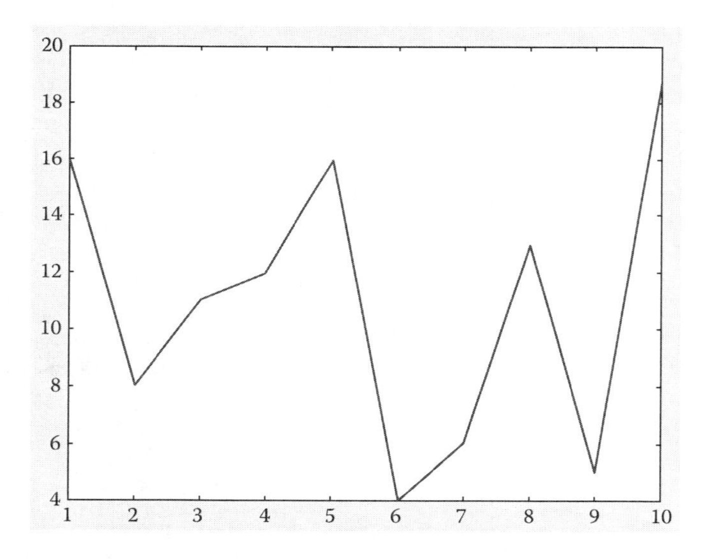

FIGURE 8.17: Number of states versus time in HMM for N = 20.

algorithm. During the testing phase, estimate the posterior probability of the model. Then find the number of data that are properly recognized with respect to the total number of data in the database. This value has the predominant performance rate of the classifier. The second type of classifier here is the SVM for pattern recognition in hand gestures. It follows several steps: (a) it classifies the row of feature vector points from the hand image; (b) once the feature vector points are labeled, it creates a matrix in such a way that a row represents the replicate data and a column represents the feature points; and (c) this leads to an equal amount of the sample data and the training data. Here, the group of gesture patterns indicates the row of sample data are assigned and feature vector points indicate the column of sample data as shown in Figure 8.20. The recognition rate is defined as the ratio of number of feature points classified to the total number of feature points of sampled

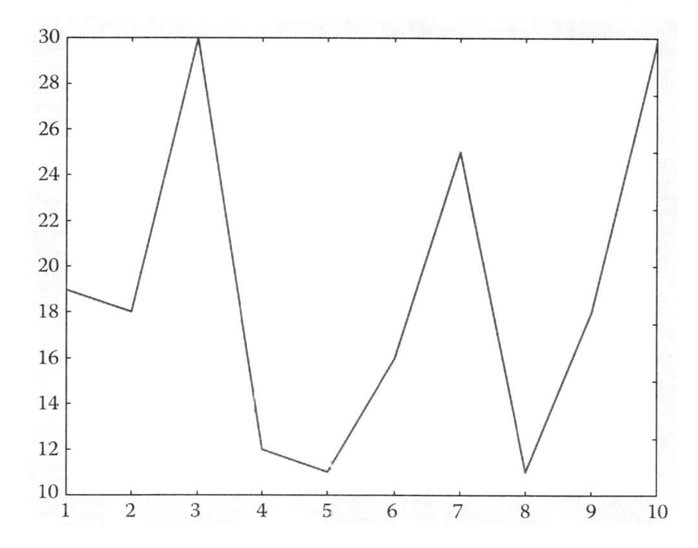

FIGURE 8.18: Number of states versus time in HMM for N = 30.

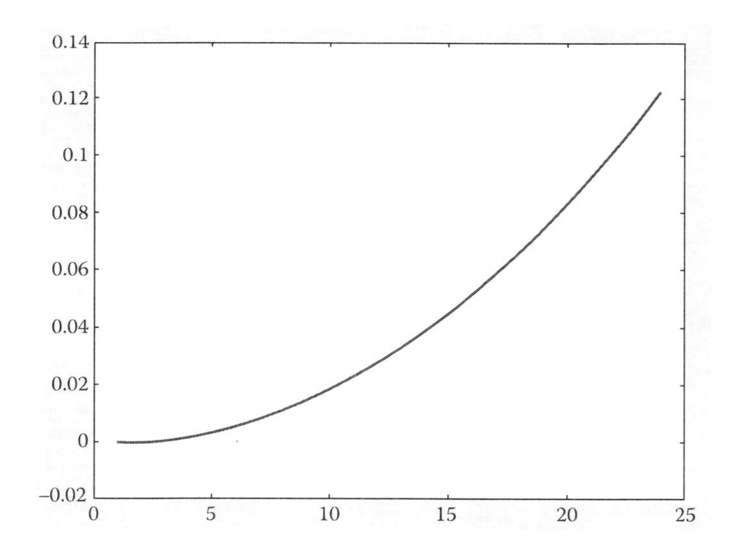

FIGURE 8.19: Updated value of forward-backward scheme in 2D HMM.

data. The third type of classifier here used in the ANN. After labeling the data, create the neural network model for classifying the pattern. To configure the network adjust the weight and bias value of the single layer perceptron. Assign the input and target value by defining the values of the hidden layer in the network. Then validate the process to calculate the performance measure with some iteration. In this scenario, 115 iterations have been taken to validate the mean square error (MSE) value of 3.7959×10^{14} to test the trained data given in the network. Also find the fitting function for one case, that of the hand gesture moving from the center to the left position. This produces maximum regression between input and target values with minimum error rate as shown in Figures 8.21 and 8.22. Similarly, this applies to all classes of hand gestures in the database. This helps to calculate the MSE

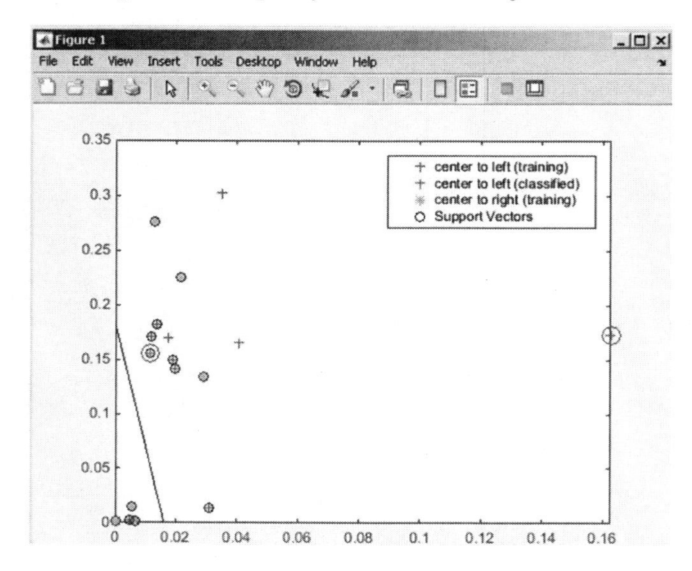

FIGURE 8.20: Classification of hand gesture using support vector machine (SVM).

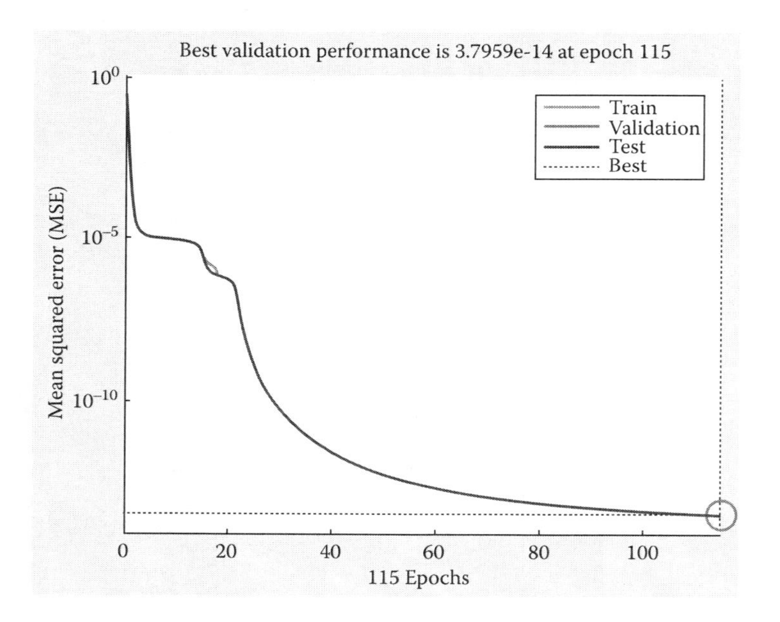

FIGURE 8.21: ANN classification, mean squared error.

and fitting value for all cases in the database. Then test the trained data with the target data to calculate the recognition rate. Table 8.1 shows the comparative analysis of various classifiers without optimization in terms of mean square error (MSE) and regression for training, validation and testing process is shown in Table 8.2.

The last type of classifier used is the random tree decision method. A regression tree has a set of coefficients which can be used to predict the target values from the past value or parent nodes. This can provide the system with very sophisticated decision making. Even then, the model parameters are good enough with the system metrics; it produces a lower recognition rate and accuracy than 2D HMM. Table 8.1 shows the comparison analysis of

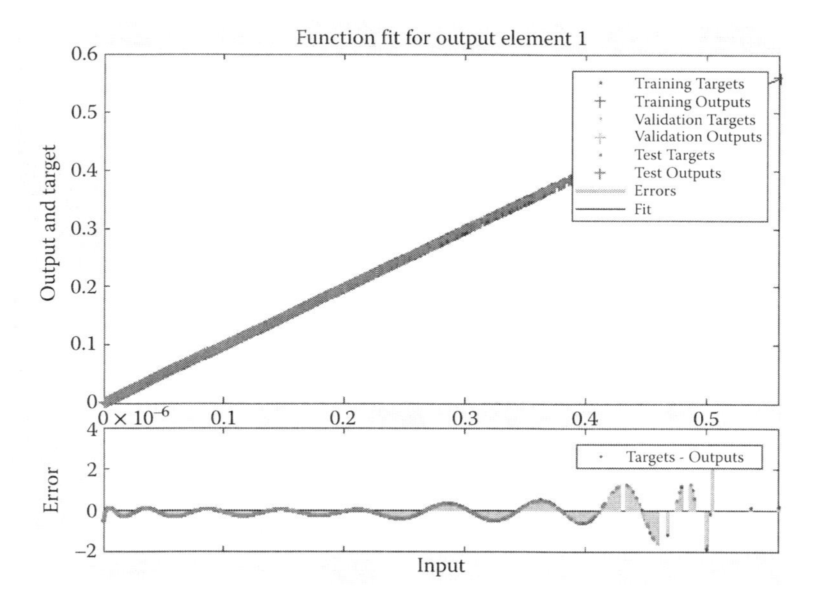

FIGURE 8.22: ANN classification, fitting function.

TABLE 8.1: Comparison analysis of various classifiers without optimization

Method	Recognition rate (percentage)	Estimated time (s)
2D HMM	97.65	2.001
SVM	96.85	2.003
DTW	96.67	1.954
ANN	96.78	2.204
Random tree decision	95.5	2.506

TABLE 8.2: Result analysis for ANN in terms of samples in MSE and regression

Data	Samples	Mean squared error	Regression
Training	11468	3.55777e-14	9.99999e-1
Validation	2458	3.79589e-14	9.99999e-1
Testing	2458	3.49857e-14	9.99999e-1

various classifiers. 2D HMM provides a better recognition rate with less time consumption compared to other classifiers used in this work.

8.7.4 Result of optimization techniques

The last stage in the hand motion recognition system is optimization. This step is required to improve the performance measure of the system. In this proposed work ABC optimization is used, which is compared with ant colony learning, PSO, and SMO. There are some issues in the system such as (a) scalability, (b) memory storage, (c) unwanted transition path in state sequence, and (d) looping to the same state. This work is focused on reducing the computational complexity of the HMM model. For that number of states the

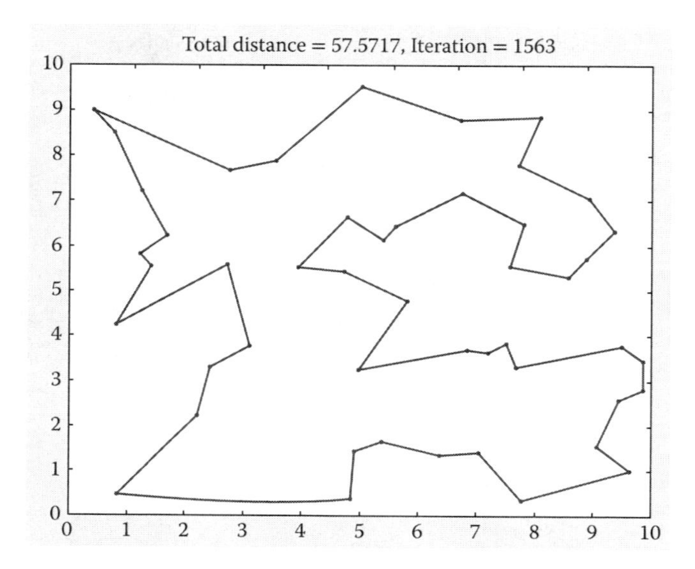

FIGURE 8.23: ABC optimization of state sequence; total distance = 57.57.

sequence that has been taken has a model parameter for this optimization part. The primary objective of this technique is to reduce the unwanted transition path between the states and also avoid the self-looping in the same state. First, the optimization technique used in this work is the ABC algorithm for hand motion recognition system. Among optimization techniques, ABC and ALA work similarly. In ABC, initially bees are deployed in the colony, represented as the number of states in HMM. Bees that collect the food and store it in the nest are represented as the starting point and terminal point of the system respectively. Bees should decide the path from source point to the target food point. Then they take the shortest path to get the food: nothing but the best cost function for the system as shown in Figure 8.23. However, bees fly faster than the ants move. If the system chooses this condition, then it achieves the better solution with less power consumption than ant colony optimization (ACO), as shown in Figure 8.24. Table 8.3 shows the comparative analysis of various classifiers with ABC algorithm. The model parameters of each classifier are applied in this optimization technique. Based on the cost function provided by this approach, again we tested the target data with trained features with the individual classifier and checked their performance measures in terms of recognition rate and estimation time. This result ultimately improves the performance metric of the system. A third optimization technique used in this work is PSO. First, it determines the local minima points in the classifier model by search method. Then it finds the optimum state variables exponentially as shown in Figure 8.25. The cluster of state variables is represented as a set of particles in the model. The limiting factor of this optimization is time consumption, stability, and sensitivity. Due to this issue, performance is lower than the proposed approach. The last optimization technique used in this work is SMO. In the pattern recognition application, it is a very challenging task and highly recommended. Risk represents mean-variance of the state sequence from one state to another state of finite state sequences in the HMM model. Expected returns represent the maximum possible state transition probability. Figure 8.26 shows the probability of the transition state sequence at a maximum of 0.0425 in which there is a risk factor of 2.2×10^{-3}. Due to the risk factor, the system has drastically lowered the performance rate of the system. To solve a QP problem by using SMO with equality condition, it further divides into the smallest QP possible. This is optimized with a pair of

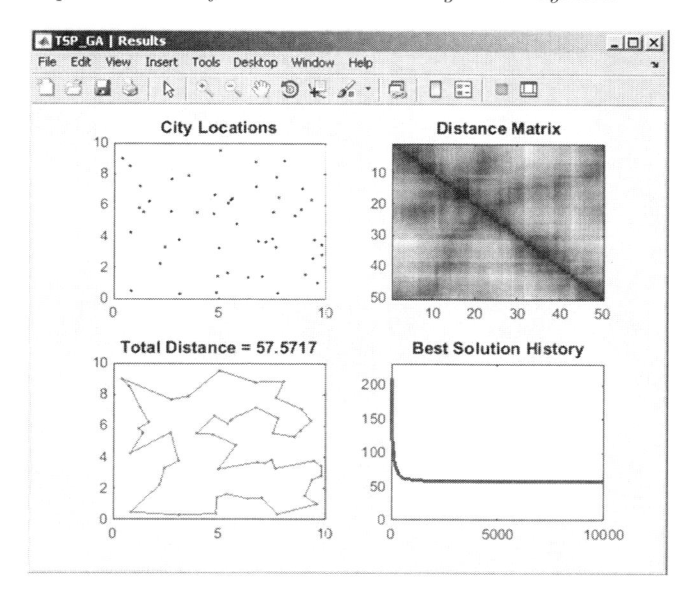

FIGURE 8.24: ABC optimization of state sequence best solution.

TABLE 8.3: Comparison analysis of various classifiers with ABC optimization

Method	Recognition rate (percentage)	Estimated time (s)
2D HMM	98.35	1.987
SVM	97.05	1.999
DTW	96.87	1.905
ANN	96.99	2.012
Random tree decision	96.15	2.125

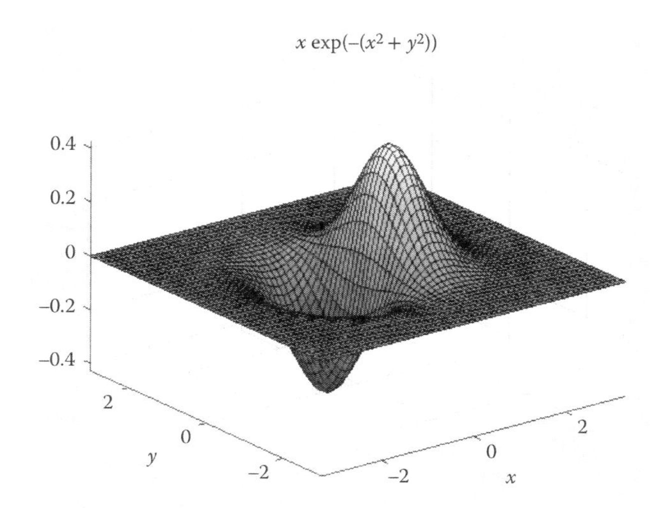

FIGURE 8.25: Particle swarm optimization for exponential condition of state sequence.

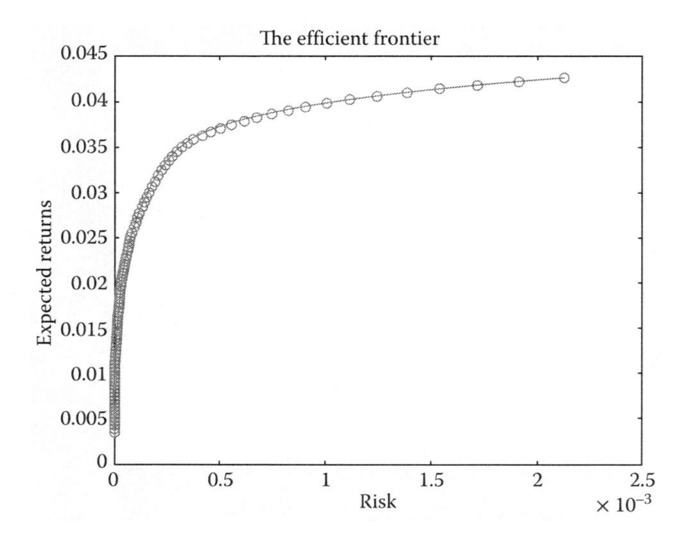

FIGURE 8.26: Sequential minimal optimization: expected returns versus risk.

Lagrange multipliers. It takes less time to compute the inner loop of QP. But it takes more computational time for the entire process of the system. In this work, SMO optimization provides a poorer performance rate than other optimization techniques.

8.8 Summary and Conclusions

Currently, the development of virtual games using hand motion is very important to all users. It is easier to access the machine with a less complex step to initiate the game through hand gestures, without any wired connection between human and machine. The hand motion recognition system consists of various stages: image acquisition, preprocessing, feature extraction, classification, and optimization. Hand images are acquired with five different classes in a fixed lighting condition. In each class, 60 different frames of hand images have been stored separately and labeled. Preprocess the input data by using the edge detection technique. Here edge detection, done by the Sobel operator, segments perfectly, unlike other operators. It provides a high tolerance rate for true detection. Among the various feature extraction techniques used in this work, the histogram of oriented gradient (HOG) and Zernike moment provide the optimum feature point that can predict what that object should be for the next stage of the system. These feature points are trained before the testing and validation process. The trained feature points are labeled and analyzed with the stochastic model.

In this work, a comparative analysis was made for the classifiers such as 2D HMM, SVM, DTW, ANN, and random tree decision method for hand motion recognition. Among the various classifiers, 2D HMM produces a high recognition rate with less computational time than other classifiers. Even then the classifier has a complex structure; the primary factor is the power consumption and efficiency of the system; This can achieved by reducing the parametric coefficient of the system, the various optimization techniques, such as artificial bee colony (ABC), ant learning algorithm (ALA), particle swarm optimization (PSO), and sequential minimal optimization (SMO) are used in this work. All the optimization techniques provide the best solution but ABC and ALA provide the best cost function in less iterative time. Additionally, the analysis was made for the classification process with ABC optimization, which provides better performance measures than the classifier without the optimization technique.

This chapter will give an idea of how to implement virtual reality applications in real time, as well as how to develop the system with less computational complexity, time complexity, and power consumption.

References

Alon, J. et al. (2009, September). A unified framework for gesture recognition and spatiotemporal gesture segmentation. *IEEE Transactions on Pattern Analysis and Machine Intelligence 31* (9), 1685–1699.

Amorim, R. C., Mirkin, B. (2012). Minkowki metric feature weighting and anomalous cluster initialization in k-mean clustering. *Journal of Pattern Recognition 45* (3), 1061–1075.

Ben-Hur, A. et al. (2001). Support vector clustering. *Journal of Machine Learning Research 2*, 125–137.

Blum, C. (2005). Ant colony optimization: Introduction and recent trends. *Journal of Physics of Life Reviews 2* (4), 353–373.

Bobulski, J. (2015). Comparison of the effectiveness of 1-D and 2-D HMM in the pattern recognition. *Journal on Image Processing & Communication 19* (1), 1–2222.

Boudaren, M. E. Y. et al. (2012). Unsupervised segmentation of random discrete data hidden with switching noise distribution. *IEEE Signal Processing Letter 19* (10), 619–622.

Boyali, A., Kavakli, M. (2012). A robust gesture recognition algorithm based on sparse representation, random projections and compressed sensing. In *7th IEEE Conference on Industrial Electronics and Applications (ICIEA)*, 243–249.

Burdick, H. E. (1997). *Digital Imaging: Theory and Applicationse.* McGraw-Hill, Arlington, TX.

Chih-wei, C.-J. L. (2002). A comparison of methods for multiclass support vector machine. *IEEE Transaction on Neural Network 13* (2), 415–425.

Coates, A. et al. (2011). An analysis of single-layer network in unsupervised feature learning. In *Proceeding of 14th International Conference on Artificial Intelligence and Statistics, JMLR: W & CP15 15.*

Dalal, N., Triggs, B. (2005). Histograms of oriented gradients for human detection. In *IEEE Computer Society Conference on Computer Vision and Pattern Recognition*, 1063–6919.

Dominic, S. et al. (1991). Comparison of feed-forward neural network training algorithms for oscillometric blood pressure estimation. In *4th International Workshop on Soft Computing Application (SOFA)A.*

Drake, Jonathan and G. Hamerly (2012). Accelerated k-mean with adaptive distance bound. In *5th NIPS workshop on Optimization for Machine Learning*, 2012.

Elkan, C. (2003). Using the triangle inequality to accelerate k-mean. In *Proceeding of 12th International Conference on Machine Learning*, Washington.

Elmezain, M. et al. (2008). A hidden Markov model based isolated and meaningful hand gesture recognition. In *Proceedings of World Academy of Science, Engineering and Technology 31*, 394–401.

El-Zaart, A., Ghosn, A. A. (2013). MRI images thresholding for Alzheimer detection. *International Journal of Computer Science and Information Technology*. DOI: 10.5121/csit.2013.3310, 95-104.

Forouzanfar, M. et al. (2010, July). Genetic reinforcement learning for neural network. In *International Joint Conference on Neural Network*, Washington.

Ganapathi, V. et al. (2010, June). Real time motion capture using a single time-of-flight camera. In *2010 IEEE Computer Society Conference on Computer Vision and Pattern Recognition*, 755–762.

Gaonkar, B., Davatzikos, C. (2013). Analytic estimation of statistical significance maps for support vector machine based multi-variate image analysis. *Journal of Neuro Image 78*, 270–283.

Gonzales, R. C., Wood, R. E. (2008). *Digital Image Processing*, 3rd edn. Prentice Hall, Upper Saddle River, NJ.

Goussies, N. A. et al. (2014). Transfer learning decision forest for gesture recognition. *Journal of Machine Learning Research 15*, 3847–3870.

Graves, A., Schmidhuber, J. (2009). Offline handwriting recognition with multidimensional recurrent neural networks. In Koller, D et al. (Eds.), *Advances in Neural Information Processing Systems*, vol. 21, pp. 545–552. Curran Associates, Red Hook, NY.

Gu, L., Su, J. (2008, May). Natural hand posture recognition based on Zernike moments and hierarchical classifier. In *2008 IEEE International Conference on Robotics and Automation*, 3088–3093.

Gupta, L. et al. (1996, April). Nonlinear alignment and averaging for estimating the evoked potential. *IEEE Transaction on Biomedical Engineering 43* (4), 348–356.

Hanegan, K. (2010). Unpivoting and pivoting your data to make it suitable for analysis. http://spot_re.tibco.com/community/blogs/tips/archive/2010/02/19/unpivoting-and-pivoting-your-data-to-make-it-suitableforanalysis.aspx.

Keogh, E., Ratanamahatana, C. A. (2004). Exact indexing of dynamic time warping. *Journal of Knowledge and Information System 7* (3). doi:10.1007/s10115-004-0154-9. 358–386.

Kumar, D., Kumar, B. (2010, October). Optimization of benchmark functions using artificial bee colony (ABC) algorithm. *IOSR Journal of Engineering 3* (10), 9–14.

Lanchantin, P., Wojciech, P. (2005). Unsupervised restoration of hidden nonstationary Markov chains using evidential priors. *IEEE Transaction on Signal Processing 53* (8), 3091–3098.

Lee, Y. et al. (2004). Multi-category support vector machines: Theory and application to the classification of microarray data and satellite radiance data. *Journal of American Statistical Association 99* (465), 67–81.

Lemire, D. (2009). Faster retrieval with a two-pass dynamic time warping lower bound. *Journal of Pattern Recognition 42* (9), 2169–2180.

Lionnie, R. et al. (2012). Performance comparison of several pre-processing methods in a hand gesture recognition system based on nearest neighbor for different background conditions. *ITB Journal of Information and Communication Technology 6* (3), 183–194.

Li, T.-H. S. et al. (2016, January). Recognition system for home-service-related sign language using entropy-based K-means algorithm and ABC-based HMM. *IEEE Transactions on Systems, Man and Cybernetics: Systems 46* (1), 150–162.

Liu, F. et al. (2015, January). Hand recognition based on finger contour and PSO. In *International Conference on Intelligent Computing and Internet of Things (ICIT)*, 35–39.

Liwicki, M. et al. (2009). A novel connectionist system for unconstrained handwriting recognition. *IEEE Transaction on Pattern Analysis and Machine Intelligence 31* (5), 855–868.

Ma, A. J. et al. (2013). Supervised spatio-temporal neighborhood topology learning for action recognition. *IEEE Transactions on Circuits and Systems for Video Technology 23* (8). 1447–1460.

Mantecon, T. et al. (2015a, August). Enhanced gesture-based human computer interaction through a compressive sensing reduction scheme of very large and efficient depth feature descriptors. In *12th IEEE International Conference on Advanced Video and Signal Based Surveillance (AVSS)*, 1–6.

Mantecon, T. et al. (2015b, June). Hand gesture-based human machine interface system using compressive sensing. In *International Symposium on Consumer Electronics (ISCE)*, 1–2.

Miya, W. (2008). Condition monitoring of oil-impregnated paper bushings using extension neural network, Gaussian mixture and hidden Markov models. In *IEEE International Conference on Systems Man and Cybernetics.*

Nyirarugira, C. et al. (2016). Hand gesture recognition using particle swarm movement. *Journal of Mathematical Problems in Engineering 2016.* article ID 1919824. 1–8.

Otsu, N. (1979). A threshold selection method from gray-level histogram. *IEEE Transaction on System Man Cybernetics 9* (1), 62–66.

Panda, C. S., Patnaik, S. (2010). Better edge gap in grayscale image using Gaussian method. *International Journal of Computational and Applied Mathematics 5* (1). 53–65.

Pandita, S., Narote, S. P. (2013). Hand gesture recognition using SIFT. *International Journal of Engineering Research and Technology 2* (1). 1–4.

Platt, J. C. (1998). Sequential minimal optimization: A fast algorithm for training support vector machine. *Microsoft Research, Technical Report.* MSR-TR-98-14.

Rabiner, L. R. (1989). A tutorial on hidden Markov model and selected applications in speech recognition. *Proceedings of IEEE Conference 77* (2), 257–286.

Raheja, J. L. et al. (2015, June). Robust gesture recognition using kinect: A comparison between DTW and HMM. *International Journal on Light and Electron Optics 126* (11–12), 1098–1104.

Rautaray, S. S., Agrawal, A. (2011, December). Interaction with virtual game through hand gesture recognition. In *2011 International Conference on Multimedia, Signal Processing and Communication Technologies*, 244–247.

Ren, Z. et al. (2013). Robust part based hand gesture recognition using kinect sensor. *IEEE Transaction on Multimedia 15* (5), 1110–1120.

Rumelhart, D. E. et al (1986). *Parallel Distributed Processing. Explorations in the Microstructure of Cognition: Foundations*, vol. 1. MIT Press. Cambridge, USA.

Sathis, L., Gururaj, B. I. (2003, April). Use of hidden Markov models for partial discharge pattern classification. *IEEE Transaction on Electrical Insulation 28* (2). 172–182.

Shalev-Shwartz, S. et al. (2011). Pegasos: Primal estimated sub-gradient solves for SVM. *Mathematical Programming 127* (1), 3–30.

Sharp, T. (2008). Implementing decision tree and forests on a GPU. In *Proceeding of the European Conference on Computer Vision*, 595–608.

Sheenu, G. J. et al. (2015, September). A multi-class hand gesture recognition in complex background using sequential minimal optimization. In *2015 International Conference on Signal Processing, Computing and Control (ISPCC)*, 92–96.

Sheta, B. et al. (2012, October). Assessments of different speeded up robust feature (surf) algorithm resolution for pose estimation of UAV. *International Journal of Computer Science and Engineering Survey 3* (5). 15–41.

Song, S. et al. (2013). An ant learning algorithm for gesture recognition with one-instant training. In *IEEE Congress on Evolutionary Computation* (ISSN: 1089-778X), 2956–2963.

Vala, H. J., Baxi, A. (2013, February). A review on Otsu image segmentation algorithm. *International Journal of Advanced Research in Computer Engineering and Technology 2* (2). 387–389.

Wang, X. et al. (2013). Experimental comparison of representation methods and distance measures for time series data. *Data Mining and Knowledge Discovery 26* (2), 275–309.

Yoon, H. et al (2009). Image contrast enhancement based sub-histogram equalization technique without over-equalization noise. In *Proceedings of the International Conference on Control, Automation and System Engineering.*

Zyda, M. (2005, September). From visual simulation to virtual reality to games. *Computer 38* (9), 25–32.

Chapter 9

Smart Visual Surveillance Technique for Better Crime Management

Narendra Rao T. J. and Jeny Rajan

9.1 Introduction

Recently, visual surveillance has become the need of the day due to rising crimes and terrorism, hence, the feeling of vulnerability in the society. In this modern era, with great advancement in technology, there is an emphasis on sophisticated systems which not only carry out surveillance but also automatically detect anomalous behaviors which can have devastating outcomes and alert the concerned authorities. Advanced video surveillance applied to detecting suspicious behaviors can also be a powerful tool for pre-event deterrence. Intelligent video analysis forms the underlying logic of such surveillance systems. They learn the normal events occurring in the context, and any behaviors deviating from this learned set of actions are inferred to be anomalous or suspicious.

9.1.1 Visual sensor networks for surveillance

In the last few years, Visual Sensor Networks (VSNs) consisting of surveillance cameras are in wide use due to their highly effective monitoring ability which is beyond human capacity. They have been positioned in mammoth proportion in common areas like airports, subways, shopping centers, sports centers, residential localities, etc., with an objective to serve as a tool for crime detection, reduction, and risk management [1].

9.1.2 Automated visual surveillance

Usually visual surveillance systems are operated manually to a large extent to monitor activities, detect any unusual incidents such as the behaviors observed in Figure 9.1a through c [3–5] and to bring them to the attention of the authorities. Unfortunately, many possible threats may go undetected with such a manual system due to obvious limitations of relying solely on human operators. Missdetections could be caused by an unmanageable number of video screens to monitor (Figure 9.1d), loss of interest and exhaustion due to prolonged monitoring, ignorance regarding the events to be deemed anomalous, and distraction by

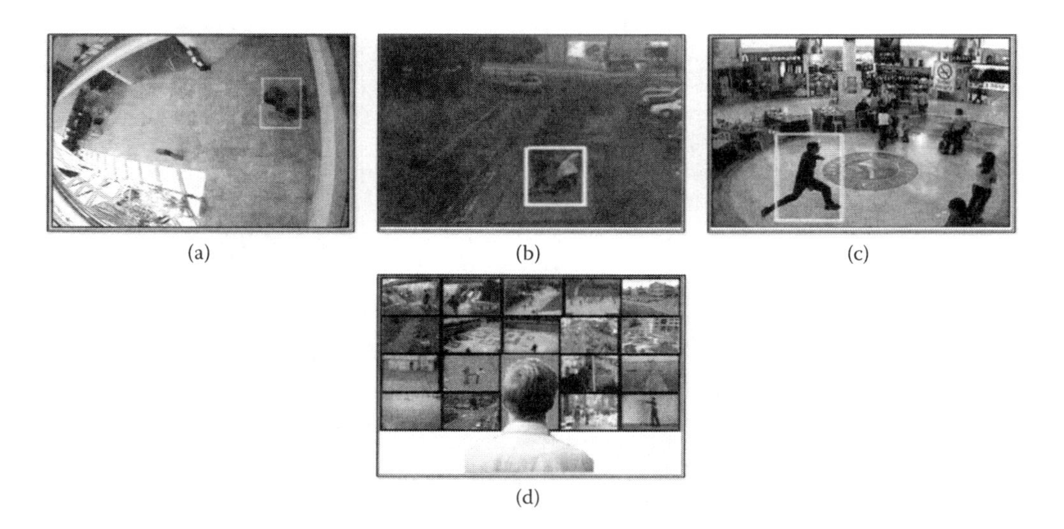

FIGURE 9.1: Representative images of (a) the act of abandoning an object, (b) the act of violence, (c) the act of running suspiciously in a mall, and (d) a human operator monitoring a large number of video screens. (The bounding box is used to indicate the anomaly.) ((a,b) Image reprinted from the open source CAVIAR (Caviar: Context-aware vision using image-based active recognition. http://homepages.inf.ed.ac.uk/rbf/CAVIAR/caviar.htm, 2011) and BEHAVE (R Fisher. Behave: Computer-assisted prescreening of video streams for unusual activities. http://homepages.inf.ed.ac.uk/rbf/BEHAVE/, 2011) datasets respectively, (c) image reprinted with permission from the first author of (A Adam et al., *IEEE Transactions on Pattern Analysis and Machine Intelligence*, 30, 555–560, 2008.))

other assigned jobs. Consequently, the captured surveillance videos are not useful for crime prevention, instead they often end up as passive records or as evidence for post-event investigations. Failing to detect anomalous activities can turn out to be hazardous in crucial scenarios such as border security, airports, train stations, etc. For example, the perpetrators of the bombing at the Luton train station in London in 2005 and during the marathon in Boston in 2013 were captured on CCTV on 7 July 2005 and 15 April 2013 respectively. The Nissan Pathfinder vehicle employed during the attempted attack on Times Square, New York in 2010 was also captured on CCTV on 1 May 2010. However, these images were unfortunately used only for post-event investigations [2].

So, technology providers and end-users have recognized the inadequacy of the manual surveillance process alone in meeting the needs of screening and timely detection of interesting events from an exhaustively enormous quantity of visual information that is captured by the increasing number of visual sensors [1,2]. To accomplish this, in the first years of the third millennium, studies began to shift the visual data evaluation technology from a completely human-oriented one to a computer-aided or a fully digital automated design of video-based surveillance systems, ranging from the sensor level up to the presentation of adequate visual information to the operators.

Finally, an advanced man-machine interface (MMI) assisted in drawing the operator's focus towards only the events of interest and probable threats. The design of upcoming embedded digital visual sensors is focused towards processing the captured digital data locally at sensor level itself. The final outcome of a wireless network with such advanced visual sensors is the transmission of only the interesting events from the individual sensors to the control center after processing the captured frames locally in real time by the sensors [6]. When it comes to surveillance systems, these interesting events can be termed unusual/ anomalous/ abnormal/ irregular/ suspicious events. Such systems perform the task of detecting the anomalous/suspicious or actual crime events when they occur. Some of the applications of a surveillance system involving unusual event detection include the following:

- **Intruder detection** often implies fence trespassing detection when a person is caught intruding a prohibited perimeter. This application is useful in ensuring perimeter security of critical and prohibited areas like borders, controlled-access buildings, military facilities, defense research centers, etc. (Figure 9.2a [7] and b).
- **Unattended object detection** aims to detect the object that has been placed in an environment for an extended time period. This may be useful to alert the security personnel regarding the suspicious object which may turn out to be a threat to the public (Figure 9.2c [8]).
- **Loitering activity detection** is used to identify people staying in the context for a protracted time without any apparent lawful purpose. This has an application in detecting the suspicious behavior before a real security violation occurs (Figure 9.2d and e [5]).
- **Tailgating detection** is the detection of unauthorized follow-through behavior at entry checkpoints. An alarm is produced if more than one person enter a prohibited location when actually only one person is permitted.
- **Crowd management** is used to avoid overcrowding situations at transport infrastructures, malls, etc. (Figure 9.2f [9]), where the statistics on the crowd volume are collected by dedicated software.

However, designing such surveillance systems that serve their true purpose to the full extent is quite challenging. To develop an intelligent and efficient unusual-event-detection algorithm capable of detecting genuine anomalies in any given context is the primary goal of such design, and it has been an active research area in recent years.

FIGURE 9.2: Representative images of the different applications of an automated surveillance system: (a,b) intruder detection, (c) unattended object detection, (d,e) loitering detection, and (f) crowd management. (The bounding box is used to indicate the anomaly.) ((a) Image reprinted with permission from IPS Intelligent Video Analytics, (b) image reprinted from open source ViSOR dataset (R Vezzani and R Cucchiara et al., *Multimedia Tools and Applications*, 50, 359–380, 2010), (c) image reprinted from open source PETS dataset (Computer Vision Pattern Recognition (CVPR). Performance evaluation of tracking and surveillance (PETS). http://www.cvg.rdg.ac.uk/PETS2001, http://www.cvg.rdg.ac.uk/PETS2006, 2009), (d) image reprinted with permission from IPS Intelligent Video Analytics, (e) image reprinted with permission from the first author of (A Adam et al., *IEEE Transactions on Pattern Analysis and Machine Intelligence*, 30, 555–560, 2008), (f) image reprinted from the open source UCF Crowd Counting dataset (H Idrees et al., Multi-source multiscale counting in extremely dense crowd images. In *Proceedings of the IEEE Conference on Computer Vision and Pattern Recognition*, pp. 2547–2554, 2013.))

9.1.3 Anomalous event and its detection

By definition, an anomaly is something that deviates from the usual. From the technical perspective, they are patterns deviating spatiotemporally from the frequently observed patterns. But the unusual events cannot be generalized, because, a particular behavior which is suspicious in a context may be usual in another, [10] i.e., the context helps in correctly understanding whether or not an activity in that context is usual. For example, loitering in the context of a park is usual whereas it is an anomaly near a military storage facility. In fact, there is also psychological support suggesting that contextual information is necessary for scene understanding from human perception [11,12]. Since unusual events cannot be defined explicitly, the surveillance systems need to rely on the fact that the unusual events occur rarely in comparison to the usual events, and to consider the events that occur occasionally as potentially suspicious and hence anomalous [10]. Such events are rare, difficult to describe, hard to predict, and can be subtle. However, given a large number of observations, it is relatively easy to verify if an event is indeed unusual [13]. Hence, this easy-to-identify, hard-to-describe property of anomalous events is reflected in their very low probability of occurrence when compared to normal events. These facts form the basis for the unusual event detection methodologies.

In the literature, the approaches for unusual behavior detection have been broadly classified into rule-based methods and statistical methods [10,14]. In the former methods, the normal/abnormal behaviors are predefined by a set of rules for detection in the form of a training set. As a result, the rules have to be defined for a new or changing context every time. In the latter case, they do not assume any rules. Instead, they try to automatically learn the notion of normal behavior from the given data and thus detect the abnormal behaviors. Hence, statistical methods are considered to be more promising.

Object tracking has been the basis in many of the works of anomaly detection [15–23]. Here, a trajectory is obtained by tracking the path of a moving object (as a blob) (Figure 9.3a [7] and b [3]) [24] and an anomaly will be identified through the deviation from the learned trajectories. Such approaches have the drawbacks that they are not suitable for detecting abnormal patterns in a crowded or complex scene and that the composition of the blob needs to be known for human behavior analysis.

On the other hand, non-tracking approaches which focus on local spatiotemporal anomalies in videos also exist. These methods rely mainly on extracting and analyzing local low-level visual features, such as motion and texture [5,13,25–31] (Figure 9.3c). The main drawback associated with these methods is that only local temporal changes are captured which cannot be used for behavioral understanding due to the absence of contextual information.

Adam et al. [5] presented a novel algorithm for detection of certain types of unusual events from videos of real world scenarios. Multiple, local, low-level feature monitors which extract specific low-level measurements/observations are utilized to monitor the scene and to detect unusual events. The observations extracted are actually the optical flow direction

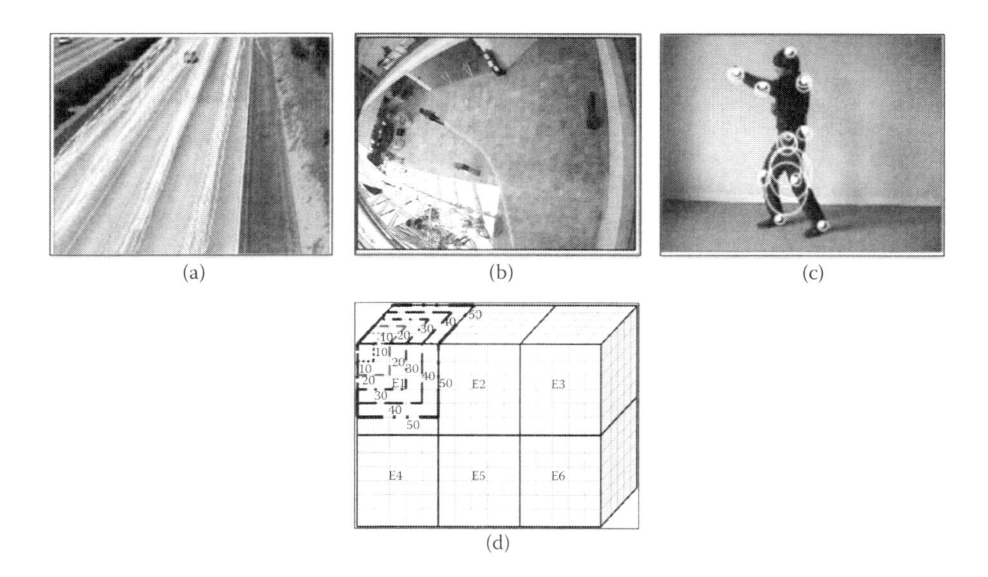

FIGURE 9.3: Representative images of object tracking (a,b). Non-tracking methods based on spatiotemporal feature points selection (c) and spatiotemporal volume construction (d). ((a,b) Modified images from the opens source ViSOR (R Vezzani and R Cucchiara et al., *Multimedia Tools and Applications*, 50, 359–380, 2010) and CAVIAR (Caviar: Context aware vision using image-based active recognition. http://homepages.inf.ed.ac.uk/rbf/CAVIAR/caviar.htm, 2011) datasets respectively, (c) modified image from the open source KTH dataset (C Schuldt et al., Recognizing human actions: A local SVM approach. In *Proceedings of the 17th International Conference on Pattern Recognition, 2004 (ICPR 2004)*, vol. 3, pp. 32–36. IEEE, 2004.)).

at the monitor's location. The extracted observations are stored in the cyclic buffer of the monitor. Given a new observation, each monitor computes the likelihood of this observation when compared to the likelihood of the observations currently stored in the buffer. If the likelihood result falls below a threshold, then the monitor outputs an alert. The algorithm uses an integration rule to integrate the decisions from all the monitors to decide whether to produce an alert or not. The method's has the limitation is that it fails to infer activities as anomalous that consisted of unusual sequences of short-term normal actions as it does not consider contextual information for anomaly detection.

The approach proposed by Mahadevan et al. [25] is called mixture of dynamic texture (MDT), where appearance and dynamics of the scene are jointly modeled to determine anomalies in crowded scenarios, which are composed of highly dynamic backgrounds, by using the ability to detect temporal as well as spatial irregularities. For temporal abnormality detection, spatiotemporal patches are extracted from the cells formed and the MDT is learned during training. The patches with low probability are considered to be anomalies. For spatial abnormality detection, a spatial abnormality map is created using the center-surround saliency technique to identify dynamic textures that are unusual by comparison to neighboring image patches. The map is finalized by estimating the likelihood of each patch with respect to the adjacent blocks using a single MDT for entire patch collection. The final spatiotemporal abnormality map is obtained by the summation of temporal and spatial abnormality maps which is thresholded to identify low probability textures within the video. Being computationally intensive is the limitation of MDT.

Xu et al. [28] proposed an approach for anomalous event detection in crowded scenes using dynamic textures described by local binary patterns (LBP) from three orthogonal planes (LBP-TOP). This algorithm involved three parts: feature extraction using LBP-TOP, model training, and detection. LBPs are defined using the spatial texture and extended to volume LBP (VLBP) by combining temporal information to model dynamic textures using latent Dirichlet allocation (LDA) as a bag of LBPs. During the training phase, LBP-TOP features are extracted from the spatiotemporal patches of sub-regions formed in the region of interest (ROI) in the scene. The LDA models are trained using the outputs of the LBP-TOP descriptors. In the detection process, for each patch, LBP-TOP features are extracted. The LDA model that corresponds to the patch is used to calculate the log likelihood of the patch. If the likelihood value is less than a threshold, then an alarm is generated at the location of the patch. The method is an unsupervised approach and does not involve object tracking and background subtraction. However, the method does not consider the spatiotemporal order of the LBP-TOP patterns in a spatiotemporal patch.

Huang and Lee [30] developed a low-cost distributed smart-camera system to detect abnormal events by analyzing the sequential behavior of a crowd of people and by applying collaborative strategy. The method captures some frames in a particular interval and histogram of oriented gradients (HOG) is analyzed to determine the presence of any person in a frame. The height of the region of interest (ROI) drawn around the detected person is the feature extracted to represent the person and a feature vector is formed from the changes in the ROI for event recognition. The system then has to convert the height of each ROI to its original size in order to construct the complete feature vector for each person. The authors used an indirect encoding scheme where each feature vector is reduced to include three new features–the maximal change of the ROI height, the number of image frames within which the maximal change happens, and the frequency of the considerable ROI change. A SVM classifier was trained and dispatched to all the smart cameras. Then, the recognition phase operated on each camera and the identified behavior sequences were shared with the nearest cameras to ensure the completeness of each feature vector. The transmission happened only if an unusual event was detected, to reduce the transmission load on the system.

A method for detecting both local and global anomalies using the hierarchical spatio-temporal interest point (STIP) feature representation and Gaussian process regression was proposed by Cheng et al. [31] in 2015. The authors used two level hierarchical representations for the events and their interactions. Under low-level representation, the STIP features of an event are extracted and are defined using a suitable descriptor. These descriptors of the normal events are quantized into a visual vocabulary using the K-means algorithm based on Euclidean distance. The local anomaly detection is achieved by measuring the k-nearest neighbors (k-NN) distance of a test cuboid against the visual vocabulary. The possible interactions in the video are acquired by extracting the ensembles of the nearby STIP features. In order to find the frequent geometric relations of the nearby STIP features from training videos, the ensembles are clustered using a bottom-up greedy clustering approach, into a high level codebook of interaction templates. Each template in the codebook is formulated into a k-NN regression problem and a model is constructed using Gaussian process regression (GPR) for learning and inferring. The likelihood based on semantic and structural similarities of a test ensemble with respect to the GPR models of normal events is calculated using the global negative log likelihood (GNLL) for global anomaly detection. The GPR method can detect both local and global anomalies through hierarchical event representation. It is adaptive and can learn new interactions while individually locating anomalous events. Also, the method is robust to noise.

Using spatiotemporal video volumes in the context of bag of video words (BOV) models is the recent trend in anomaly detection [10,14,29,32,33]. Their computationally simple nature and their capacity to detect anomalous activity in highly crowded contexts make them popular [32]. However, classical BOV approaches ignore the relationships between the volumes (contextual information) during the process of grouping similar volumes even if they are critical for anomalous event detection. It has been quoted in [10] that anomaly detection based on spatiotemporal volumes without taking into account their volume arrangement will not yield acceptable results.

Recently, attempts have been made to consider contextual information for the purpose of anomaly detection (Figure 9.3d). In [33] the similarity of a particular video volume to its eight neighboring volumes is determined to get the volume composition. Those volumes that are not similar to all others in this set are marked as being anomalous. In [32], video volumes are represented using 3D Gaussian distributions of the spatio-temporal gradient. The temporal relationship between these distributions is modeled using HMMs.

Boiman et al. presented an alternative approach based on the spatiotemporal composition of a large number of volumes. They considered anomalies as irregularities and addressed this issue by a method named inference by composition (IBC) [14]. Contextual information about an event obtained by considering the arrangement of volumes in a larger area was employed in this method. This involves the use of the spatiotemporal patches extracted from previous visual examples (database) to compose the new observed visual data which can be an image or a video sequence (query). The regions in the query which can be constructed using large contiguous chunks of data extracted from the database are considered to be very likely or normal, whereas the regions which cannot be composed, or which can be composed using only small fragments from the database, are regarded to be unlikely or suspicious. To account for local non-rigid deformations in the chunks of data, the large chunks are broken down to ensembles of many small patches at different scales with their relative geometric positions considered. Now, during the inference process a search is made for ensemble of patches in the database which have similar configuration (both in descriptors and relative arrangement) as in the ensemble of patches considered from the query. Since IBC uses every spatiotemporal patch for reconstruction, it is computationally intensive. Computationally intensive methods have higher time complexity and hence cannot work in real time. Real-time anomaly detection plays a major role in the timely management of a post-event situation or in probable prevention of such events.

An unsupervised framework of dynamic sparse coding and online reconstructibility has been presented by Zhao et al. [29] to detect unusual events in videos. In this approach, given a video sequence, the sliding window employed scans it along the spatial and temporal axes and divides the video into a set of events, each represented by group of spatiotemporal cuboids. Thus, the task of unusual event detection is formulated as finding the unusual group of cuboids present in the same sliding window. Initially, a dictionary is learned from video using sparse coding and later updated as more data becomes available in an online manner. The input signal to sparse coding in case of unusual event detection is a group of weight vectors with both spatial and temporal location information representing an event. The relationship between these vectors represented by the arrangement of the cuboids is also considered. Given the learned dictionary of usual events, a reconstruction weight vector is learned for query event and then a normality measure is computed from the reconstruction weight vector. During the detection of unusual events, an event is unusual if it is either not reconstructible from the dictionary of usual events with small error or, if constructible, it may involve a combination of a large number of bases from the usual event dictionary. The method is fully unsupervised and is capable of automatic construction of the normal event dictionary from the initial portion of the video sequence itself and updating itself using each newly observed event.

Roshthakhari et al. [10] proposed a spatio-temporal compositions (STC) approach similar to the IBC method for anomaly detection to overcome the drawback of IBC. The method is based on spatiotemporal video volume formation through dense sampling at various spatial and temporal scales and considering local and global volume composition. A codebook is constructed by grouping redundant and similar volumes, and each video volume is assigned a codeword based on similarity. Then, the relative compositions /arrangements of the volumes inside each ensemble (group of volumes) are modeled probabilistically. Finally, an inference mechanism is developed to make decision about whether any region in the newly observed sample is anomalous. A similarity map is constructed for all frames by calculating its similarity to all previous observations. If the similarity of a region is less than a threshold it contains unusual behavior(s). The contextual information is obtained through ensembles as in [14]. The use of probability distribution of the codewords instead of actual volumes greatly simplifies the computation.

Another important issue is the use of a supervised technique such as in [14–16,21–23,25,28,30–32,34–38] or a semi-supervised technique such as in [39,40] to learn the behavior model. Both these types require a large amount of training data representing the possible normal behaviors. But, it is said that it is almost impossible to have a training set for all the normal behaviors in all contexts, as the behaviors may change with the changing contexts and over time [41]. Hence, in case of anomaly detection, unsupervised learning is inevitably required in order to incrementally update the learned model for both normal and abnormal events [5,10,29,33,41].

It can be noted from the above discussion that the STC method [10] is real-time, has unsupervised learning ability, considers contextual information, and also claims to outperform the state-of-the-art methods found in the literature. The method relies on a probabilistic framework to detect anomalous events. The principle of using spatiotemporal volumes and ensembles form the basis of obtaining contextual information. A codebook is constructed to reduce the number of volumes stored in the dataset in order to limit the search time and reduce the memory requirement, thus making the algorithm fast. The unsupervised learning is initiated through an incremental updation of probability distribution functions of the normal events. Since the process of learning is fast, it requires a small training set to initiate the online learning process. Apart from this, the algorithm is also able to detect multiple complex activities in crowded environments. But the algorithm generates volumes in large number which are overlapping due to dense sampling. Also, inferencing is based on a single threshold value determined experimentally. It is mentioned that improvement in terms of performance of the method could be achieved with an adaptive threshold. Inspired by this, Rao et al. [42] proposed a modified approach for anomaly detection referred to as the MSTC method which forms the focus of this chapter.

9.1.4 Related issues in visual surveillance

Video data captured from public spaces is often intrinsically ill-conditioned due to variations in image quality, resolution, noise, diversity of pose and appearance, and due to occlusion in crowded scenes [1]. As a result of such poorly conditioned footage, the normal events may be missinterpreted as abnormal, resulting in false alarms or vice-versa, leading to no alarm in genuinely suspicious cases. Thus, in the long run, the installed expensive video analytics systems may be abandoned or otherwise infrequently used. So, for a VSN to function as an active, reliable and robust real-time surveillance system, it must be able to capture the events in high resolution so as to give a more clear and useful footage. This requires the sensor nodes to be up and running all the time at a high-resolution capture mode. But, the lifetime of battery-operated camera nodes is limited by their energy consumption, which is proportional to the amount of work a VSN node performs, i.e., energy required for sensing, processing, and transmitting the data [43]. Given the large amount of images captured by the camera nodes, both processing as well as transmitting them are quite costly in terms of energy as it requires large bandwidth for transmission data.

There are two promising solutions for this. One is to allow the sensors to monitor at low/normal resolution and switchover to high resolution mode to capture the events of interest and transmit only those frames. Another option is to acquire frames in low resolution itself (i.e., no switching over) and apply some enhancement techniques, such as super resolution, to get their high resolution versions [44–46].

9.1.5 Overview of the chapter

In this chapter, research on the development of a smart visual surveillance approach is presented. The above mentioned MSTC method [42], which is a contextual information based anomalous activity detection algorithm, has been discussed in detail as a primary aspect of the approach. Other features like event-driven processing of query videos, the aspect of anomaly-driven transmission of only the anomalous frames for energy conservation, and visual enhancement of the frames having the detected anomalous events to enable clear viewing of the activities of interest [42] have been discussed as well.

9.1.6 Organization of the chapter

The rest of the chapter is organized as follows:

- Section 9.2 explains the design of the approach for visual surveillance in detail, including the method employed for anomaly detection and localization along with the related algorithms.
- Section 9.3 provides the extensive information on experiments carried out for performance evaluation of the MSTC method along with the analysis of the results.
- Section 9.4 summarizes and concludes the chapter along with the scope for future research.

9.2 Design of the Visual Surveillance Approach

The approach for smart visual surveillance discussed here is a collection of algorithms which work coherently toward providing intelligent visual surveillance. The MSTC algorithm [42] for anomaly detection and localization forms the core of this design. It is a modified version of the STC concept [10]. The MSTC algorithm is intended to be incorporated in the sensor node itself. It provides a way to conserve the energy of a battery operated sensor through event-driven processing where a query will be processed only if there exists any event to be classified as an anomaly or normal and more importantly by transmitting only those frames containing anomalous events instead of sending all the captured frames (anomaly-driven transmission). The visual enhancement of the anomalous frames received at the centralized location helps to better understanding of the events.

Figure 9.4 represents the functioning of the surveillance approach discussed in this chapter. The component $'a'$ symbolizes the scene with usual and unusual events in the form of frames which will be captured by the visual sensor node $'b'$. The MSTC algorithm intended to be in the node performs the event-driven processing of the frames, detects the anomalous event (in the second set of frames in $'a'$), and localizes the anomaly. The anomaly-driven transmission algorithm in the node transmits the anomalous frames denoted by $'c'$ to the receiver end $'d'$. The receiver reads the frames, enhances them, and writes them into a video a file. These algorithms are discussed in detail in the following subsections.

9.2.1 Anomaly detection and localization algorithm

The MSTC method [42] in principle is based on the concept of STC [10]. Nevertheless, the implementation has been deviated, as discussed in the following subsection.

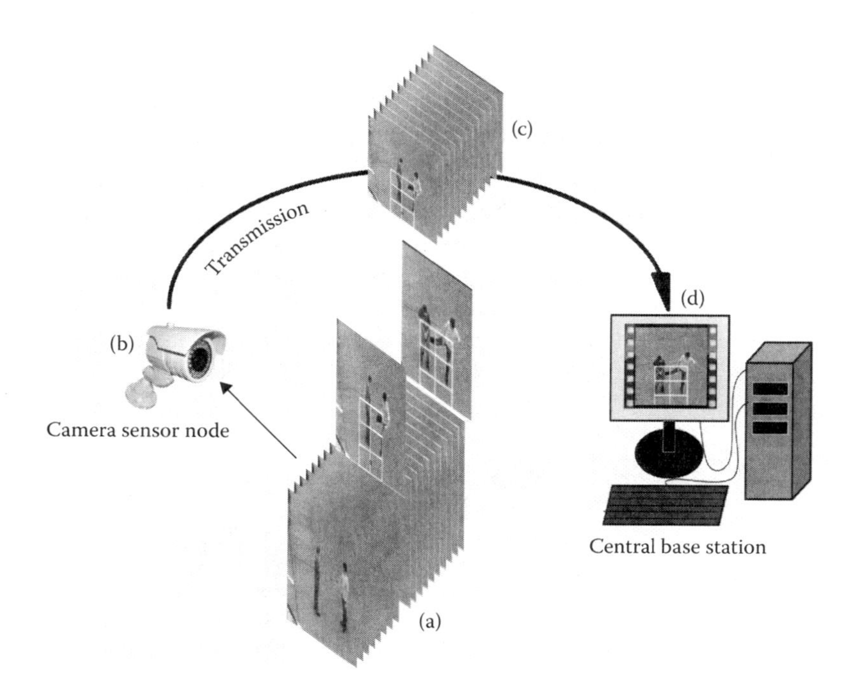

FIGURE 9.4: Representative image of the surveillance system framework where, the camera sensor node (b) captures the scene shown as the set of frames (a) and transmits only the anomalous frames (c) to the base (d).

9.2.1.1 Underlying principle

An event in a video is defined by the changing pixel intensities in time, across the frames. The changes may be caused due to the movement of a living being or any object. The underlying principle of the anomalous event detection approach (MSTC) described in this chapter is that an activity which is rare, having a very low frequency of occurrence among the majority of events, is considered anomalous. As already mentioned, the context corresponding to an event is also crucial in determining an anomalous event. For obtaining the contextual information, the entire video sequence is initially divided into blocks both spatially and temporally which are termed ensembles. The ensembles containing a partial or a whole activity represent the context of the activity as well. Moreover, the ensembles also help in localizing the anomalous events upon their detection. In surveillance applications, pixel-wise localization is beneficial and is preferred over frame-level anomaly detection. Understanding the event and the context of its occurrence requires knowing the pixel arrangement within the ensemble. For this purpose, the ensemble is further divided into smaller blocks called spatiotemporal volumes at different scales in spatial and temporal domains and the topology is determined. Through a probabilistic modeling framework, the probability or likelihood of occurrence of this volume composition and hence the probability of occurrence of the ensemble as a whole is determined.

This likelihood of occurrence is the key to separate the anomalous ensembles, having very low likelihood, from their normal counterparts. So, for a new query, a comparison of each of its ensemble's composition is made with the ensembles of the training, in terms of similarity, probabilistically. If the similarity is less than a threshold for an ensemble, then that ensemble is labeled to have the event(s) of interest. The method discussed in this chapter is carried out in two stages – the initialization stage and the anomalous events detection stage as illustrated in Figure 9.5.

9.2.1.2 Initialization with a training video sequence

The initialization phase requires a short video sequence describing the given context, including the normal activities involved in it. This process of initialization consists of the following three steps.

Step 1: Formation of spatiotemporal blocks and construction of volumes

The training sequence with regular activities is divided into ensembles which are blocks of equal size in both spatial and temporal domains, after initial down-sampling. The down-sampling eases the processing and the corresponding reduction of spatial dimensions of the video, reduces the number of volumes that can possibly be constructed.

Even though resizing generally causes loss of information, the experimental results in Section 9.3 show that it has only negligibly affected the anomalous event detection process, while requiring only minimum time for volume construction. The ensembles formed give the contextual information aiding in the anomaly detection.

Each ensemble is further subdivided to generate a collection of non-overlapping spatiotemporal volumes at different spatiotemporal scales, such as, of size $10 \times 10 \times 10$ and in the increments of ten, up to the size of the ensemble as shown in Figure 9.6. The smaller volumes capture minute changes in space and time while the larger volumes represent larger spatiotemporal changes, corresponding to the activities. The algorithm for ensemble and volume construction is given in Algorithm 9.1.

The volume construction in a non-overlapping manner at each scale generates approximately one volume for every 10 pixels, thereby reducing many iterations when compared to volume construction around every pixel. Generally, it is not possible to reproduce

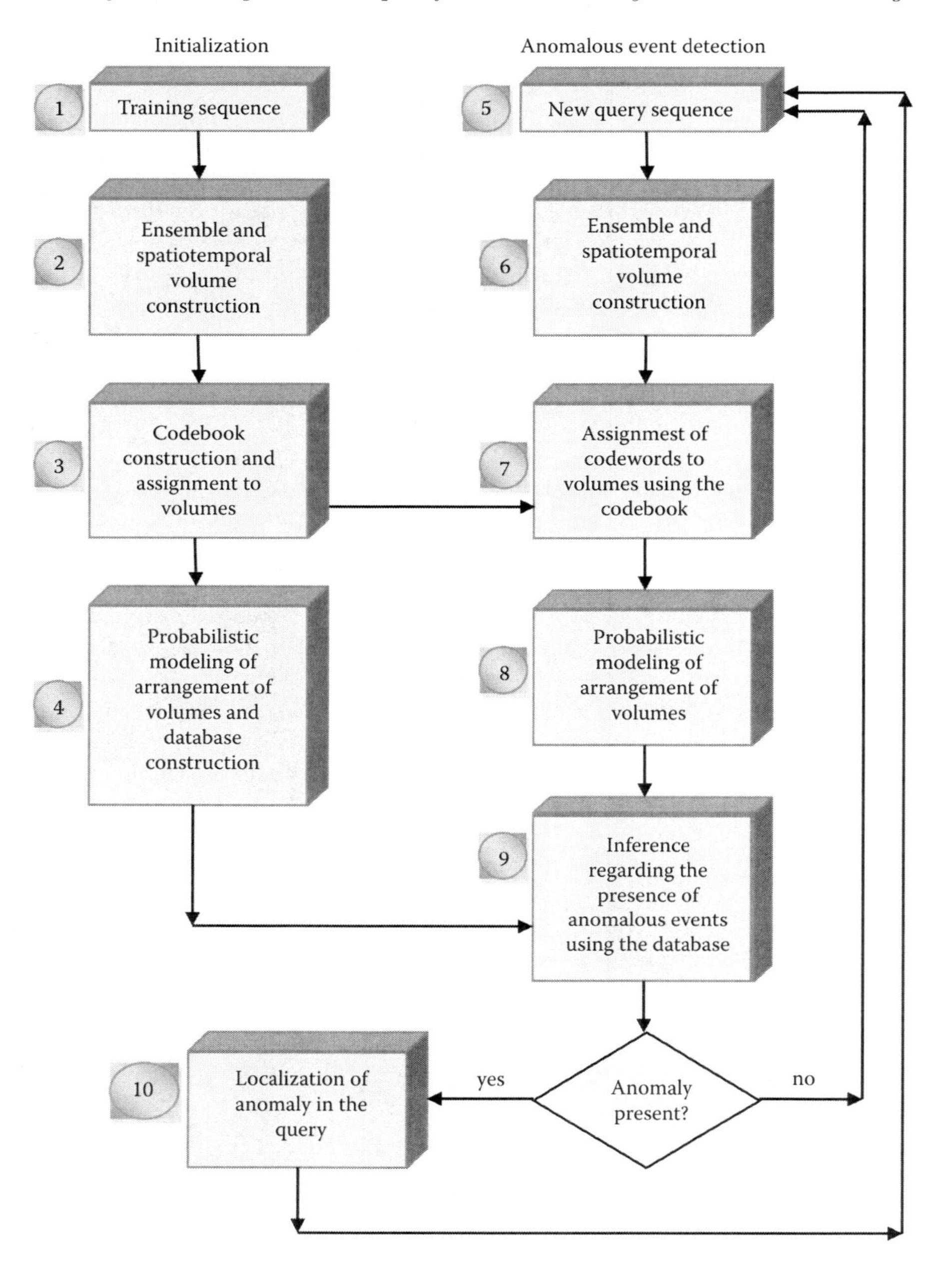

FIGURE 9.5: Flow chart of the two stage working of the MSTC method, starting from the initialization till the detection and localization of the unusual events.

human activities and any other natural spatial structures [10,14]. Therefore, these local misalignments of the events in space and time are not to be ignored or missed. This is achieved by sampling at different scales. This varies from [10] where volumes and ensembles construction is carried out around each pixel. This kind of dense sampling generates a very huge number of overlapping ensembles and volumes. For example, in [10], a one minute video yielded 10^6 video volumes. Processing this many volumes increases the time complexity of subsequent steps of anomaly detection.

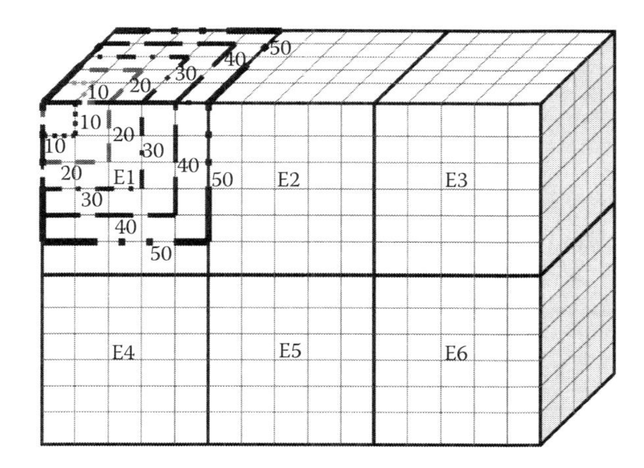

FIGURE 9.6: Ensemble formation and volume construction at different spatial and temporal scales.

A set of descriptors is defined for each spatiotemporal volume v_i and these temporal descriptors (f_z) represent the volumes from this point on. It is determined as the gradient (∇) along the temporal axis and is given by, [10,42]

$$[f_x, f_y, f_z] = abs\left(\nabla\left(v_i\right)\right), \forall v_i \tag{9.1}$$

The gradient is computed using the first order gradient function of MATLAB®.

The changes in the scene that would have occurred during the course of time is captured by the temporal gradient (f_z). All these volume descriptors at different scales are reduced into vectors and stacked into a compact form after normalizing to unit length so as to reduce the computational complexity of the overall anomaly detection process.

Step 2: Codebook construction and codeword assignment to volumes

The resulting volumes from the volume construction process above are large in number requiring large memory space for storage. For example, approximately 4×10^4 volumes are formed from a video with a frame rate of 30 frames/sec and a length one minute. As the volumes constructed is non-overlapping in nature, they are mostly non-identical, thus complicating the computation of the probability of occurrence with regard to the arrangement of volumes in the next step. However, because of the possibility of many similar volumes existing, grouping them limits their number without considerable loss of information, while easing the probability calculation. The combining of similar volumes results in a codeword which represents all the volumes of the group. The set of all the codewords together form the codebook.

The procedure for codebook construction discussed here is much simpler and avoids the pruning process of codewords as well when compared to the codebook procedure in [10]. The Euclidean distance is used as a measure of similarity to group the volumes. The procedure starts by considering the first volume descriptor v_1 from the set of volumes V as the current volume. The similarity between each of the remaining volumes v_i and v_1 is calculated in the form of Euclidean distance. The volumes having the distance from v_1 lesser than a threshold ϵ (mean of the distances of all the volumes v_i relative to v_1 in V) are grouped together into a set of similar volumes. All these volume descriptors are similar enough to v_1 and are eligible to be grouped together. The first codeword c_1 is formed from this set of similar volumes by taking the average of the all those volume descriptors. These volumes are removed from V

Algorithm 9.1 Algorithm for initialization, given a training video sequence

Input: A video sequence containing normal behaviors in a context.

Output: Database P_{E^T} containing posterior probabilities of all the ensembles of the training sequence.

Let n be the number of frames of usual events for training.

Ensembles and spatiotemporal volumes construction:

- Down-sample the n frames.
- Let $E^T = \{E_1^T, E_2^T, ..., E_i^T\}$ be a set of ensembles formed, where i=($n \times 6$)/50 (i.e., six ensembles per 50 frames).
- Divide each ensemble to obtain volumes $V = \{v_1, v_2, ..., v_k\}$ of size $10 \times 10 \times 10$, up to $50 \times 50 \times 50$ making a total of N volumes for the training video.
- Smooth each volume with a Gaussian filter and obtain its gradient descriptor.

Codebook construction and assignment:

Let N be the number of volumes represented by their descriptor vectors, $V = \{v_i\}_{i=1}^N$

$codes\_set \;=\; V$

$j \;=\; 0$

while *codes_set* is not empty **do**

  for all volumes in *codes_set* **do**

    if $euclidean\_distance(v_1, v_i) \; < \epsilon$ **then**

      $move\ v_i\ from\ codes\_set\ to\ set_{j+1}$

    end if

  end for

  $j \;=\; j + 1$

  create new codeword, $c_j = mean(set_j)$

end while

Let M be the number of codewords in $C = \{c_j\}_{j=1}^M$

for all volumes in V **do**

  for all codewords C **do**

    compute the weight $w_{i,j}$

  end for

  assign c_j *corresponding to* $max(w_{i,j})$ *to* v_i

end for

Training sequence probability database construction:

- Calculate volume posterior probability $P(c, c'|v_k, v_i, p)$ for each volume $v_i, i = 1$ to k in each ensemble $E_i^T \in E^T$
- Calculate ensemble posterior probabilities, $P_{E^T} = \left\{P_{E_1^T}, P_{E_2^T}, P_{E_3^T}, ..., P_{E_i^T}\right\}$ of each ensemble $E_i^T \in E^T$.

Source: TJ Narendra Rao et al., An improved contextual information based approach for anomaly detection via adaptive inference for surveillance application. In *Proceedings of International Conference on Computer Vision and Image Processing*, pp. 133–147. Springer, 2017.

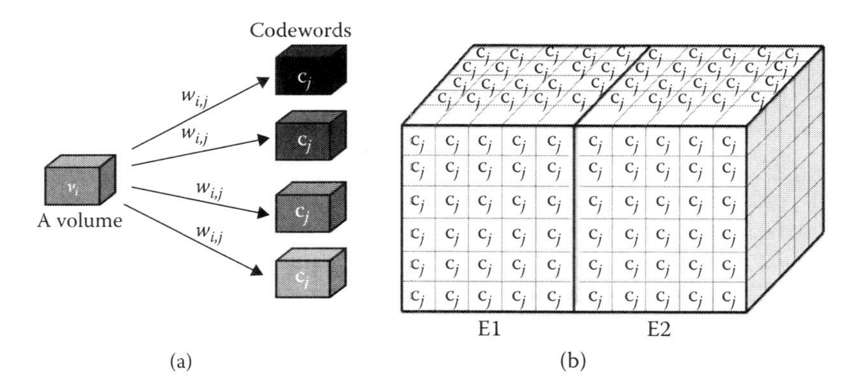

FIGURE 9.7: (a) Representation of a volume v_i with its similarity weight $w_{i,j}$ for the different codewords c_j. (b) Codeword c_j with the highest similarity representing each volume v_i in the ensembles.

and the process is repeated for the remaining volumes. All these codewords together form a codebook, $C = \{c_j\}_{j=1}^{M}$. It must be observed that the number of codewords created, M, is far smaller than the number of volumes generated, N.

After the codebook is constructed, each volume is to be assigned a codeword so as to avoid the use of the volumes during further processing. This reduces the computational complexity of the subsequent steps. The codeword c_j chosen for assignment must be most similar to the volume v_i. So, the similarity weight between each volume and each of the codewords is computed as given by, [10,42]

$$w_{i,j} = \frac{1}{\sum_j \frac{1}{distance(v_i, c_j)}} \times \frac{1}{distance\,(v_i, c_j)} \tag{9.2}$$

Each volume is assigned a codeword which has the highest degree of similarity with it as in Figure 9.7. The algorithm for this step is given in Algorithm 9.1.

Step 3: Non-parametric probabilistic modeling of volume configuration in an ensemble

Once the spatiotemporal volumes are generated and the codebook is constructed out of them, the volume configurations of the ensembles are to be modeled to represent and realize the usual activities observed in the context of the training sequence. A probabilistic framework has been employed in [10] for this purpose. The probability of a normal event in an ensemble can be known by the likelihood of occurrence of the volume arrangement in that ensemble. Likewise, the volume arrangement probabilities of all the ensembles define the probability of occurrence of all the usual behaviors in the training video. An unusual ensemble will be the one having very low probability for its composition.

In [42], it is considered that each volume within an ensemble has a definite position, as in Figure 9.8a, and these position numbers are utilized to know the arrangement of volumes. After the volumes are assigned with codewords, each ensemble may be considered as a set of codewords. Thus, the probabilistic modeling of volume arrangement in an ensemble turns into codeword composition modeling around a central codeword. For a volume v_k at position p in an ensemble, let the codeword labeled be c and let the central volume v_i be represented by codeword c'. Then the probability of finding (c, c') given the observed video volumes (v_k, v_i, p) is termed the volume posterior probability [10] and is given by $P(c, c'|v_k, v_i, p)$ and is

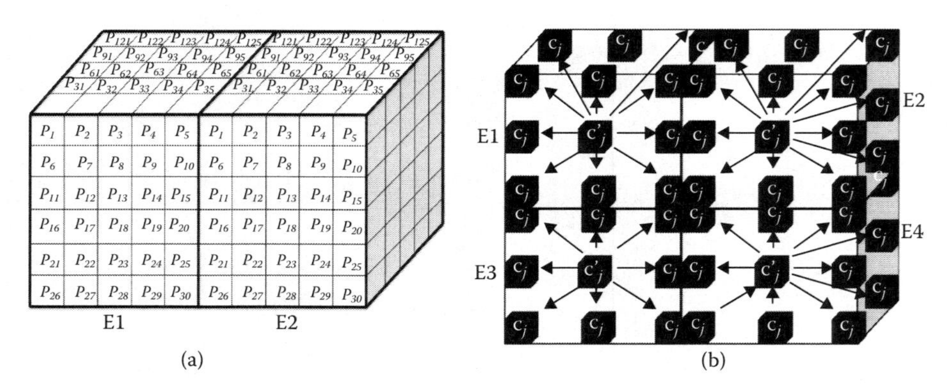

FIGURE 9.8: (a) Pictorial representation of two ensembles where each volume v_i has a distinct position p_i. (b) Shows four ensembles (E1, E2, E3, E4) with the arrangement of certain volumes relative to their central codewords c'_j which is used for the computation of the first term of Equation 9.5.

represented using the principle of conditional probability as, [42]

$$P(c, c'|v_k, v_i, p) = P(c|c', v_k, v_i, p)P(c'|v_k, v_i, p) \tag{9.3}$$

Since each v_i is represented by a codeword c', v_i can be removed from the first term of Equation 9.3. Also, since v_i and v_k are independent, v_k along with its position p can be eliminated from the second term of Equation 9.3.

$$P(c, c'|v_k, v_i, p) = P(c|c', v_k, p)P(c'|v_i) \tag{9.4}$$

The first term of Equation 9.4 involves determining the probability of codeword c occurring at every position p in an ensemble with the central codeword c' considering all the other ensembles and also determining the probability of observing c given v_k within an ensemble. Hence Equation 9.4 becomes,

$$P(c, c'|v_k, v_i, p) = P(c|c', p)P(c|v_k)P(c'|v_i) \tag{9.5}$$

The first term of Equation 9.5 is assessed non-parametrically by taking into account all the ensembles pertaining to the training video and determining the probability of codeword c occurring at every position p in an ensemble relative the central codeword c', as depicted in Figure 9.8b. Unlike as in [10] where Gaussians Mixture Model is used for this estimation, here a non-parametric approach is employed to determine the true distributions. The second and third terms of Equation 9.5 are computed by applying Bayes' theorem of conditional probability. The posterior probability of the whole ensemble is the product of posterior probabilities of all its volumes. All the ensemble posterior probabilities thus obtained for the training sequence together constitute the database P_{ET}.

9.2.1.3 Anomalous events detection in a query video

Step 1: Event-driven processing

The new query video observed is sampled to construct ensembles and volumes as explained for training sequence under initialization and similarly the descriptors of the volumes are also obtained. To make the algorithm power aware and efficient, unnecessary wastage of processing energy is avoided by further processing the query only if there exists an event at all. The existence of a non-zero gradient descriptor indicates the presence of an event. If none exists, the codeword assignment, probabilistic modeling and inferring processes are ignored. Else the query is further processed. The probability calculation is similar to that of training and the posterior probabilities of all the ensembles of the query sequence are obtained. This event-driven processing of a query video is explained in Algorithm 9.2.

The MSTC method has unsupervised learning ability for understanding the newly observed normal behaviors. For this purpose, the frequency of the codewords in different positions of the ensembles in the training is incremented, updated and used to obtain the first term values of Equation 9.5, when a query is processed. For the next query the frequency values from previous query are utilized and are carried forward to the subsequent queries, so as to learn. This characteristic is advantageous in contexts where the normal events pattern varies quite often.

Step 2: Inferencing and localization

A predetermined threshold is required for the decision making regarding the presence of anomalous events. In [42] three inference mechanisms have been developed which are adaptive with regard to fixing the threshold based on the context, as opposed to the employment of a single experimentally obtained threshold value as in [10]. The three mechanisms are derived to suit the contexts encountered normally under surveillance. Adaptive inferencing can be considered superior to a single-value based decision making for the reason that the probability values of the video sequences and in turn the decision-making threshold vary with the contexts. Hence, this approach can be expected to produce the desired detection rate for all contexts.

Inference mechanism for crowded/non-crowded real-world contexts *(IM 1)* The underlying principle of *IM 1* is that an anomalous event in a real-world context has a very low frequency of occurrence and is a very short length in time in comparison to the usual activities in the scene, and that the range of probabilities in P_{ET} represent usual events. During new query processing, the ensemble posterior probabilities corresponding to normal actions are expected to be within the range of P_{ET} or higher. Therefore, threshold (T) for decision making regarding the presence of anomalous events chosen is the minimum ensemble posterior probability of P_{ET} added with a small constant deviation α. Any ensemble of the query with the probability less than this threshold is concluded to be anomalous.

Inference mechanism for non-crowded contexts consisting of a definite activity pattern *(IM 2)* If the purpose of surveillance in a context is the detection of the activity in the query deviating from the regular activity found in the training: e.g., when boxing replaces normal walking, exhibiting a pattern change, then this mechanism *IM 2* is suitable. The pattern here refers to the number of ensembles and their arrangement corresponding to the activity (ignoring the background ensembles), determined based on the ensemble posterior probability values. Thus, the anomalous ensembles in the query show a mismatch in their ensemble posterior probability pattern when compared to the pattern of the training.

Inference mechanism for contexts involving object detection *(IM 3)* The Euclidean distance similarity measure forms the basis for *IM 3* and is computed between ensemble probabilities P_{ET} of the training and ensemble probabilities P_{EU} of a usual sequence. This distance is fixed as the threshold (T). The presence of an anomaly (object of interest) in the query is indicated by the similarity value less than T (or distance greater than T) when the distance is computed with the training video. The anomalous object in the query is localized to the ensemble with minimum posterior probability.

Once the anomalous events are detected, those ensembles containing them are highlighted with a bounding box to localize the anomaly.

Algorithm 9.2 Anomalous event detection procedure, given a query

Input: A query video sequence containing normal or anomalous behaviors in a context.
Output: Inference regarding whether the given query is usual or anomalous.
Query processing:

- Form the ensembles $E^Q = \left\{E_1^Q, E_2^Q, ..., E_i^Q\right\}$, construct volumes $V = \{q_1, q_2, ..., q_k\}$, compute gradient descriptors like in Algorithm 1.
- Every volume is assigned a codeword as in Algorithm 1 from the codebook only if a non-zero gradient descriptor exists.
- Compute the posterior probability of each ensemble $P_{E^Q} = \left\{P_{E_1^Q}, P_{E_2^Q}, P_{E_3^Q}, ..., P_{E_i^Q}\right\}$ as in Section 9.2.1.2.

Inference mechanisms:

IM 1:
$T = min(P_{E^T}) + \alpha$
for all volumes in P_{E^Q} **do**
  if $P_{E_i^Q} < T$ **then**
    $P_{E_i^Q}$ *is anomalous*
  end if
end for

IM 2:
for all values in P_{E^Q} **do**
  if the pattern of P_{E^Q} != the pattern of P_{E^T} **then**
    $P_{E_i^Q}$ *deviating from pattern of* P_{E^T} *is anomalous*
  end if
end for

IM 3:
$T = distance(P_{E^T}, P_{E^U})$
if $distance(P_{E^T}, P_{E^Q}) > T$ **then**
  $min(P_{E^Q})$ *is anomalous*
end if

Source: TJ Narendra Rao et al., An improved contextual information based approach for anomaly detection via adaptive inference for surveillance application. In *Proceedings of International Conference on Computer Vision and Image Processing*, pp. 133–147. Springer, 2017.

9.2.2 Anomaly-driven transmission and visual enhancement

Conventionally, all the frames captured by the visual sensor camera are transmitted across to the base station, irrespective of the presence or absence of any anomalous event in them. This requires larger bandwidth which generates greater energy expenditure and also increases the probability of packet collision. Hence, it is required that some of the image processing is done locally and send the processed information only [43]. Hence, here in this surveillance approach, the unusual event detection and localization is performed at the sensor node and only the anomalous frames are transmitted to the centralized location. The transmission is through a TCP/IP-based socket communication setup using MATLAB®. The communication between the receiver (termed Server) which is the computer at the base station and the sender (termed Client) which is the MSTC algorithm in the visual sensor, occurs as follows

- The Server waits for the connection from the Client.
- Once the Client establishes the connection, it sends the anomaly localized frames to the Server and then processes the next query video.
- The Server reads the data from the socket and writes the frames into a video file.
- The Server waits for the next set of anomalous frames to be received.

The received anomalous frames are subjected to enhancement using simple Gaussian filter and edge filter before writing them into a video file. This improves the visibility of the unusual event(s) which helps with better surveillance.

9.3 Experimental Results and Discussion

The MSTC method [42] discussed in this chapter was tested on seven different standard datasets namely, the *Subway Surveillance* dataset[1], *Anomalous Walking Patterns* dataset[2] [14], the *UT-Interaction* dataset[3] [47], the *MSR Action Dataset II*[4] [48], the *KTH* dataset[5] [24], the *Weizmann* dataset[6] [49], the *Spatiotemporal Anomaly Detection* dataset (Train, Boat-River, and Canoe video sequences)[7] [27], with simple and complex activities, in both non-crowded and crowded environments and under varied illumination conditions. Four of them have been found among the top ten, in the ranking based on the number of papers that use them for benchmarking, as mentioned in [50].

The down-sampling of the training and query videos are done to a dimension of 120×160 and then the ensembles of equal size i.e., $60 \times 53 \times 50$ are formed. Furthermore, these ensembles are sampled to construct volumes and the initialization followed by anomalous event detection are carried out as explained in Section 9.2.1.

9.3.1 Details of the datasets

The *Subway Surveillance* dataset consists of real surveillance video of a subway station recorded by a camera at the exit gate. It is 43 minutes long with 64,901 frames in total containing usual events like people coming up through the turnstiles by exiting the platform and activities of walking in the wrong direction and loitering near the exit, mainly considered to be anomalous.

---

[1]obtained from the first author of [5]
[2]http://www.wisdom.weizmann.ac.il/~vision/Irregularities.html (openly available)
[3]http://cvrc.ece.utexas.edu/SDHA2010/Human_Interaction.html (openly available)
[4]http://research.microsoft.com/en-us/um/people/zliu/actionrecorsrc/ (openly available)
[5]http://www.nada.kth.se/cvap/actions/ (openly available)
[6]http://www.wisdom.weizmann.ac.il/~vision/SpaceTimeActions.html (openly available)
[7]http://www.cse.yorku.ca/vision/research/spatiotemporal-anomalous-behavior.shtml (openly available)

The *Anomalous Walking Patterns* dataset depicts the scenario with one or two persons. The video contains normal walking and jogging activities and suspicious behaviors such as abnormal walking patterns, crawling, falling down, jumping over objects, etc.

The *UT-Interaction* dataset contains a total of 20 video sequences of interactions between two persons. Each video contains around eight executions of interactions and is of approximately one minute long. Multiple actors with more than 15 types of clothing participated in the videos.

The *MSR Action Dataset II* is an extended version of the Microsoft Action Dataset. It consists of video sequences recorded in a crowded environment. Each video sequence consists of multiple actions of hand waving, clapping, and boxing. Some sequences contain actions performed by different people. All the video sequences are captured with clutter and moving backgrounds with lengths ranging from 32 to 76 seconds.

The *KTH* and *Weizmann* datasets are the standard benchmarks in the literature used for action recognition. The *KTH* dataset contains six different actions (walking, jogging, running, boxing, hand waving, and hand clapping) performed by 25 different persons in four different scenarios (outdoors, outdoors with scale variation, outdoors with different clothes, and indoors). The *Weizmann* dataset contains ten different actions performed by nine actors.

The *Spatiotemporal Anomaly Detection* dataset contains real-world videos with more complicated dynamic backgrounds with variable illumination conditions. Even though there exists significant environmental changes in this dataset, the abnormalities are actually simplistic motions (e.g., movements observed in the scene). Three video sequences from this dataset are used for experiments, which have variable illumination and dynamic backgrounds: the *Boat-River*, the *Train*, and the *Canoe* video sequences. The *Train* sequence is the most challenging one in this dataset due to drastically varying illumination and camera jitter. In this sequence, the abnormalities relate to the movement of people. In the *Boat-River* and *Canoe* video sequences, the abnormalities are the passing boat and canoe in the sea respectively.

In the *UT-Interaction*, *MSR II*, *KTH* and *Weizmann* datasets, the behaviors which are frequently encountered or which appear to be usual in that context are grouped to be normal and those which deviate from these are considered to be anomalous, as *Ground Truth* (GT). For the remaining datasets, the unusual events ground truth has been provided. Table 9.1 gives the details of the activities considered for experimentation from the different datasets. The last column shows the inference mechanism applied during query processing based on the context of the datasets.

9.3.2 Performance analysis

Analysis of the experimental results of MSTC [42] is done using the parameters like detection rate, precision-recall (PR) curves, receiver operating characteristic (ROC) curves, and the equal error rate. Comparative analysis of the performance of the MSTC method has been done against the state-of-the-art STC [10], inference by composition (IBC) [14], MDT [25], local optical flow (LOF) [5], spatiotemporal oriented energy filters (STEF) [27], space-time Markov random fields (ST-MRF) [51], and dynamic sparse coding [29] methods for the common datasets. The experimental results given in Table 9.2 show the performance of the MSTC method on the above datasets. It should be noted from the second column of Table 9.2 that the training video used for initialization for each of the datasets is made up of a very small number of frames (in comparison to the entire length of the video), ranging from 50 to 600 frames, just so as to contain the normal actions in the context. The algorithm processes 100 frames of a query video at a time to detect any anomaly if present. The number of unusual events rightly inferred as anomalous (true positives) is given in the third column and the number of false alarms (false positives) generated for

TABLE 9.1: Details of the activities classified as usual and anomalous and the inference mechanism applied based on the context

Datasets	Activities classified as usual	Activities classified as anomalous	Inference mechanism applied
Subway Surveillance dataset (Exit video)	Persons moving out of the platform	Entering through the exit gate, walking in the wrong directions, loitering	*IM 1*
Anomalous Walking Patterns dataset	Walking	Stealth walking, crawling, falling, jumping over obstacles	*IM 1*
UT-Interaction dataset	Participants in the scenario performing no physical interactions	Punching, kicking, pushing	*IM 1*
MSR II dataset	Usual behaviors in a dining area	Hand waving, clapping, punching	*IM 1*
KTH dataset	Walking	Boxing, hand waving, clapping, running	*IM 2*
Weizmann dataset	Walking	Bending, jump jacking, jumping, hopping, sideways galloping, skipping, one and two hands waving	*IM 2*
Spatiotemporal anomaly detection dataset (Train, Boat-River, and Canoe videos)	No movement of passengers inside the train cabin, no boat in the river	People moving in and out of the cabin, boat moving across the river	*IM 1* for Train video sequence, *IM 3* for others

the events containing normal behaviors is given in the fourth column of Table 9.2. The number of misdetections produced is given in the fifth column The sixth column shows the detection rate i.e., ((True positives/ (True positives + False negatives)) $\times$ 100) obtained for the different datasets.

In all, the MSTC method exhibited appreciable results, with an anomalous event detection rate of 92.03%, with only six false alarms. The possible causes for the few misdetections could be that the movements by the participants were not being very noticeable at a distance from camera, poor illumination leading to least gradient changes in *Subway Surveillance*, *UT-Interaction* and, *Train* and *Boat-River* videos of *Spatiotemporal Anomaly Detection* datasets and the high similarity of the activities in the queries and training video in the case of the *KTH* and *Weizmann* datasets.

TABLE 9.2: Consolidated experimental results for the different datasets

Datasets		No. of training frames	True Positives	False positives	False negatives	Detection rate
Subway Surveillance dataset (Exit video)		600	43	02	02	43/45 (95.5%)
Anomalous Walking Patterns dataset		100	08	00	00	8/8 (100%)
UT-Interaction dataset		100	19	00	02	19/21 (90.47%)
MSR II dataset		50	16	02	00	16/16 (100%)
KTH dataset		100	77	01	03	77/80 (96.25%)
Weizmann dataset		50	12	00	06	12/18 (66.6%)
Spatio-temporal anomaly detection dataset	Train video	100	05	00	02	5/7 (71%)
	Boat-River video	50	02	01	01	2/3 (66.6%)
	Canoe video	100	03	00	00	3/3 (100%)
	Total		**185**	**06**	**16**	**185/201 (92.03%)**

PR curve is the plot of precision (the fraction of detected unusual queries that are truly unusual) versus recall (the fraction of truly unusual queries detected) at varied decision-making thresholds. Precision is a measure of quality (exactness) whereas recall is a measure of quantity (completeness) and are given by Equations 9.6 and 9.7 respectively. The threshold corresponding to highest precision at highest recall is considered optimum. Figure 9.9a shows the PR curve of the discussed method for the *Train* video of *Spatiotemporal Anomaly Detection* dataset. It reveals that at the optimum threshold obtained through *IM 1*, a maximum precision of 1.0 was observed at a recall of 0.43. For the same dataset, at this recall the STC, the Spatiotemporal TEF, the local optical flow methods exhibited precision values lower than 1.0 with reference to the graph in [10]. Also, at their highest precision points, each of the above methods corresponded to lower recall values. Similarly, at their highest recall points the above methods showed lower precision values. The same is case for the *Subway Surveillance* dataset. Figure 9.9c shows the PR curve of the MSTC method for the *Subway Surveillance* dataset. The exact precision-recall values of the above methods for these two datasets are provided in Table 9.3. Figures 9.9b, d, e show the PR curves plotted for the, *UT-Interaction*, *MSR II*, *Anomalous Walking Patterns* datasets respectively. In all these graphs, the optimum threshold observed at highest precision and highest recall was found to be in agreement with the threshold derived through the inference mechanism applied to these datasets based on their context.

$$Precision = \frac{True\ positives}{True\ positives\ +\ False\ positives} \tag{9.6}$$

$$Recall = \frac{True\ positives}{True\ positives\ +\ False\ negatives} \tag{9.7}$$

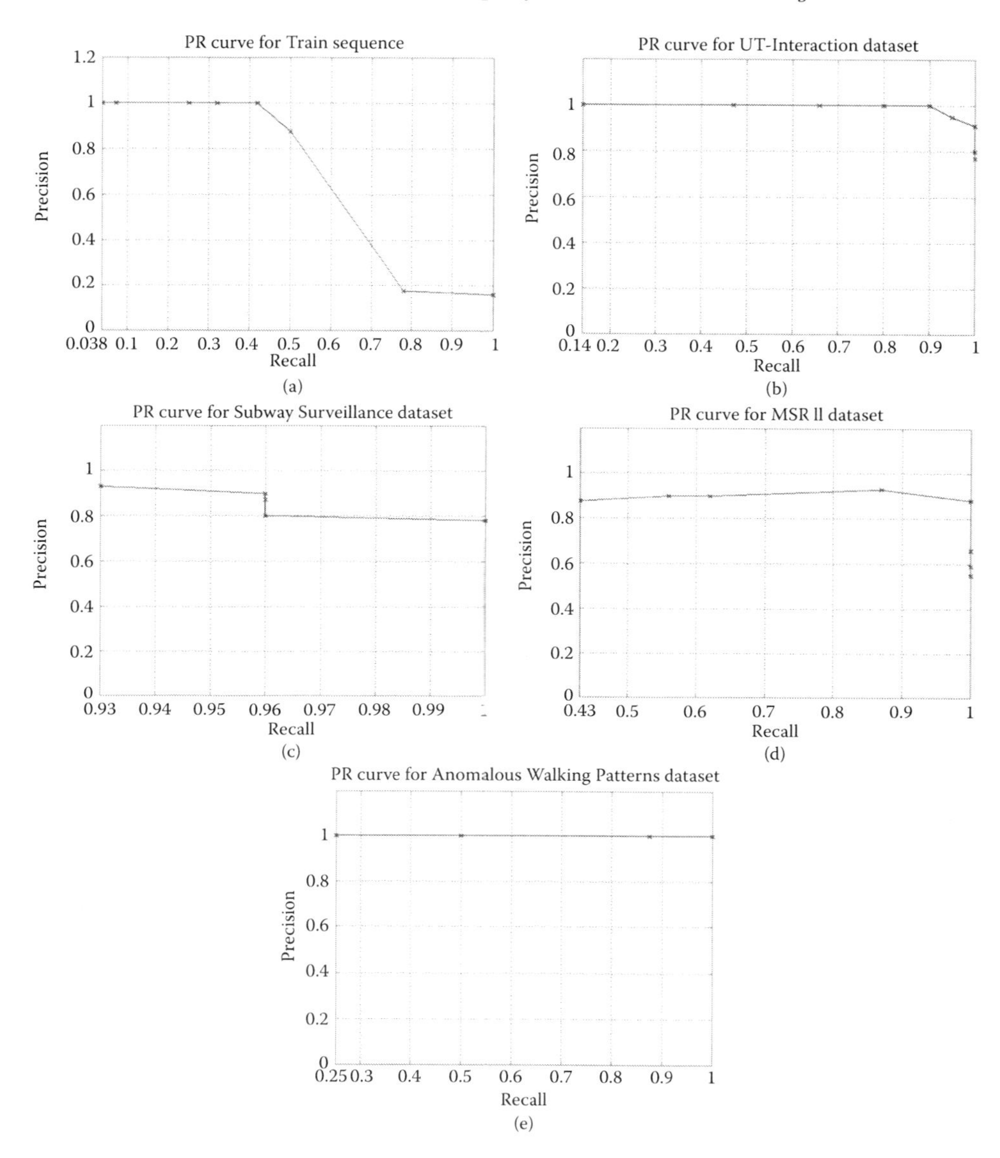

FIGURE 9.9: PR curves of the MSTC method for (a) the *Train sequence*, (b) *UT-Interaction dataset*, (c) *Subway Surveillance* dataset, (d) *MSR II* dataset, and (e) *Anomalous Walking Patterns* dataset.

Figure 9.10 shows the ROC curves of the MSTC method for the different real-world datasets. ROC curve is a plot of true positive rate (TPR) versus false positive rate (FPR) at varied decision-making thresholds. TPR measures the proportion of actual positives (unusual queries) which are correctly identified as such and is given by Equation 9.8. FPR measures the proportion of detected positives which are wrongly identified as positives and is given by Equation 9.9. The threshold corresponding to highest TPR at lowest FPR is considered optimum. With reference to the ROC plot in [10] for *Anomalous Walking Patterns* dataset, it may be noted that the highest TPR value of 1.0 and FPR values ranging from 0.47 to 0.73 were produced by the STC, the IBC, the MDT, the STEF, and the local optical flow methods. However, the MSTC method achieved a TPR of 0.84 at 0.0

TABLE 9.3: Comparative analysis of the performance of the MSTC method against other methods in terms of precision-recall values derived from the graphs in [10] for the train video sequence and subway surveillance dataset

Datasets	Method	Precision and recall at optimum threshold		Recall at maximum precision		Precision at maximum recall	
		Recall	Precision	Recall	Precision	Recall	Precision
Train video sequence	STC [10]	0.43	0.80	0.0	1.0	1.0	0.15
	STEF [27]	0.43	0.85	0.0	1.0	1.0	0.15
	LOF [5]	0.43	0.40	0.025	0.87	1.0	0.11
	MSTC [42]	0.43	1.0	0.038	1.0	1.0	0.15
Subway Surveillance dataset (Exit video)	STC [10]	0.96	0.66	0.0	1.0	1.0	0.64
	STEF [27]	0.96	0.53	0.0	1.0	1.0	0.50
	LOF [5]	0.96	0.32	0.0	1.0	1.0	0.28
	MSTC [42]	0.96	0.9	0.93	0.93	1.0	0.79

FPR at the threshold obtained through *IM 1* for the dataset as shown in the ROC curve in Figure 9.10. The exact TPR and FPR values of the above methods are provided in Table 9.4. In all these curves, the optimum threshold observed at highest TPR and lowest FPR corresponded to the threshold derived through the inference mechanism applied to these datasets based on their context.

$$TPR = \frac{True\ positives}{True\ positives\ +\ False\ negatives} \tag{9.8}$$

$$FPR = \frac{False\ positives}{False\ positives\ +\ True\ negatives} \tag{9.9}$$

The same graph in Figure 9.10 is used to compute the frame level equal error rate (EER) for the different datasets. EER is the percentage of misclassified frames (either misdetected frames or false positive frames) when the FPR is equal to the misdetection rate [10,33]. Hence, lower EER values are preferred. As given in Table 9.5, the EER value ranges from 7% to 25% across the datasets, indicating good performance of the algorithm. Also, the detection rate of the MSTC algorithm in case of *Subway Surveillance (Exit video)* dataset is comparable to those of STC [10], LOF [5], ST-MRF [51] and dynamic sparse Coding [29] methods (results taken from corresponding references). The dataset provider [5] has considered people coming up from the platform facing the camera (exiting the platform) as usual behavior and has mentioned that the choice of anomalous events in the dataset is subjective. Hence, the number of total unusual events considered varies in the above works. Going by the definition of the provider, the action of people exiting the platform is considered normal and movement in any other direction i.e., going down through the turnstiles, going from left to right, vice versa and the loitering events were considered as anomalies (total of 45 events). The STC, ST-MRF, and dynamic sparse coding methods detected 19 out of 19 anomalous events while LOF detected nine out of the nine events. The MSTC method detected 18 out of 19 anomalous events common across other methods considered for comparison and it detected 43 out of 45 unusual events considered in total for the experiments. Figures 9.11 through 9.17 show a few samples of the anomalous events detected by the MSTC method.

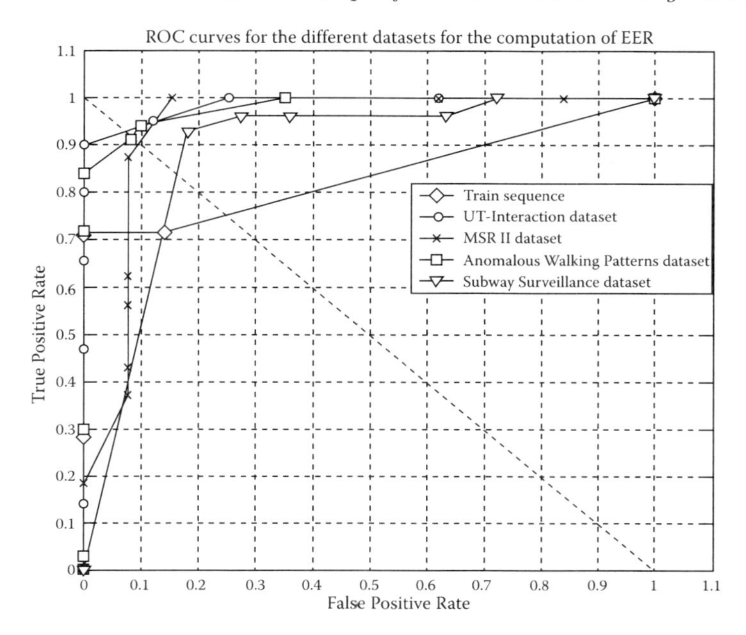

FIGURE 9.10: ROC curves of the MSTC method for the different real-world datasets for the computation of EER. The dashed diagonal is the EER line.

TABLE 9.4: Comparative analysis of the performance of the MSTC method against other methods in terms of TPR and FPR values derived from the graph in [10]

Dataset	Method	TPR	FPR
Anomalous Walking Patterns dataset	STC [10]	1.0	0.48
	IBC [14]	1.0	0.47
	MDT [25]	1.0	0.56
	LOF [5]	1.0	0.73
	STEF [27]	1.0	0.67
	MSTC [42]	0.84	0.0

TABLE 9.5: EER values for frame level anomaly detection in the different datasets

Datasets	Frame level EER (%)
Subway Surveillance dataset (Exit video)	16.25
Anomalous Walking Patterns dataset	08.5
UT-Interaction dataset	07.0
MSR II dataset	09.5
Train video sequence	25.0

9.3.3 Computation complexity analysis

As quoted in [10], the STC approach employs dense sampling around every pixel during volume construction and hence produces a large number of volumes (10^6 volumes for 1 minute video). The time complexity of the codebook construction is $O(V)$ for V number of volumes and the complexity of codeword assignment is $O(K \times M)$ if an ensemble has K number of volumes and there are M codewords. The posterior probability computation process has a complexity of $O(K \times M \times M)$.

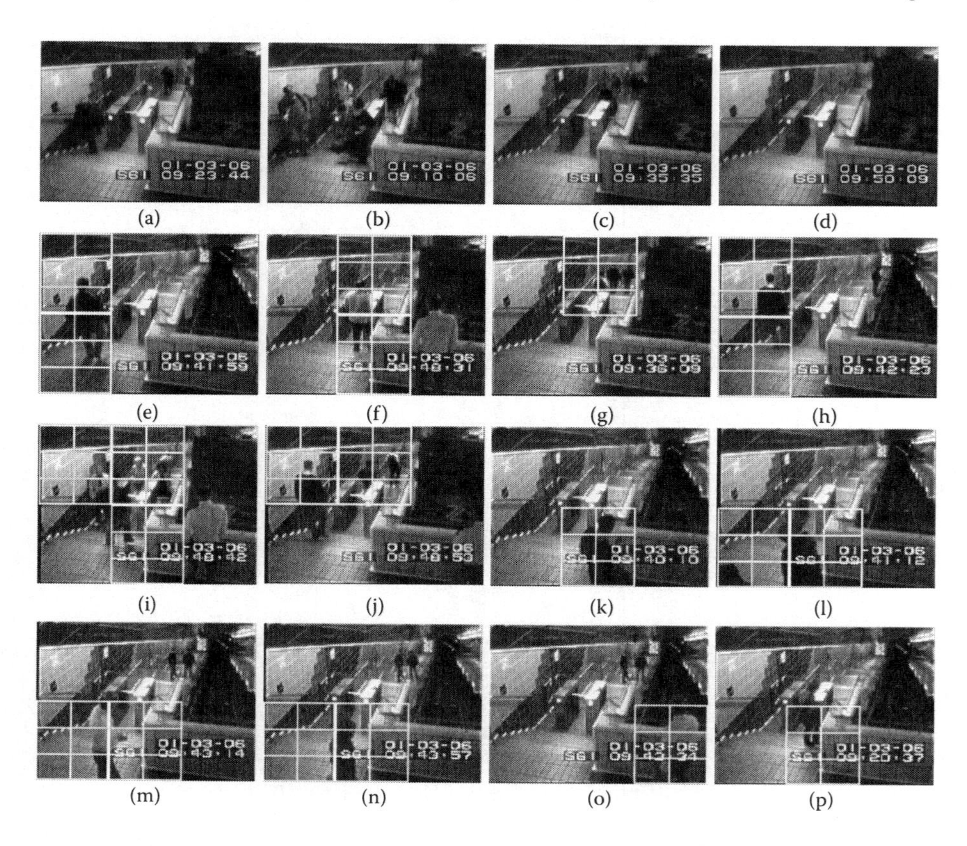

FIGURE 9.11: Anomaly detection in *Subway Surveillance* dataset: (a,b,c,d) Sample frames of normal actions of passengers exiting from the platform and of an empty platform. The anomalies detected are shown as follows: (e,f,g,h,i,j) a person is entering through the exit gate, (k,l,m,n) person moving in the wrong directions, (o) person loitering, and (p) janitor cleaning.

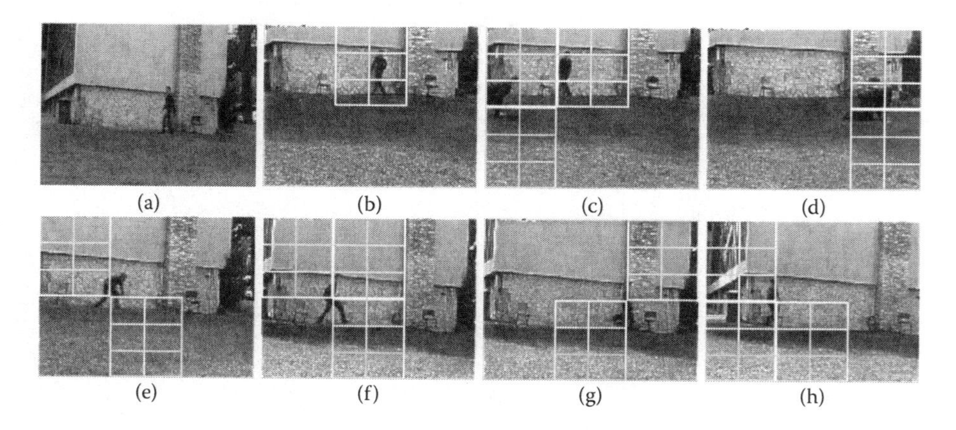

FIGURE 9.12: Anomaly detection in the *Anomalous Walking Patterns* dataset: (a) shows valid action of walking. The anomalies detected are shown as follows: (b) person walking in abnormal manner; (c) person walking in abnormal manner, while another person is crawling; (d) person crawling; (e) person falling; (f,g) person running; and (h) person lifting the chair.

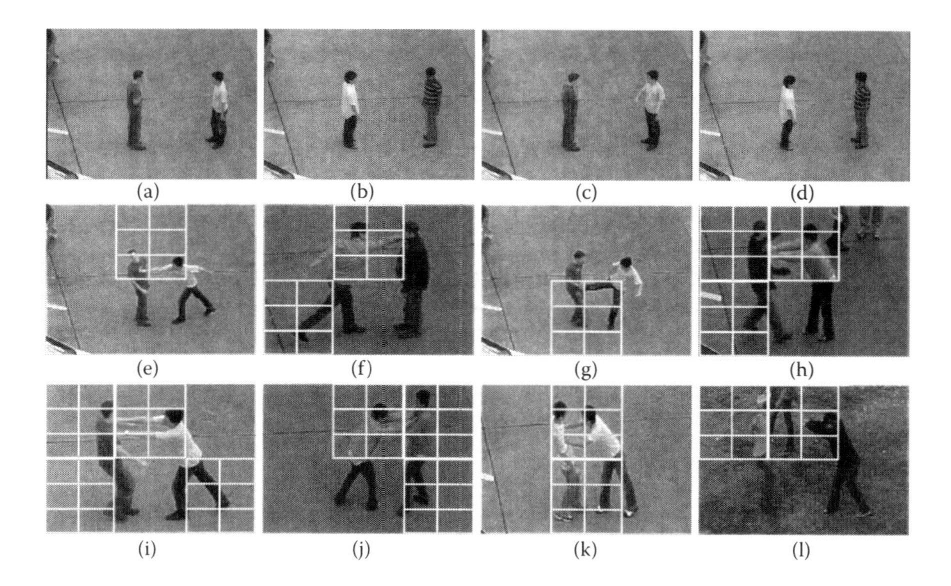

FIGURE 9.13: Anomaly detection in *UT-Interaction* dataset. (a,b,c,d) Sample frames of normal actions of persons in the scene having no physical interactions. The anomalies are shown as follows: (e,f) performing punching action, (g) performing kicking action, and (h,i,j,k,l) performing pushing action.

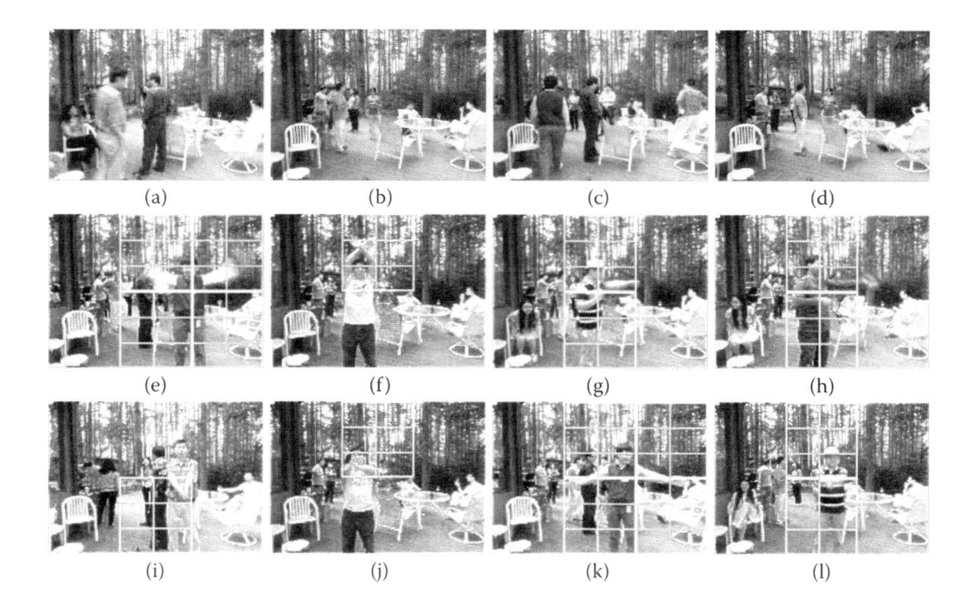

FIGURE 9.14: Anomaly detection in *MSR II* dataset. (a,b,c,d) Sample frames of persons performing normal actions in a dining area. The anomalies are shown as follows: (e,f) performing hand waving action, (g,h) performing boxing action, and (i,j,k,l) performing hand clapping action.

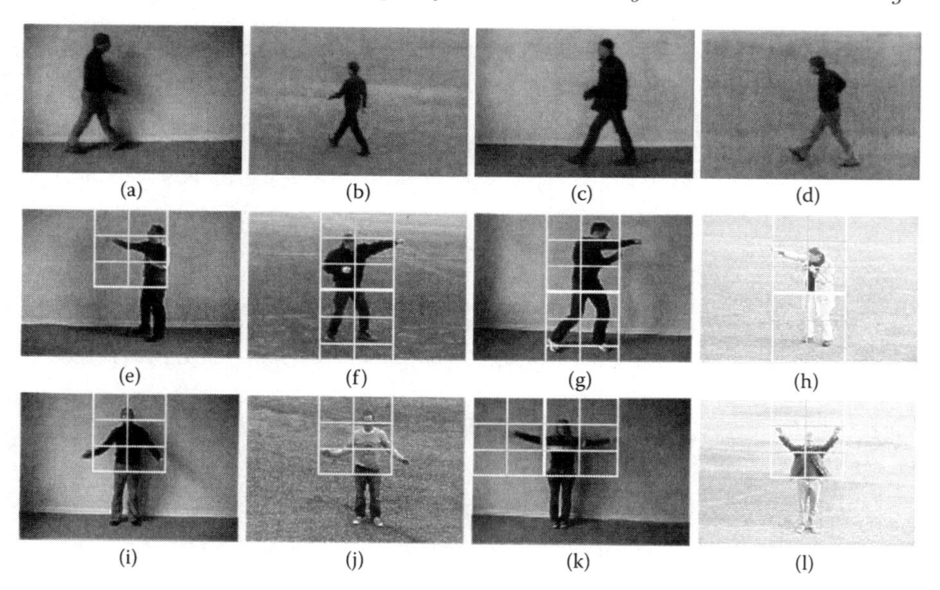

FIGURE 9.15: Anomaly detection in *KTH* dataset: (a,b,c,d) the valid action of walking. The anomalies are shown as follows: (e,f,g,h) performing boxing action, (i,j) performing hand clapping action, and (k,l) performing hand waving action. These anomalous events are detected in both indoor and outdoor environments and with participants wearing similar and contrast-colored clothes relative to background.

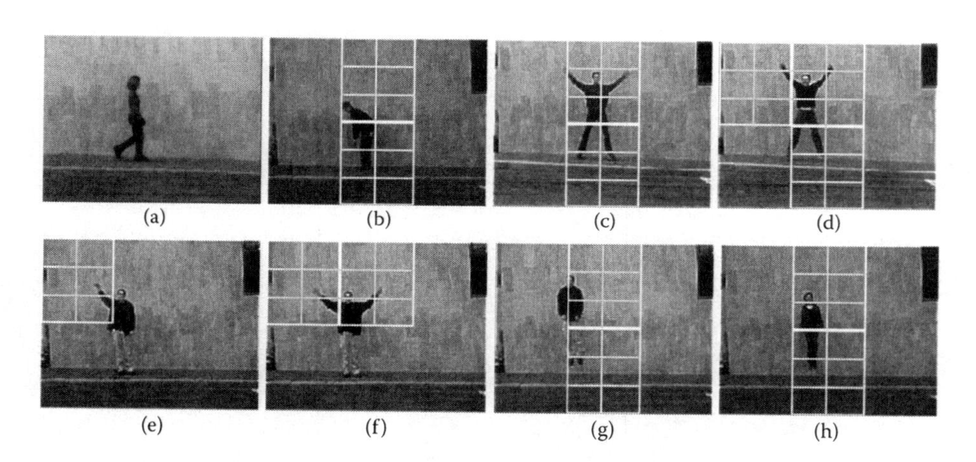

FIGURE 9.16: Anomaly detection in *Weizmann* dataset: (a) the valid action of walking. The anomalies are shown as follows: (b) performing bending action, (c,d) performing jumping jack action, (e) performing hand waving action, (f) performing two hands waving action, and (g,h) performing jumping action. These anomalous events are detected with participants wearing similar and contrast-colored clothes relative to background.

In case of MSTC method [42], a relatively low number of volumes are created due to the non-overlapping volume construction i.e., only about 4×10^4 volumes are formed when a 1 minute video of 120×160 frame dimension is processed. Consequently, codebook construction and codeword assignment both involve a lower number of iterations. Though the complexity of the probability calculation in the MSTC method is $O(K \times M \times K)$,

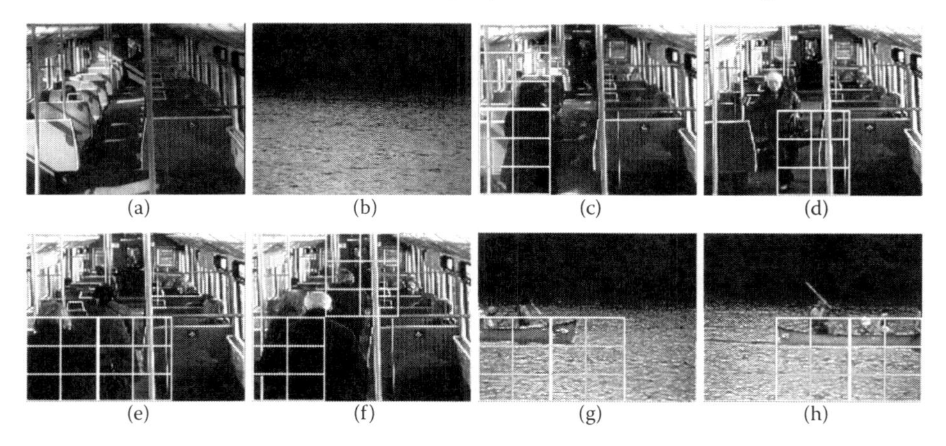

FIGURE 9.17: Anomaly detection in *Spatiotemporal Anomaly Detection* dataset: (a,b) the valid frames from Train and Canoe video sequences. The anomalies are shown as follows: (c,d,e,f) passengers moving in the cabin and (g,h) canoe moving across the river.

TABLE 9.6: Average computational time (in seconds) required per frame of a query from the different datasets (Average of three trials considered)

Datasets		Computational time (in seconds) per frame
Subway Surveillance dataset (Exit video)		0.172
Anomalous Walking Patterns dataset		0.107
UT-Interaction dataset		0.572
MSR II dataset		0.142
KTH dataset		0.151
Weizmann dataset		0.154
Spatiotemporal Anomaly detection dataset	Train video	0.228
	Boat-River video	0.424
	Canoe video	0.301

it still requires much fewer iterations because of smaller K and M, resulting in reduced computational complexity in comparison to that of [10].

The MSTC is implemented and experimented using the versatile MATLAB® R2012a (version: 7.14.0.739) software on a PC with 2.93GHz Intel Core(TM) i3 CPU and 4 GB RAM. Table 9.6 shows the processing time in seconds required by the anomaly detection algorithm to process a single frame of a query. The computational time is the sum of the time required for volume construction, for codeword assignment, for ensemble posterior probability calculation, for inferring and for localization of the anomaly in the query. It can be observed from Table 9.6 that for the majority of datasets, it takes only a fraction of a second to process each frame, thus making anomaly detection in real time. It should be noted that the computational time would have been much quicker if a compiled version of the MATLAB® codes were to be used, especially for the *UT-Interaction* dataset, *Boat-River* and *Canoe* video sequences.

9.3.4 Effect of codebook size and performance

Since the codebook construction process plays a highly significant role in the working of the method, it is necessary to analyze the effect of the codebook size on the performance.

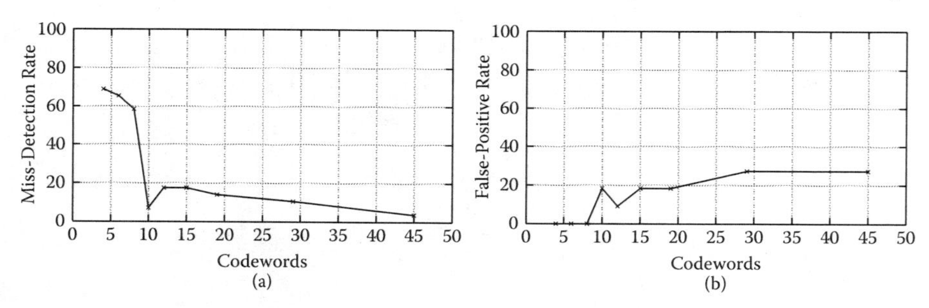

FIGURE 9.18: Effect of codebook size on the performance: (a) shows the variation in the miss-detection rate (%) relative to the increase in number of codewords; (b) shows the variation in the false alarm rate (%) relative to the increase in the codebook size.

This is done by varying the threshold ϵ which affects the volume grouping process, thus, resulting in a different number of codewords formed. The results shown in Section 9.3.2 are obtained by choosing ϵ as the mean of the distances of all the volumes v_i relative to v_1 in the set of volumes V generated for the training video sequence as explained in Section 9.2.1.2. The variations in the misdetection rate and the false alarms at different codebook sizes were analyzed for the *Subway Surveillance (Exit video)* dataset and the result has been shown in Figure 9.18. The observation is that the higher threshold values produce a lower number of codewords resulting from improper grouping of volumes leading to loss of information. It can be seen from the graph that this drastically increases the number of misdetections which is due to inadvertent grouping of anomalous events with the normal behaviors. On the other hand an expected decrease in the false positives also can be seen. Lower threshold values result in larger codebooks. Though these increased number of codewords retain more information and hence improve performance, they need more memory space and also increase the computational time of the method.

In brief, from the graph, it can be concluded that the misdetection rate greatly decreases with the increase in the number of codewords up to a point and then further decreases very slightly, but at the cost of increased number of codewords. The false positive rate correspondingly increases with the increase in codebook size. So, although there is a trade-off between codebook size and performance, using a compact sized codebook of ten codewords produces acceptable result for this dataset. This proves the reliability of the codebook construction process of MSTC.

9.4 Summary and Conclusion

This chapter presents a detailed discussion on an intelligent visual surveillance approach, the main component of which being the improved method for anomalous event detection and localization (MSTC algorithm) based on the principle of STC method [10]. The complexity is reduced and the performance is improved through specific deviations incorporated. The algorithm works at two stages: the initialization and the anomalous events detection. Under initialization, the method is trained with a short video sequence of few seconds (termed the training sequence) composed of the normal behaviors observed in the context of deployment which are learned and modeled. This process involves dividing the training video into ensembles and sampling them to generate spatiotemporal volumes at different spatiotemporal scales. Similar volumes are grouped into codewords to reduce the memory needed to store the volumes and also to aid in the probabilistic modeling process. The topology of the ensembles is probabilistically modeled through a non-parametric approach. The anomaly detection is based on detecting the events in the query that are not similar enough to the usual events of the training sequence. For this purpose, the probabilities of volume arrangement in the ensembles of the query and training videos are compared, and if an unusual event is detected in an ensemble with the probability less than a threshold, then that ensemble is inferred to have the event(s) of interest and the event(s) is localized. The decision making regarding the presence of anomaly is done depending on the context by the three adaptive inference mechanisms.

Experimentation on seven standard video datasets with simple and complex activities, in both non-crowded and crowded environments and in scenes with varied illumination conditions proved to be effective with a detection rate of 92.03% across all datasets, with only six false alarms. It should be noted that appreciable results were produced at reduced complexity for the standard datasets in comparison to STC and other state-of-the-art methods.

This surveillance approach is made power aware and energy efficient through event-driven processing and anomaly-driven transmission incorporated into the algorithm which is intended to be executed in the camera sensor node. Also, the visual enhancement of the anomalous frames aids in the surveillance application. All the software solutions work together towards providing smart and intelligent visual surveillance.

9.4.1 Characteristic features of MSTC and benefits of the surveillance approach

The characteristic features of the MSTC algorithm include the following:

- Non-overlapping volumes and ensembles which simplify the codebook construction and probabilistic modeling processes.
- Simple yet reliable codebook construction process.
- Non-parametric probabilistic modeling to obtain the true distribution of codewords in the ensembles. This is to aid in the desirable detection of the unusual events.
- Three adaptive context-based inference mechanisms to adaptively determine the decision-making threshold.
- Event-driven processing where a query will be processed only if there exists any event to be classified as an anomaly or normal.
- The feature of anomaly-driven transmission of only the anomalous frames is also added as a part of MSTC.

The benefits of the above approach for visual surveillance include the following

- The method avoids the employment of tracking, background subtraction or other processes such as foreground segmentation.
- The initialization is very fast due to the use of very short training sequence. It can also be attributed to the use of non-overlapping ensembles and volumes, and to the simple yet reliable codebook construction.
- The event-driven query processing eliminates computational energy wastage and the anomaly-driven transmission reduces bandwidth and in turn image transmission energy requirement.
- The anomalous event detection is in real time due to the use of non-overlapping ensembles and volumes.

- The adaptive inference mechanisms increase the robustness of the method to changing contexts.
- The visual enhancement of the anomalous frames improves the visibility of the anomalous activities, which helps towards better surveillance.

9.4.2 Future research directions

The performance and robustness of the MSTC method can be improved by introducing a unified inference technique which is adaptive to any context. Also, utilizing partially overlapped volumes and ensembles and applying a more robust descriptor for the volumes may further increase the efficiency of the method.

Transmission of only those frames containing the unusual events instead of all the captured frames has a risk that the video clips belonging to misdetected events will be lost, as those frames will not be transmitted to the base. Hence, the scope of improvement lies in increasing the unusual event detection accuracy to maximum possible level. In some cases, however, it will be required to access the frames caught by the sensor node for some purpose. One possible solution is to buffer the frames captured in a time span (like a week) and then replace the older ones with the newly captured frames.

References

1. S Gong, CC Loy, and T Xiang. Security and surveillance. In *Visual Analysis of Humans*, pp. 455–472. Springer, 2011.

2. N Mould, JL Regens, CJ Jensen III, and DN Edger. Video surveillance and counterterrorism: The application of suspicious activity recognition in visual surveillance systems to counterterrorism. *Journal of Policing, Intelligence and Counter Terrorism*, 9(2):151–175, 2014.

3. Caviar. Context aware vision using image-based active recognition. http://homepages.inf.ed.ac.uk/rbf/CAVIAR/caviar.htm, 2011.

4. R Fisher. Behave: Computer-assisted prescreening of video streams for unusual activities. http://homepages.inf.ed.ac.uk/rbf/BEHAVE/, 2011.

5. A Adam, E Rivlin, I Shimshoni, and D Reinitz. Robust real-time unusual event detection using multiple fixed-location monitors. *IEEE Transactions on Pattern Analysis and Machine Intelligence*, 30(3):555–560, 2008.

6. GL Foresti, C Micheloni, C Piciarelli, and L Snidaro. Visual sensor technology for advanced surveillance systems: Historical view, technological aspects and research activities in italy. *Sensors*, 9(4):2252–2270, 2009.

7. R Vezzani and R Cucchiara. Video surveillance online repository (visor): An integrated framework. *Multimedia Tools and Applications*, 50(2):359–380, 2010.

8. Computer Vision Pattern Recognition (CVPR). Performance evaluation of tracking and surveillance (PETS). http://www.cvg.rdg.ac.uk/PETS2001, http://www.cvg.rdg.ac.uk/PETS2006, 2009.

9. H Idrees, I Saleemi, C Seibert, and M Shah. Multi-source multiscale counting in extremely dense crowd images. In *Proceedings of the IEEE Conference on Computer Vision and Pattern Recognition*, Portland, OR, pp. 2547–2554, 2013.

10. MJ Roshtkhari and MD Levine. An on-line, real-time learning method for detecting anomalies in videos using spatio-temporal compositions. *Computer Vision and Image Understanding*, 117(10):1436–1452, 2013.

11. M Bar. Visual objects in context. *Nature Reviews Neuroscience*, 5(8):617–629, 2004.

12. O Schwartz, A Hsu, and P Dayan. Space and time in visual context. *Nature Reviews Neuroscience*, 8(7):522–535, 2007.

13. H Zhong, J Shi, and M Visontai. Detecting unusual activity in video. In *Proceedings of the 2004 IEEE Computer Society Conference on Computer Vision and Pattern Recognition, 2004 (CVPR 2004)*, Washington, DC, vol. 2, pp. II–819. IEEE, 2004.

14. O Boiman and M Irani. Detecting irregularities in images and in video. *International Journal of Computer Vision*, 74(1):17–31, 2007.

15. F Jiang, Y Wu, and AK Katsaggelos. Abnormal event detection from surveillance video by dynamic hierarchical clustering. In *2007 IEEE International Conference on Image Processing*, San Antonio, TX, vol. 5, pp. 145–1485. IEEE, 2007.

16. I Ivanov, F Dufaux, TM Ha, and T Ebrahimi. Towards generic detection of unusual events in video surveillance. In *Sixth IEEE International Conference on Advanced Video and Signal Based Surveillance, 2009 (AVSS'09)*, Genova, Italy, pp. 61–66. IEEE, 2009.

17. F Tung, JS Zelek, and DA Clausi. Goal-based trajectory analysis for unusual behaviour detection in intelligent surveillance. *Image and Vision Computing*, 29(4):230–240, 2011.

18. S Calderara, U Heinemann, A Prati, R Cucchiara, and N Tishby. Detecting anomalies in people's trajectories using spectral graph analysis. *Computer Vision and Image Understanding*, 115(8):1099–1111, 2011.

19. C Piciarelli and GL Foresti. On-line trajectory clustering for anomalous events detection. *Pattern Recognition Letters*, 27(15):1835–1842, 2006.

20. C Piciarelli, C Micheloni, and GL Foresti. Trajectory-based anomalous event detection. *IEEE Transactions on Circuits and Systems for Video Technology*, 18(11):1544–1554, 2008.

21. J Xu, S Denman, C Fookes, and S Sridharan. Detecting rare events using Kullback-Leibler divergence. In *2015 IEEE International Conference on Acoustics, Speech and Signal Processing (ICASSP)*, Brisbane, Australia, pp. 1305–1309. IEEE, 2015.

22. S Zhou, W Shen, D Zeng, and Z Zhang. Unusual event detection in crowded scenes by trajectory analysis. In *2015 IEEE International Conference on Acoustics, Speech and Signal Processing (ICASSP)*, Brisbane, Australia, pp. 1300–1304. IEEE, 2015.

23. H Guo, X Wu, H Wang, L Chen, Y Ou, and W Feng. A novel approach for global abnormal event detection in multi-camera surveillance system. In *2015 IEEE International Conference on Information and Automation*, Lijiang, China, pp. 73–78. IEEE, 2015.

24. C Schuldt, I Laptev, and B Caputo. Recognizing human actions: A local SVM approach. In *Proceedings of the 17th International Conference on Pattern Recognition, 2004 (ICPR 2004)*, Cambridge, UK, vol. 3, pp. 32–36. IEEE, 2004.

25. V Mahadevan, W Li, V Bhalodia, and N Vasconcelos. Anomaly detection in crowded scenes. In *2010 IEEE Conference on Computer Vision and Pattern Recognition (CVPR)*, San Francisco, CA, vol. 249, pp. 250, 2010.

26. R Mehran, A Oyama, and M Shah. Abnormal crowd behavior detection using social force model. In *IEEE Conference on Computer Vision and Pattern Recognition, 2009 (CVPR 2009)*, Miami, FL, pp. 935–942. IEEE, 2009.

27. A Zaharescu and R Wildes. Anomalous behaviour detection using spatiotemporal oriented energies, subset inclusion histogram comparison and event-driven processing. In *European Conference on Computer Vision*, Crete, Greece, pp. 563–576. Springer, 2010.

28. J Xu, S Denman, C Fookes, and S Sridharan. Unusual event detection in crowded scenes using bag of LBPS in spatio-temporal patches. In *2011 International Conference on Digital Image Computing Techniques and Applications (DICTA)*, Noosa, Australia, pp. 549–554. IEEE, 2011.

29. B Zhao, L Fei-Fei, and EP Xing. Online detection of unusual events in videos via dynamic sparse coding. In *IEEE Conference on Computer Vision and Pattern Recognition (CVPR), 2011*, Colorado Springs, CO, pp. 3313–3320. IEEE, 2011.

30. JY Huang and WP Lee. A smart camera network with SVM classifiers for crowd event recognition. In *Proceedings of the World Congress on Engineering, 1*, London, UK, 2014.

31. K-W Cheng, Y-T Chen, and W-H Fang. Video anomaly detection and localization using hierarchical feature representation and gaussian process regression. In *IEEE Conference on Computer Vision and Pattern Recognition*, Boston, MA, pp. 2909–2917. IEEE, 2015.

32. L Kratz and K Nishino. Anomaly detection in extremely crowded scenes using spatio-temporal motion pattern models. In *IEEE Conference on Computer Vision and Pattern Recognition, 2009 (CVPR 2009)*, Miami, FL, pp. 1446–1453. IEEE, 2009.

33. M Bertini, AD Bimbo, and L Seidenari. Multi-scale and real-time non-parametric approach for anomaly detection and localization. *Computer Vision and Image Understanding*, 116(3):320–329, 2012.

34. H Zhou and D Kimber. Unusual event detection via multi-camera video mining. In *18th International Conference on Pattern Recognition (ICPR'06)*, Hong Kong, China, vol. 3, pp. 1161–1166. IEEE, 2006.

35. P Antonakaki, D Kosmopoulos, and SJ Perantonis. Detecting abnormal human behaviour using multiple cameras. *Signal Processing*, 89(9):1723–1738, 2009.

36. H Dee and D Hogg. Detecting inexplicable behaviour. In *The British Machine Vision Association*, pp. 477–486, 2004.

37. CC Loy, T Xiang, and S Gong. Detecting and discriminating behavioural anomalies. *Pattern Recognition*, 44(1):117–132, 2011.

38. C Piciarelli and GL Foresti. Surveillance-oriented event detection in video streams. *IEEE Intelligent Systems*, 26(3):32–41, 2011.

39. D Zhang, D Gatica-Perez, S Bengio, and I McCowan. Semi-supervised adapted HMMs for unusual event detection. In *2005 IEEE Computer Society Conference on Computer Vision and Pattern Recognition (CVPR'05)*, San Diego, CA, vol. 1, pp. 611–618. IEEE, 2005.

40. RR Sillito and RB Fisher. Semi-supervised learning for anomalous trajectory detection. In *Proceedings of the BMVC*, Leeds, UK, vol. 1, pp. 1035–1044, 2008.

41. A Wiliem, V Madasu, W Boles, and P Yarlagadda. A suspicious behaviour detection using a context space model for smart surveillance systems. *Computer Vision and Image Understanding*, 116(2):194–209, 2012.

42. TJ Narendra Rao, GN Girish, and J Rajan. An improved contextual information based approach for anomaly detection via adaptive inference for surveillance application. In *Proceedings of International Conference on Computer Vision and Image Processing*, Roorkee, India, pp. 133–147. Springer, 2017.

43. A Marcus and O Marques. An eye on visual sensor networks. *IEEE Potentials*, 31(2):38–43, 2012.

44. D Glasner, S Bagon, and M Irani. Super-resolution from a single image. In *2009 IEEE 12th International Conference on Computer Vision*, Kyoto, Japan, pp. 349–356. IEEE, 2009.

45. MM Islam, VK Asari, MN Islam, and MA Karim. Super-resolution enhancement technique for low resolution video. *IEEE Transactions on Consumer Electronics*, 56(2):919–924, 2010.

46. O Shahar, A Faktor, and M Irani. Space-time super-resolution from a single video. In *2011 IEEE Conference on Computer Vision and Pattern Recognition (CVPR),* Colorado Springs, CO, pp. 3353–3360. IEEE, 2011.

47. MS Ryoo and J Aggarwal. UT-Interaction dataset, ICPR contest on Semantic Description of Human Activities (SDHA). In *2010 IEEE Conference on Pattern Recognition Workshops,* p. 4. IEEE, 2010.

48. J Yuan, Z Liu, and Y Wu. Discriminative video pattern search for efficient action detection. *IEEE Transactions on Pattern Analysis and Machine Intelligence*, 33(9): 1728–1743, 2011.

49. L Gorelick, M Blank, E Shechtman, M Irani, and R Basri. Actions as space-time shapes. http://www.wisdom.weizmann.ac.il/vision/SpaceTimeActions.html, 2011.

50. JM Chaquet, EJ Carmona, and A Fernández-Caballero. A survey of video datasets for human action and activity recognition. *Computer Vision and Image Understanding*, 117(6):633–659, 2013.

51. J Kim and K Grauman. Observe locally, infer globally: A space-time MRF for detecting abnormal activities with incremental updates. In *IEEE Conference on Computer Vision and Pattern Recognition, 2009 (CVPR 2009)*, Miami, FL, pp. 2921–2928. IEEE, 2009.

Chapter 10

Watershed Image Analysis Using a PSO-CA Hybrid Approach

Kalyan Mahata, Rajib Das, Subhasish Das, and Anasua Sarkar

10.1 Introduction

In 1999, by Cogalton and Green introduced remote sensing a method for obtaining knowledge of any object with no direct physical contact with it [1]. For grouping pixels among predefined classes (like turbid water or an urban area) in satellite images, a vast set of approaches exists.

Remote sensing to interpret features for considering geospatial data, objects, and classes on Earth's surface is a method that doesn't involve actual contact. There exists a panoply of methods for segmenting pixels in remote sensing images (for example, turbid water or an urban area) In theory, we can define a set for the remote sensing dataset, as shown in the following equation,

$$P = \{p_{ijk} | 1 \leq i \leq r, 1 \leq j \leq s, 1 \leq k \leq n\}(1) \tag{10.1}$$

of $r \times s \times n$ dimensions in pixels, where

$$p_{ij} \in \{p_{ij1}, p_{ij2}, ...p_{ijk}\} \tag{10.2}$$

denotes the values of n spectral bands for (i, j)th pixel. Among overlapping segments, we partition this image by our hybrid unsupervised algorithm over the chosen spatial image to find similar regions.

Let P (usually R^n or Z^n) represent the remote sensing image space; x, y denote points (pixels) in P as spatial variables consequently. In remote sensing, let $d_P(x, y)$ define a distance norm between two $\{x, y\} \in P$ pixels; d_P as the Euclidean distance is taken on the existing work [2,3]. In the domain of spatial images, a crisp object C may be projected as a subset of $P, C \subseteq P$. Utilizing the particle swarm movements over FC-Means (FCM) clustering solution is the scope of this chapter. This work is the extension of the author's earlier paper, using fuzzy-PSO (FPSO) and other similar approaches over similar remote sensing datasets [4,5].

Among unsupervised methods of classifications, clustering is based on similarity maximum within classes and minimum similarity among classes. State-of-the-art methods of which are used several times for remote sensing image segmentation are the self-organizing map algorithm [6], K-means algorithm [7], and to recognize arbitrarily shaped land-cover regions efficiently in remote sensing images.

Our proposed hybrid approach to classify pixels in the chosen Landsat image on the catchment of Barkar River. We also evaluate quantitatively using external validity measures, our experimental results with solutions obtained from K-means, FCM, and our implementation of FPSO algorithms over the Barakar catchment area including Tilaya Dam. The existing two internal validity indices indicate the quantitative evaluation of efficiency of our proposed hybrid PSO (particle swarm optimization) and cellular automata (CA)–based approach (FPSOCA) for detecting land-cover clusters. Next, our generated clustered regions are verified with the ground truth knowledge. The statistical evaluation over both datasets also denotes the significance of the proposed hybrid FPSOCA approach compared with K-means, FCM, and FPSO methods to detect imprecise clusters.

10.2 Existing Works

From the panoply of existing works on remote sensing image analysis, exploiting hybrid intelligences to find out significant accuracies in solutions recently has become a trend in challenging tasks.

Jacobsen [8] identifies hybrid intelligent systems of four categories:

For considering real-world problems, the semi-infinite problems of linear programming can be remodeled as a fuzzy-based problem with an infinite number of fuzzy constraints [9]. It is a hybrid cutting-plane algorithm, solving sub-problems using Zimmerman methods.

Hybrid intelligent systems have been researched more than incomplete information systems recently. Xiao et al. [10] develop three different kinds of approaches for incomplete knowledge engineering. They combine the variable precision method with hybrid methods of multi-granulation in one approach on a rough set. In the two other approaches, they experiment with variable precision with the hybrid multi-granulation-based rough set on similarity relation and limited relation of tolerance, respectively.

Several other hybrid intelligent approaches in theoretical analysis in computer science fields are also enhancing new dimensions for applicable computational approaches recently [11–14].

Raouzaiou et al. [15] exploit the facial expression detection problem from the view of a combined AI approach on appropriateness of sub-symbolic to symbolic mapping.

Furkan et al. [16] implement a hybrid fuzzy support vector regression (HF-SVR) model for fuzzy regression modeling in the linear and nonlinear domains. They utilize combined SVM and least squares principle for parameter estimates.Varying different kernel functions, different learning machines for fuzzy regression can be constructed nonlinearly.

Multidisciplinary approaches to utilize artificial intelligence in real-life applications have been explored earlier in several works. Goel [17] proposes a land cover method for feature extraction using hybrid swarm intelligence methods, which are a combination of the ACO2/BBO classification, over a training dataset of a 7-Band Alwar remote sensing image. In this hybrid approach, he utilizes the rough set approaches from the Rosetta tool to discretize each of the 20 clusters using the semi-native algorithm.

Similarly, Singh et al. [18] experiment to estimate the area of agricultural land in India using a new hybrid intelligence approach of the neuro-fuzzy method. As widely termed, the neuro-fuzzy hybrid method is used to create fuzzy neural network (FNN) as well as neuro-fuzzy system (NFS), from the study of existing works. A neuro-fuzzy model is defined as a fuzzy model which includes a learning approach, which is trained and derived from neural networks (NNs).

The application of hybrid intelligence for pattern recognition is a wide area of exploration. Recently, Teh et al. [19] propose a system to combine SOM (self-organizing maps) with the kMER (kernel-based maximum-entropy learning-rule) approach and PNN (probablistic neural-network) methods for fault detection and analysis. According to Teh, SOM never achieves optimum usage for its resources. To overcome this problem, they use a topographic map formation algorithm, namely kMER, to generate an equiprobablistic map with the maximum-entropy learning-rule. However, the kMER algorithm lacks for its computational efficiency. This drawback is covered by the SOM method without using the RF (receptive field) region. Therefore, in this hybrid intelligent system, they overcome the limitations and the drawbacks of individual methods in a combined manner. This is the benefit of hybrid inetelligent techniques in real-world applications.

Biomedical applications of hybrid intelligence systems also provide several improvements in clinical diagnosis treatment methodologies in recent years. [20] also works with hybrid intelligent methods in cardiotocographic signals which is a hybrid tightly coupled model. They propose CAFE (computer aided foetal evaluator) [20] as an intelligent, tightly coupled hybrid system to overcome the difficulties inherent in CTG (cardiotocograph) analysis.

Villar et al. [21] experiment with a hybrid implementation of the intelligent recognition models based on algorithms for detection of strokes in earlier stages. Their approach combines two wearable devices to monitor movement data in phases: a human activity recognition (HAR) device and a device to generate alarms. Hybridization of two HAR methods, one using genetic-fuzzy finite-state machines and the other using time-series (TS) analysis, has been done with symbolic aggregate approximation (SAX) TS representation for the alarm generation. Villar et al. evaluate their intelligent system over data gathered from healthy individuals.

Park et al. [22] develop a method for filtering quantum noise from medically applied X-ray CT images. They propose a hybrid NN filter method combining one bilateral filter (BF) and many neural-edge-enhancer(s) (NEE) with a neural filter (NF). This hybrid filter can be applied on different modality images, without considering filter selection for fusion.

Kumar et al. [23] implement a novel approach to hybridization based on bacterial foraging optimization (BFO) and PSO methods of feature vectors of the best facial selection. This selection enhances the identification accuracy of the individual recognition. This hybrid approach reduces facial features that are irrelevent in the feature space, and thus enhances accuracy.

Kasabov et al. in [24] define another dynamic neural-fuzzy inference evolving system, online and offline learning adaptively, and apply it for predicting dynamic time-series. It uses the most activated dynamically chosen fuzzy rules based on fuzzy inference system.

Negnevitsky [25] proposed a method on short-term wind power forecasting using a hybrid intelligent system. Abraham et al. [26] propose a soft approach hybridized to automatically forecast stock markets and trend analysis. Principal-component analysis has been used to preprocess inputs and then fed to the artificial NN to forecast stocks. Results from stock predictions are analyzed using NFS to detect market trends.

In a similar research domain, Araujo et al. [27] solve the problem in the financial time series forecasting analysis using a new concept of time phase adjustment by random walk dilema (RWD). They develop a system, namely increasing-decreasing-linear-neuron (IDLN) in stock market forecasting with high frequency. Hybridization is implemented using another descending gradient-based method, which can automatically perform time-phase adjustments. They demonstrate and test their work over a set of different markets with high-frequency financial time series: for example, the Brazilian stock market.

In their other work, Araujo et al. [27] develop the morphological-linear system, namely dilation-erosion-linear-perceptron (DELP), to forecast in finance. This hybrid system is based on morphological operators using lattice theory with linear operators. They incorporate one gradient-based approach with the next function for automatical fixing of phases to observe the adjusted time-phase distortions within financial phenomena. They analyze the S&P 500 Index, with five existing performance norms to evaluate the forecasting model.

Some recent developments in this field are those proposed by Khemaja [28] in his hybrid intelligent tutoring system. He develops a knapsack-based system to utilize intelligent decision-making for e-training activities in the information and communication technology (ICT) domain.

Dennis et al. [29] propose a blended method for a plug-in hybrid electric vehicle (PHEV). It implements a method to recognize driving patterns in real-time with control adaptation. They use the k-nearest neighbor (KNN) algorithm based on a driving pattern recognion module. The PHEV acommodates all necesary parameters, such as associate speed pattern recognition for trip-length, to form a blended model.

Bagher [30] proposes a method which describes hybridizing an artificial hormones controller with a type-two fuzzy controller. He explores three possible cases to fully hybridize two controllers. Several metrics for controlling features, like settling-time, steady-state error, overshoot, integral-absolute-error (IAE), integral-squared-error (ISE), integral time-weighted-absolute-error (ITAE), and integral of time multiplied by the squared error (ITSE) are estimated for his work. He also demonstrates that the proposed hybrid controller consumes energy 20%–30% less than the allowable base-controller.

Masoud et al. [31] propose three meta-heuristic algorithms using the genetic algorithm method by heterogeneous fleet for the waste collection problem of vehicles with multiple separated compartments. In their approach, they combine the multiple depot carriers over the problems of vehicle routing as well as on mixed close-open model on this problem. They propose a new mathematical mixed-integer programming model for minimizing the cost of servicing with respect to the customer's demands with the customer's available constraints.

To develop a hybrid intelligent transportation system, Lopez-Garcia [32] develops a system of genetic algorithms (GA) with their full application with the cross-entropy (CE) method in a hybridization. He refers to this method as GACE. This method optimizes the elements of a hierarchy of a fuzzy-rule-based system (FRBS) by passing through different weights for both GA and CE combined techniques. This methods can predict short-term traffic congestion.

Hybrid intelligence has also recently been applied to vision-based navigation methods. Lima et al. [33] develop a hybrid control approach for autonomous robotic automobiles in urban environments. This is a combined approach of the line-following visual-serving (VS) controller for road lane as a deliberative control and a dynamic-window method based on images (IDWA) as a reactive control to avoid obstacles. This method considers car kinematics to guarantee safe movement of a car in urban environments.

Hybrid intelligent systems can be utilized to improve energy management system (EMS). Qi et al. [34] propose the method of a contemporary evolutionary algorithm (EA), a self-adaptive state-of-charge (SOC) controlling method used for and applied to plug-in hybrid electric-vehicles (PHEVs). This system cannot detect common errors which are very frequently proclaiming a power demand in real-time, in predicting propulsion appropriate for online, easy-to-handle implementation.

Ming et al. [35] develop a hybrid system of a tracked bulldozer using wavelet transform-based energy management strategy; where an application of fuzzy control is proposed, EMS is evaluated with HILS (hardware-in-loop-simulation) both to increase the engine's fuel efficiency and effectiveness and to prolong the working life of the used battery pack.

Singla et al. [36] applied hybrid swarm intelligence techniques to analyze the land cover regions in Alwar in India. They applied hybrid ACO (ant colony optimization)/BBO (biogeography-based optimization), hybrid CS (cuckoo search)/PSO; and other methods in their experiments.

Ayerdi et al. [37] propose a spectral spatial processing method with spectral images and anticipative hybrid extreme rotation operations on hyperspectral image analysis using forest classification. Developing a spectral clasifier, namely anticipative-hybrid-extreme-rotation-forest (AHERF) with a spatial-spectral semi-supervised method for their analysis.

The batch scheduling method independently minimizes the problem of make span and flow time [38]. These two implement methods differ in main control and cooperation between both methods. The hierarchic and simultaneous optimization modes both are experimented with for this bi-objective scheduling approach.

10.3 Algorithm of PSO

The PSO algorithm optimizes a problem by iteratively considering a chosen norm for quality evaluation. PSO is a population-based algorithm that uses the best position within a search space. In PSO, the movement of the particle in the search space is dependent on the local best position but is also moved in the direction of the previous best position of swarm. Let the swarm size be N, and it is an M-dimensional search space; then the position of the p-th particle is defined as

$$X_p = \{x_{p1}, x_{p2}, ..., x_{pM}\} \tag{10.3}$$

The velocity of p-th particle is defined as

$$V_p = \{v_{p1}, v_{p2}, ..., v_{pM}\} \tag{10.4}$$

The best previous position of the p-th particle is defined as

$$P_p = \{p_{p1}, p_{p2}, ..., p_{pM}\} \tag{10.5}$$

Consequently, the best previous position of whole swarm is defined as

$$P_S = \{p_{S1}, p_{S2}, ..., p_{SM}\} \tag{10.6}$$

Let us assume the present iteration as t. As already said, the measured value [39–41],

$$v_{im}^{(t+1)} = \omega^t * v_{im}^t + (rand() * c_1 * (p_{im} - x_{im}^t))/\delta\tau + (rand() * c_2 * (p_{sm} - x_{im}^t))/\delta\tau \quad (10.7)$$

$$x_{im}^{(t+1)} = x_{im}^t + v_{im}^t * \delta\tau \quad (10.8)$$

$$\omega^t = \omega_{max} - (t * (\omega_{max} - \omega_{min}))/T \quad (10.9)$$

where $1 \leq t \leq T, 1 \leq m \leq M$, and $rand()$ generates the random number within uniform distribution $U(0,1)$; c_1 and c_2 define coefficients for accelerations; ω defines inertia weight, with ω_{max} and ω_{min} being the maximum and minimum values. For the initial matrix,

$$X = \begin{bmatrix} x_{11} & x_{12} & x_{1M} \\ x_{21} & x_{22} & x_{2M} \\ x_{n1} & x_{n2} & x_{NM} \end{bmatrix},$$

the equation to generate particle value is

$$x_{initial} = x_{im} = x_{min} + (x_{max} - x_{min}) * rand(), \forall m = \{1, ..., M, n = 1, ..., N\}. \quad (10.10)$$

Then the boundary constraints for $x_{im}^{(t+1)}$ and $v_{im}^{(t+1)}$ are defined in the following equations,

$$x_{im}^{(t+1)} = \{(x_{im}^{(t+1)}, \quad x_{min} \leq x_{im}^{(t+1)} \leq x_{max} \\ x_{initial}, x_{im}^{(t+1)} \quad > x_{max} \\ x_{initial}, x_{im}^{(t+1)} \quad < x_{min} \quad (10.11)$$

$$v_{im}^{(t+1)} = \{(v_{im}^{(t+1)}, \quad -v_{max} \leq v_{im}^{(t+1)} \leq v_{max} \\ v_{max}, v_{im}^{(t+1)} > \quad v_{max} \\ -v_{max}, v_{im}^{(t+1)} < \quad -v_{max} \quad (10.12)$$

where $\{v_{max}, v_{min}\}$ *and* $\{x_{max}, x_{min}\}$ are maximum and minimum values of v and x.

10.4 Cellular Automata Method

The cellular automaton system (plural - cellular automata, abbreviation - CA) was recently experimented with computational applications, mathematical applications, applications of physics, complexity science, and theoretical biology. This is a discrete model for microstructure modeling. It is used for dynamical systems to study physical system models [42]. It evolves the computational devices in time as well as in discrete space. A CA can be seeded in any state, with states with a single 1 and all 0s at separate positions, which results in unique patterns of fixed numbers.

Stephen Wolfram [43] proposes cellular automation in a simple model of cells consisting of a spatial latice. The discrete variables are stored in cells. At time t, the discrete variable denotes the present state for that cell. The next state at $(t+1)$ time is dependent on present as well as neighborhood states in time t. If the three-neighborhood CA with self, left, and

TABLE 10.1: Cellular automata rule thirty look up table

Present state	111	110	101	100	011	010	001	000
Rule	(7)	(6)	(5)	(4)	(3)	(2)	(1)	(0)
Next state	0	0	0	1	1	1	1	0

right neighbors is considered, with each containing two states, either 0 or 1, then the next state of a cell is computed as

$$S_i^{t+1} = f(S_{i-1}^t, S_i^t, S_{i+1}^t) \tag{10.13}$$

where f represents a transition function to next state. S_{i-1}^t, S_{i+1}^t and S_i^t are the present states for left position and right position neighbors and self of i-th cell at time t. The f is shown in the lookup table in Table 10.1. Rule R_i [44] denotes the decimal equivalent of eight outputs as shown in Table 10.1. In the two-state three-neighborhood CA, there are 256 rules.

We utilize this pattern generation property in our scheme.

In the periodic boundary CA, the immediate leftmost and rightmost cells are neighbors. Alternatively, it is called a null boundary CA. Further classification of CA defines deterministic and probabilistic CA, which are also stochastic. Previously defined CA are of the deterministic type.

The self-generating automaton which von Neumann developed is defined as M_c, which is a 2D cellular space with its specific CA rule and a specific initial state configurations embedded in it. There are 29 possible states per cell.

Here, we have used Codd's eight-state five-neighborhood (von Neumann) CA for the process. The initial state is any one state from 0,1,2,4,5,6, i.e., $q_0=\{0,1,2,4,5,6\} \in Q$ and $q_f=3 \in Q$.

10.5 Hybrid FPSOCA Clustering Algorithm

The proposed hybrid method, the FPSOCA algorithm, contains two steps. In the initial step, the FCM clustering method in the initial phase is used to allocate pixel memberships in the remote sensing image as chosen to random clusters. These membership values are given as input to the PSO algorithm. In the final fine-tuning step, the CA-based method of neighborhood priority correction is implemented. The two steps are depicted in Figure 10.1.

Initially random clusters have been assigned to pixels in the FCM algorithm, as mentioned in Section 10.2. The initial cluster centroids $C^{(0)}$ and membership matrix $U^{(0)}$ have been computed from initial pixel allocations. The mean-square-error objective function value ϵ for stopping iterations is 1E-05.

After the initial membership matrix generation using the FCM algorithm, we utilize $U^{(q+1)}$ after q iterations for $M = (C)$ number of fuzzy classes for input to PSO approach. The value $x_{initial} = U^{(q+1)}$ is set. The constraints $\{-v_{max}, v_{max}\} = \{0, 1\}$ for $V_{initial}$ in our PSO implementation is computed using the Rastrigrin function. The Rastrigrin function is defined as

$$f(x_i) = \sum x_i^2 - 10 * cos(2 * \pi * x_i) + 10 \tag{10.14}$$

$$x_{min} = [0, 0, ..., 0] \tag{10.15}$$

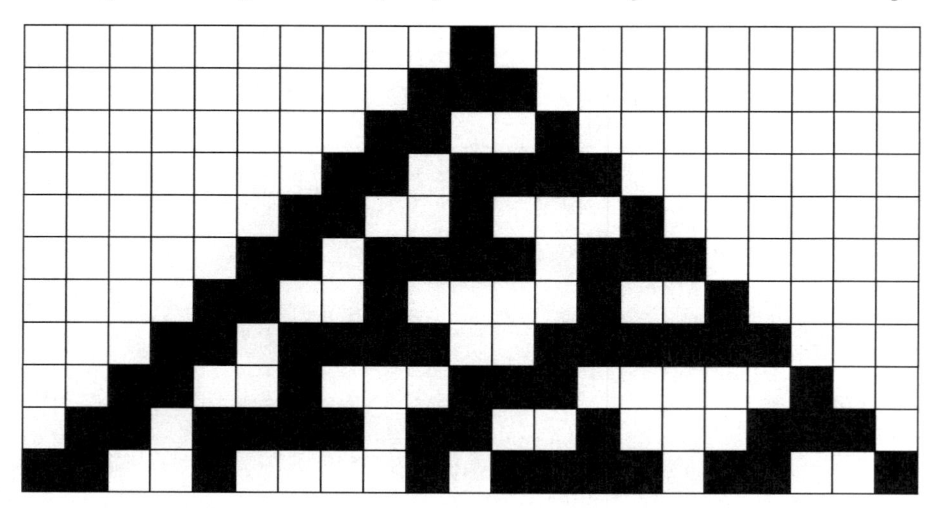

FIGURE 10.1: Example of cellular automata of Wolfram rule 30.

$$f_{x_{min}} = 0 \tag{10.16}$$

Similarly, the constraints $\{x_{max}, x_{min}\} = \{0, 1\}$ are set. Utilizing all these constraints, the new membership degree matrix $U_{SE} = x_m^{t+1}$ is generated and all pixels are reassigned to the clusters with the highest membership values. The centroids of clusters are recomputed in each iteration. The convergence criteria is set between the difference of f_{Gbest}, the previous best particle within the current population, and global minimum (GM). When this difference becomes less than ϵ (terminating threshold), the iterations stop.

After initial allocations from the combined FPSO method, our proposed CA based upgradation method on neighborhood priority is exploited among all pixels. The method is discussed in the next subsection. From finally obtained CA-corrected solutions.

10.6 CA-Based Neighborhood Priority Correction Method

Our CA model with two dimensions for neighborhood-based priority correction enables the selection of a formal algorithm to implement the scheme, which has been depicted in the flowchart in Figure 10.2.

The CA's first phase is implemented with clustering solutions from the initial well-known partitioning K-means on both sets of data: the remote sensing image and the remote sensing dataset. Thereafter, if four neighbors at left position, right position, top position, and bottom position show similar cluster values, then the current cell priority upgrades to similar cluster assignments of at least two neighbors.

Subsequently, if no two neighbors contain similar cluster assignments, the next state of the current cell remains the same. This approach reduces the number of outliers in small homogeneous regions. It also produces more efficient solutions for mixed pixels. The CA matrices are iterated to generate appropriate neighborhood upgradation among pixels on a chosen catchment region.

A particular transition function f is specified and shown to yield certain computation and construction properties discussed in the next section.

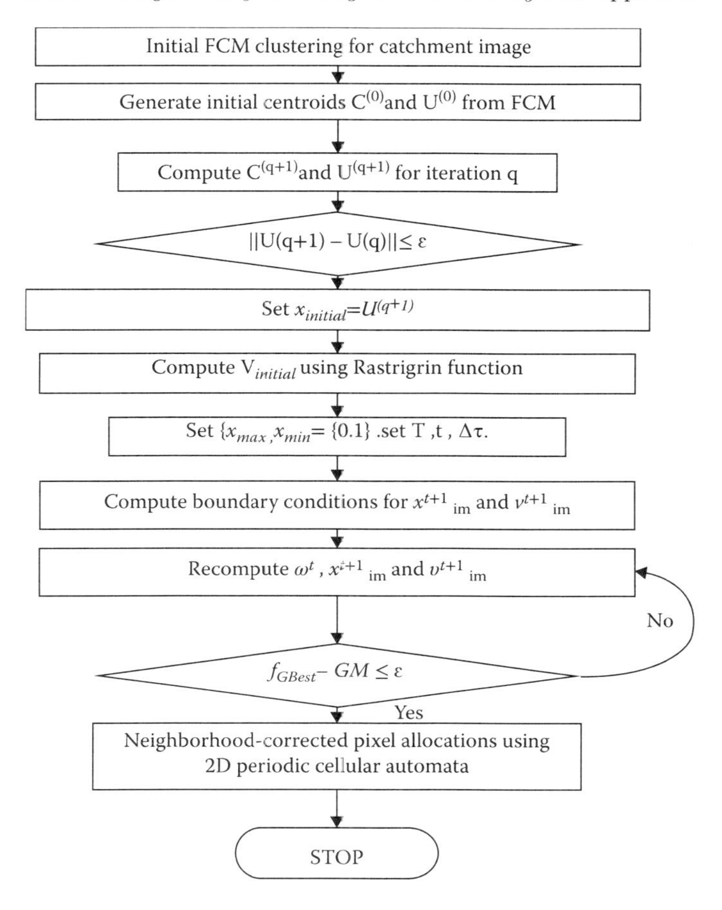

FIGURE 10.2: The flowchart of FPSOCA algorithm for Barakar River catchment segmentation.

10.7 Pixel Classification Application

The new FPSOCA method has been implemented using MATLAB® R2014a on Intel® Core™ i3-3220 processor with 3.30 GHz. For comparative evaluation, existing *K*-means, FCM, and FPSO methods are also experimented. For quantitative evaluation, Dunn [45] and Davies–Bouldin (DB) [46] internal validity indices are evaluated of clustering solutions on FPSOCA, fuzzy *C*-means, *K*-means and FPSO methods. The segmented regions of land use land cover models by FPSOCA approach is also compared with ground-truth knowledge.

The Barakar River catchment is chosen as a Landsat image in year 2015 has been extracted for research analysis. There are red, green, and blue bands available, as shown in Figure 10.3 with the original image, which shows the catchment region of the Barakar River with seven classes with histogram equalization: turbid-water (TW), pond-water (PW), concrete (Concr.), vegetation (Veg.), habitation (Hab.), open-space (OS), and roads (with bridges) (B/R).

Barakar River flows from top-left to the middle catchment region. At the top-left corner of the image, Tilaya Dam exists. Some waterbodies also exist in the middle-left area of catchment. The Barakar River seems to be a thinner white line from top-left to the

FIGURE 10.3: The original Barakar River catchment area in 2015.

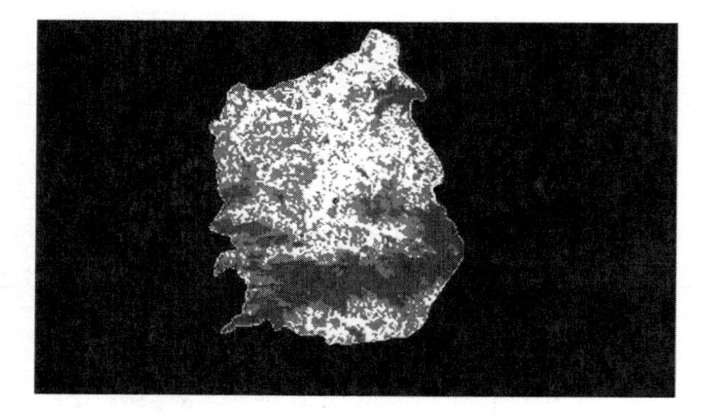

FIGURE 10.4: Barakar River catchment area in 2015 using pixel classification by K-means method (K=7).

left-middle of the catchment. The Tilaya Dam remains in deep-green color in the top-left region in Figure 10.3.

The segmentation solutions of the Barakar River catchment area as produced by K-means, FCM, and FPSO methods are depicted in Figures 10.4 through 10.6 for (K = 7). In Figure 10.4, the K-means method fails to separate the river from its catchment background in white color. The FCM method in Figure 10.5 produces better solutions, but also did not detect the lower part of the catchment area accurately. The FPSO method also segments the catchment with improper detection of the river and background as shown in Figure 10.6. Some waterbodies in the middle-left parts are shown to be mixed in K-Means as well as in fuzzy C-means clustering results as shown in Figures 10.4 and 10.5.

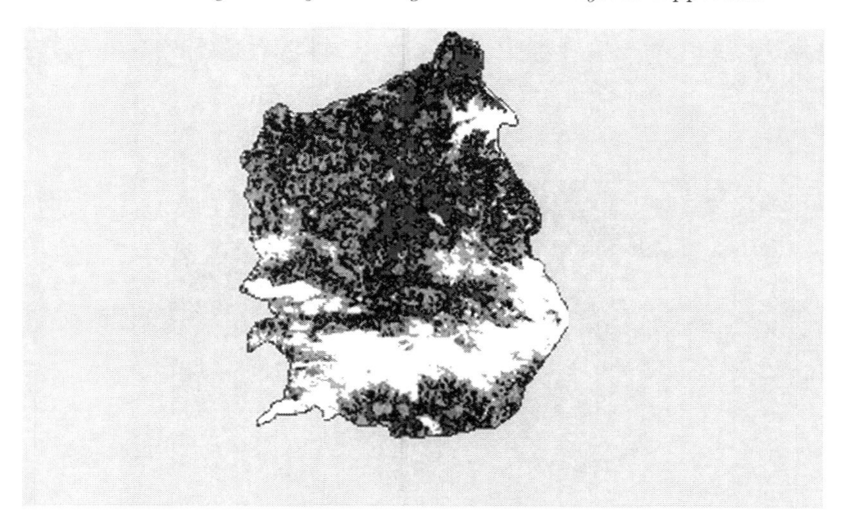

FIGURE 10.5: Catchment area in 2015 of the Barakar River produced by FCM approach (when K = 7) after pixel classification.

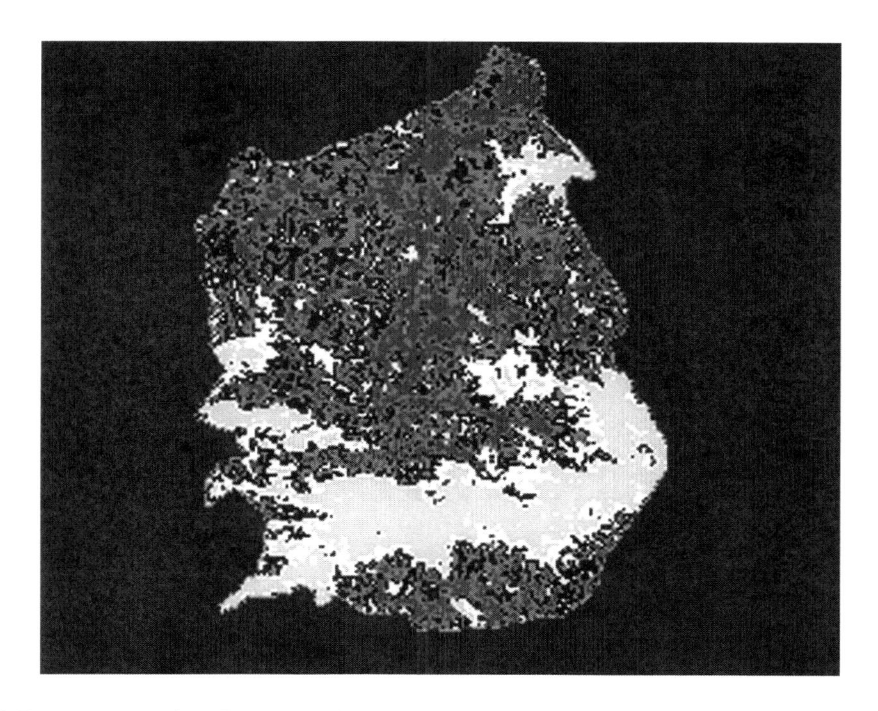

FIGURE 10.6: Barakar River catchment area in 2015 by FPSO algorithm (when K = 7) after pixel classification.

However, our new FPSOCA approach in Figure 10.7 is able to separate waterbodies at the middle-left catchment. It also detects the boundaries of the segments more prominently than the other three approaches. This new FPSOCA method also detects Tilaya Dam properly, separating it from its surrounding areas. These segmentation analyses show that our FPSOCA approach can detect the dam and catchment regions more prominently than existing K-means, FCM, and FPSO approaches.

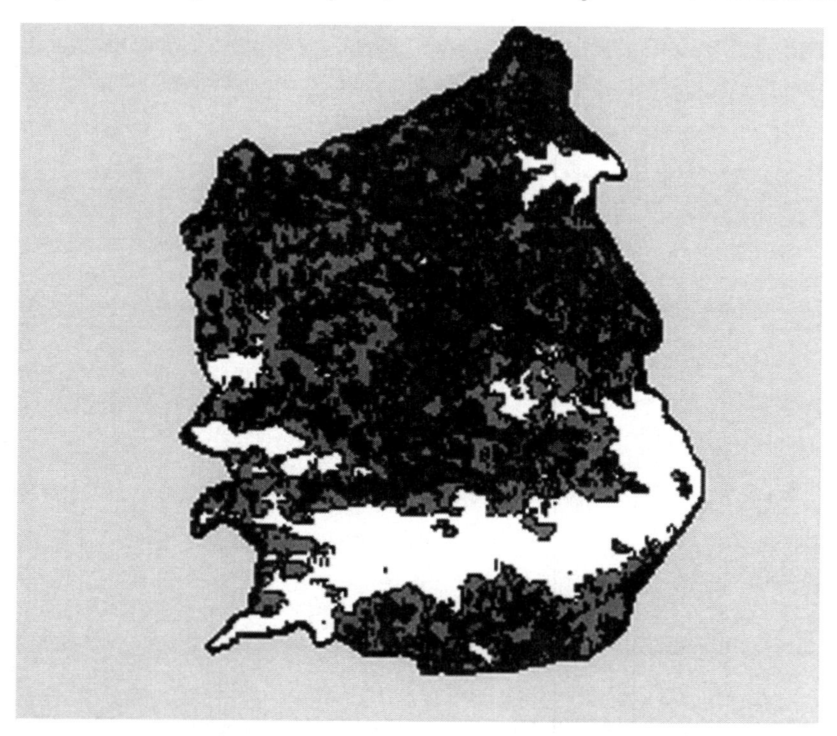

FIGURE 10.7: Pixel classification of Barakar River catchment areas in 2015 as depicted by an FPSOCA algorithm.

TABLE 10.2: Values of validity indices for the pixel classification solutions of Barakar River catchemnt region generated by K-means, fuzzy C-means, FPSO, and FPSOCA methods

Index	Barakar catchment 2015			
	***K*-means**	**FCM**	**FPSO**	**FPSOCA**
Davies–Bouldin index (MINIMIZING)	0.5538	0.4710	0.4896	0.3253
Dunn index (MAXIMIZING)	0.9948	0.6497	0.9568	1.3316

10.8 Quantitative Analysis

The clustering solutions obtained by the chosen four methods are objectively evaluated by validity norms Davies–Bouldin (DB) [46] as well as Dunn [45] indices, on the Barakar catchment area in Table 10.2. FPSOCA obtains minimum DB index value as 0.3253, and K-means exhibits a DB index value of 0.5538. FCM and FPSO methods exhibit 0.4710 and 0.4896 values respectively. Subsequently, maximum Dunn index value as obtained with FPSOCA method is 1.3316, while K-means, FCM, and FPSO methods obtain smaller values.

In Table 10.2, the clustering solutions obtained by K-means, FCM, FPSO, and proposed FPSOCA algorithms are evaluated quantitatively using the internal validity DB and Dunn indices. For this dataset also FPSOCA approach obtains smallest DB index value of 0.3253 and the highest Dunn index value of 1.3316 among the chosen four experimented methods.

Therefore, for this hyperspectral dataset also our proposed FPSOCA approach provides efficient clustering solutions while in comparison with the other three methods. All implications of these solutions indicate that the FPSOCA method acn optimize DB and Dunn indices more efficiently than K-means, FCM, and FPSO algorithms. Therefore, it has been shown that the FPSOCA method is comparable to K-means, FCM, and FPSO methods when considering the quality of solutions.

10.9 Statistical Analysis

The statistical non-parametrically test for significance analysis, named Wilcoxon's rank-sum, has been conducted on independent samples [47] with a 5% significance-level. Three sets are generated by 10 consecutive runs from the DB index value for the chosen three methods: K-means, FCM, FPSO, and FPSOCA algorithms on both datasets. The obtained median values on both datasets of each group are shown in Table 10.3. The results show that FPSOCA obtains smaller values than K-means and FCM methods, showing more efficiency than those methods.

The proposed scheme is modeled and checked for the satisfiability of the selection of exactly the clusters in the image at a time using MATLAB® programming. The performance analysis establishes that our proposed method is superior compared to K-means as well as FCM methods.

Table 10.4 exhibits values produced by Wilcoxon's rank-sum tests on three groups, FPSOCA-K-means and FPSOCA-FCM, and FPSOCA-FPSO over chosen catchment image. All P-values obtained show smaller than 0.005 (5% significance-level). For the Barakar River catchment area, the rank sum test P-value between FPSOCA and K-means methods is 2.00 E-003 which is very small. This determines that the clustering solutions obtained from FPSOCA show statistical significance and they do not occur by chance. Other groups for FCM and FPSO solutions also show similar significance, which also satisfy the null hypothesis. Therefore, all results satisfy the statistical significance of the land cover, and land use segments from the FPSOCA algorithm over K-means, FCM, and FPSO solutions.

TABLE 10.3: Performance comparisons of medians over ten consecutive runs on DB indices using chosen methods

Data	Algorithms			
	***K*-means**	**FCM**	**FPSO**	**FPSOCA**
Barakar River catchment region	0.5538	0.4710	0.4896	0.3266

TABLE 10.4: Comparing FPSOCA with K-means, FCM, and FPSO, respectively

Algorithm	Comparison with FPSOCA	
	H	**P-value**
K-means	1	2.00E-003
FCM	1	3.13E-003
FPSO	1	1.17E-003

10.10 Future Research Directions

The von Neumann CA are efficient in universal computation as well as construction. The CA construction has been simplified by Codd (1968) to eight states per cell, but is efficient in universal-computation and self-reproduction. In this work, we have utilized the eight-state five-neighborhood (von Neumann) CA architecture to model the queen bee problem. Cellular automata evolve in discrete space and time. We have designed a set of transition rules, required for the problem, to get the next states. We define Q=0, 1, 2, 3, 4, 5, 6, 7 as a set of local states that can be assigned to each cell. A configuration over Q is a mapping (Z2 $\rightarrow$ Q), and is called a global state of the CA. The set of all global states or configurations is denoted by Conf. (Q). Our proposed model is shown in Figure 10.2. The change in the overlapping regions over the timeline should be detected for applications. Such a classification is the general difficulty as there is no mathematical criterion is available. We have to find out an appropriate algorithm for doing that task.

10.11 Conclusions

Image segmentation algorithms are the most challenging methods used in remote sensing to help to interpret the land use, land cover models in the satellite imagery. Cellular automata model provides the changes in cell states depending on their neighborhood, which occur in discrete time form. Therefore, the cellular automata model can be used to analyze land cover segments in remotely sensed images, considering the neighborhood pixel classes.

This article contribution lies in significant improvement over detection for mixed land cover regions in satellite imagery than existing clustering algorithms. This new approach introduces a hybrid soft set based pixel segmentation with cellular automata based neighborhood upgradation in our proposed FPSOCA algorithm for clustering. The contributions primarily of this work are to implement one new clustering method on satellite images, which is initially soft PSO-based with final CA-based neighborhood corrections. The neighborhood upgradation stage helps to modify outlier allocations and overlapping clusters significantly. It verifies allocations overall considering the neighborhood, to generate more efficient land cover solutions.

The obtained performance from proposed FPSOCA approach is demonstrated on chosen satellite images on Barakar River catchment region in 2015, including Tilaya dam. Significant efficiency of the proposed FPSOCA segmentation method in comparison with existing K-means and fuzzy C-means methods is established both quantitatively and statistically. For quantitative evaluation, two internal external validity indices are analyzed over the catchment image. The ground truth knowledge verification also exhibits significant superiority for the proposed FPSOCA method over three well-known algorithms. Statistical significance tests are performed to show statistically significant FPSOCA solutions, when comparing with K-means, fuzzy C-means and FPSO methods on a remotely sensed chosen image. This scheme is highly adaptive in concurrent topological changes and is, therefore, well suited for use in any remote sensing satellite image.

Glossary

Cellular automata: A discrete and dynamic system consists of uniformly interconnected cells.

Clustering: Method to allocate similar points in one group to enhance intra-cluster similarity and decrease inter-cluster similarity.

Fuzzy C-means algorithm: Clustering approach to classify n point in C clusters, where centroids are calculated iteratively from means of all points of that cluster, using the fuzzy set theory concept.

Fuzzy set: Set of points with membership values between 0 and 1 for each cluster following fuzzy-set theory by Zadeh.

Hybrid intelligence: A computational approach to apply combined methods from artificial intelligence subfields.

Land cover: The physical coverage of forests, wetlands, impervious surfaces, agriculture, and open water regions on earth's surface.

PSO: Optimization method to move particles stochastically towards the best previous positions of self as well as for the whole swarm.

Validity index: Index to evaluate cluster compactness, to detect different clusters.

Watershed: The ridge dividing the draining region of a river.

Watershed analysis: A systematic approach to distribute ecosystem information with human, aquatic, riparian, and terrestrial features, conditions, processes, and interactions within the region of a watershed.

References

1. R. G. Congalton and K. Green. *Assessing the Accuracy of Classifications of Remotely Sensed Data: Principles and Practices.* Lewis Publications, 1999.

2. U. Maulik and A. Sarkar. Efficient parallel algorithm for pixel classification in remote sensing imagery. *Geoinformatica*, 16(2):39–407, 2012.

3. S. Bandyopadhyay. Satellite image classification using genetically guided fuzzy clustering with spatial information. *International Journal of Remote Sensing*, 26(3):57–593, 2005.

4. A. Sarkar and R. Das. Remote sensing image classification using fuzzy-PSO hybrid approach. In *Handbook of Research on Swarm Intelligence in Engineering*, 2015.

5. D. Chakraborty, A. Sarkar, and U. Maulik. A new isotropic locality improved kernel for pattern classifications in remote sensing imagery. *Spatial Statistics*, 17:7–82, 2016.

6. R. Spang. Diagnostic signatures from microarrays, a bioinformatics concept for personalized medicine. *Biosilico*, 1(2):64–68, 2003.

7. S. Tavazoie, J. Hughes, M. Campbell, R. Cho, and G. Church. Systematic determination of genetic network architecture. *Bioinformatics*, 17:405–414, 2001.

8. H.-A. Jacobsen. A generic architecture for hybrid intelligent systems. In *Proceedings of IEEE Fuzzy Systems*, 1998.

9. F. J. Alireza and K. Somayeh. A hybrid method for solving fuzzy semi-infinite linear programming problems. *Journal of Intelligent and Fuzzy Systems*, 28(2):879–884, 2015.

10. H. Lin, Q. Wang, B. Liu, B. Han, and X. Lu. Hybrid multi-granulation rough sets of variable precision based on tolerance. *Journal of Intelligent and Fuzzy Systems*, 31(2):71–725, 2016.

11. A. K. Shah and D. M. Adhyaru. HJB solution-based optimal control of hybrid dynamical systems using multiple linearized model. *Control and Intelligent Systems*, 44(2), 2016.

12. H. M. Nehi and A. Keikha. TOPSIS and Choquet integral hybrid technique for solving MAGDM problems with interval type-2 fuzzy numbers. *Journal of Intelligent and Fuzzy Systems*, 30(3):1301–1310, 2016.

13. R. T.-Moghaddam, S. Sadri, N. P. Zia, and M. Mohammadi. A hybrid fuzzy approach for the closed-loop supply chain network design under uncertainty. *Journal of Intelligent and Fuzzy Systems*, 28(6):2811–2826, 2015.

14. E. Scarlat and V. Maracine. The hybrid intelligent systems design using grey systems theory. *Grey Systems*, 5(2):19–205, 2015.

15. A. Raouzaiou, N. Tsapatsoulis, V. Tzouvaras, G. Stamou, and S. Kollias. A hybrid intelligent system for facial expression recognition. In *Proceedings of European Symposium on Intelligent Technologies, Hybrid Systems and Their Implementation on Smart Adaptive Systems*, 2002.

16. F. Baser and A. Apaydin. Hybrid fuzzy support vector regression analysis. *Journal of Intelligent and Fuzzy Systems*, 28(5):2037–2045, 2015.

17. L. Goel. Land cover feature extraction using hybrid swarm intelligence technique—A remote sensing perspective. *International Journal on Signal and Image Processing*, 14–16, 2010.

18. H. Singh, P. Maurya, K. Singh, and P. Singh. Analysis of remote sensed data using hybrid intelligence system: A case study of Bhopal region. In *Proceedings of National Conference on Future Aspects of Artificial Intelligence in Industrial Automation, 2012*, pp. 26–31, 2012.

19. C. S. Teh and C. P. Lim. A hybrid intelligent system and its application to fault detection and diagnosis. In *Applications of Soft Computing*, vol. 36. Springer, 2006.

20. B. Guijarro-Berdi-as, A. Alonso-Betanzos, and O. Fontenla-Romero. Intelligent analysis and pattern recognition in cardiotocographic signals using a tightly coupled hybrid system. *Artificial Intelligence*, 136(1):1–27, 2002.

21. J. R. Villar, C. Chira, J. Sedano, S. González, and J. M. Trejo. A hybrid intelligent recognition system for the early detection of strokes. *Integrated Computer Aided Engineering*, 22(3):21–227, 2015.

22. K. Park, H. S. Lee, and J. Lee. Hybrid filter based on neural networks for removing quantum noise in low-dose medical x-ray CT images. *International Journal of Fuzzy Logic and Intelligent Systems*, 15(2), 2015.

23. S. Kumar and S. K. Singh. Hybrid BFO and PSO swarm intelligence approach for biometric feature optimization. *International Journal of Swarm Intelligence Research*, 7(2):36–62, 2016.

24. N. Kasabov and Q. Song. DENFIS: Dynamic evolving neural-fuzzy inference system and its application for time-series prediction, 2001.

25. M. Negnevitsky, P. Johnson, and S. Santoso. Short-term wind power forecasting using hybrid intelligent systems. In *IEEE Power Engineering Society General Meeting*, 2007.

26. A. Abraham, B. Nath, and P. K. Mahanti. Hybrid intelligent systems for stock market analysis. In *ICCS '01 Proceedings of the International Conference on Computational Science—Part II*, pp. 337–345, 2001.

27. R. de A. Araújo, A. L. I. Oliveira, and S. R. L. Meira. A hybrid model for S&P 500 index forecasting. In *Proceedings of Artificial Neural Networks and Machine Learning–ICANN 2012, Part II*, pp. 57–581, 2012.

28. M. Khemaja. Using a knapsack model to optimize continuous building of a hybrid intelligent tutoring system: Application to information technology professionals. *International Journal of Human Capital and Information Technology Professionals*, 7(2):1–18, 2016.

29. N. Denis, M. R. Dubois, R. Dubé, and A. Desrochers. Blended power management strategy using pattern recognition for a plug-in hybrid electric vehicle. *International Journal of Intelligent Transportation Systems Research*, 14(2):101–114, 2016.

30. M. B. Fakhrzad and M. A. Rahdar. Optimization of hybrid robot control system using artificial hormones and fuzzy logic. *Journal of Intelligent and Fuzzy Systems*, 30(3):1403–1410, 2016.

31. M. Rabbani, H. Farrokhi-asl, and H. Rafiei. A hybrid genetic algorithm for waste collection problem by heterogeneous fleet of vehicles with multiple separated compartments. *Journal of Intelligent and Fuzzy Systems*, 30(3):181–1830, 2016.

32. P. L. García, E. Onieva, E. Osaba, A. D. Masegosa, and A. Perallos. A hybrid method for short-term traffic congestion forecasting using genetic algorithms and cross entropy. *IEEE Transactions on Intelligent Transportation Systems*, 17(2):557–569, 2016.

33. D. A. de Lima and A. C. Victorino. A hybrid controller for vision-based navigation of autonomous vehicles in urban environments. *IEEE Transactions on Intelligent Transportation Systems*, 17(8):2310–2323, 2016.

34. X. Qi, M. J. Barth, G. Wu, and K. Boriboonsomsin. Intelligent on-line energy management system for plug-in hybrid electric vehicles based on evolutionary algorithm. In T. Friedrich, F. Neumann, and A. M. Sutton, editors, *GECCO 2016 Proceedings of Genetic and Evolutionary Computation Conference*, pp. 16–168. ACM, 2016.

35. M. Pan, J. Yan, Q. Tu, and C. Jiang. Fuzzy control and wavelet transform-based energy management strategy design of a hybrid tracked bulldozer. *Journal of Intelligent and Fuzzy Systems*, 29(6):256–2574, 2015.

36. S. Singla, P. Jarial, and G. Mittal. An analytical study of the remote sensing image classification using swarm intelligence techniques. *International Journal for Research in Applied Science and Engineering Technology*, 3(6):49–497, 2015.

37. B. Ayerdi and M. G. Romay. Hyperspectral image analysis by spectral-spatial processing and anticipative hybrid extreme rotation forest classification. *IEEE Transactions on Geoscience and Remote Sensing*, 54(5):262–2639, 2016.

38. F. Xhafa, J. Kolodziej, L. Barolli, V. Kolici, R. Miho, and M. Takizawa. Evaluation of hybridization of GA and TS algorithms for independent batch scheduling in computational grids. In F. Xhafa, L. Barolli, J. Kolodziej, and S. Ullah Khan, editors, *2011 International Conference on P2P, Parallel, Grid, Cloud and Internet Computing, 3PGCIC 2011*, pp. 14–155. IEEE Computer Society, 2011.

39. R. C. Gonzalez and R. E. Woods. *Digital Image Processing.* Addison-Wesley, 1992.

40. A. Lukashin and R. Futchs. Analysis of temporal gene expression profiles, clustering by simulated annealing and determining optimal number of clusters. *Nature Genetics*, 22:28–285, 1999.

41. Y. Xu, V. Olman, and D. Xu. Clustering gene expression data using a graph theoretic approach, an application of minimum spanning trees. *Bioinformatics*, 17:309–318, 1999.

42. A.R. Smith III. Two-dimensional formal languages and pattern recognition by cellular automata. In *Proceedings of IEEE Conference Record of 12th Annual Symposium on Switching and Automata Theory*, 1971.

43. S. Wolfram. Cryptography with cellular automata. *Lecture Notes on Computer Science*, 218:42–432, 1986.

44. S. Wolfram. Statistical mechanics of cellular automata. *Reviews of Modern Physics*, 55(3):601–644, 1983.

45. J. C. Dunn. A fuzzy relative of the ISODATA process and its use in detecting compact well-separated clusters. *Journal of Cybernetics*, 3(3):3–57, 1973.

46. D. L. Davies and D. W. Bouldin. A cluster separation measure. *IEEE Transactions on Pattern Analysis and Machine Intelligence*, 1(2):224–227, 1979.

47. M. Hollander and D. Wolfe. *Nonparametric Statistical Methods.* Wiley, 1999.

Chapter 11

An Analysis of Brain MRI Segmentation

Tuhin Utsab Paul and Samir Kumar Bandhyopadhyay

11.1 Introduction

In day-to-day life, computational applications are gaining much importance. Especially, the use of the computer-aided design (CAD) systems for application in the field of computational biomedical analysis is being studied to a greater extent. In today's health care arena, detection and analysis of a brain tumor is one of the most common occurring fatalities. According to a report of the National Cancer Institute (NCI) statistics, over the last two decades, the overall cases of cancer, including brain cancer, has increased more than 10%. The National Brain Tumor Foundation (NBTF) for research in the USA had estimated that approximately 29,000 patients in the United States are diagnosed with primary brain tumors every year; moreover, out of them 13,000 patients die. Especially among kids, brain tumors cause one quarter of all deaths related to cancer. The average yearly incidence of primary brain tumors in the United States is 11 to 12 for every 100,000 people. For primary malignant brain tumors, that rate is six to seven for every 100,000. In the United Kingdom, over 4,200 people are diagnosed with brain tumors each year (2007 estimates). There are nearly 200 other categories of tumors diagnosed in the United Kingdom every year. Approximately 16 out of every 1,000 cancers detected in the United Kingdom are in the brain (or 1.6%). In India, 80,271 people are affected by various types of tumors (2007 estimates).

One of the specific medical image analysis methodologies is fully computerized brain disorder diagnosis like brain tumor detection from MRI. A brain tumor is the uncontrolled growth of the tissue cell in the brain. The cells that supply blood in the arteries are bounded tightly together which makes general laboratory tests inadequate to analyze the chemistry of the human brain. The various modalities of biomedical imaging that allow the doctor and researchers to analyze the brain anatomy by studying the brain without surgical invasion are computed tomography, magnetic resonance imaging (MRI) and positron emission tomography (PET).

MRI is a biomedical imaging technique that is used by radiologists for viewing the anatomical structures. MRI provides detailed information about soft tissue structural anatomy of human. MRI assist in diagnosis of the brain tumor. Magnetic resonance (MR) images are used for analyzing and studying the anatomy of the brain.

One of the most challenging tasks in today's medical image analysis is automated brain tumor detection from MR images [1–3]. MR produces images of the anatomy of soft human tissue. It is used to study the human anatomy without invasive surgery. Brain image segmentation, or the process of creating a partition and analyzing the image into visibly and anatomically different regions, is among the most vital and critical aspects of computer-aided clinical diagnostics of tumors or other anatomical abnormalities. Various types of noise that are found in the brain MR images are multiplicative in nature and reductions of these noises are critical. From the clinical aspect, it is very essential to ensure that the sensitive anatomical details are not removed by the noise reduction algorithms. Thus, precise segmentation of brain MR images is a challenging research area. Hence, highly precise segmentation of the MR images is very critical for proper diagnosis by computer-aided clinical tools. A wide variety of procedures for segmentation of MR images had been proposed to date.

11.2 Brain Imaging

Medical imaging is usually related to radiology or "clinical imaging" and the medical practitioner's or radiologist's responsibility for understanding as well as acquiring the images. Diagnostic radiography defines the technical details of capturing images in the biomedical domain. The radiologist is responsible for capturing medical images of clinical quality.

Biomedical imaging is the domain of pathological investigation which combines different interdisciplinary areas of technology like biomedical engineering and physics and even medicine, if required. Research in the domain of medical instruments, image acquisition (e.g., radiography), modeling, and quantification is under the purview of biomedical physics and engineering and also computer science to a great extent. Research into the analysis of biomedical images is under the purview of various multidisciplinary domains that include radiology, neuroscience, psychiatry, cardiology, psychology, and computer science as required for the investigation of the pathological condition of the patient.

The latest state-of-the-art technologies in medical imaging are described in Sections 11.2.1–11.2.5.

11.2.1 Ultrasound

High-frequency sound waves in the range of 2.0 to 10.0 megahertz are reflected by human soft tissues to varying degrees to produce a 2D image used by ultrasonography. This is used to visualize the growth and development of fetuses in pregnant women. Ultrasound is also used in imaging the abdominal organs, human heart, male sex organs, and the arteries of arms. Although it provides more anatomical detailing than other imaging techniques, such as CT (computed tomography) or MRI, it has some advantages which make this technology suitable as a first-line investigation in several situations where it studies the function and effect of moving structures in real time. Since the patient is not exposed to radiation, it is also very safe to use. Moreover, the ultrasound does not appear to cause any adverse effects, although information on this is not well documented. It is also comparatively less expensive than other imaging techniques and quick to perform. Ultrasound scanners are easily portable and can be taken to critically ill patients in intensive care units, thereby avoiding the risk of transferring the patient to an other radiology department. The instantaneous movement of fluid and body organs can be captured as images, which is essential for performing drainage of fluids or a biopsy. Color Doppler capabilities on modern ultrasound scanners allow the flow of blood in arteries and veins to be analyzed.

11.2.2 Projection radiography (X-rays)

Radiographs, generally termed as X-rays, are mostly used to diagnose the type, position, and extent of a bone fracture and for diagnosing pathological changes in lungs. Radiography also uses radio-opaque contrast media, such as barium, to analyze the anatomy of stomach and intestines, which helps to diagnose ulcers and different varieties of colon cancer.

11.2.3 Computed tomography (CT)

A CT scan is helical in structure and conventionally generates a 2D image of human anatomy. Repeated scans are limited due to ionization. In CT, X-rays must be obstructed by any type of high-density tissue to generate an image, and hence the image quality is bad.

11.2.4 Magnetic resonance imaging (MRI)

An magnetic resonance imaging instrument (MRI scanner), one of the most advanced imaging technologies, uses strong magnets to polarize and excite hydrogen nuclei (single proton) in water molecules that are present within body tissues, which produces a detectable signal which is spatially encoded that results in images of tissues. In other words, MRI uses three types of electromagnetic fields: the strong static field of the order teslas, in which the static magnetic field is used to polarize the hydrogen nuclei; the gradient field that uses a relatively weak time-varying signal of approximately 1 kHz for encoding in a spatial domain; and a weaker radio-frequency (RF) field is used to manipulate hydrogen nuclei for generating signals of good strength that are intercepted through an RF antenna. Like CT, MRI also generates a 2D image of a thin "slice" of the anatomical structure and is hence considered a tomographic imaging technology. The latest MRI machines are also eligible to generate images as 3D blocks; that is considered a generalization of the single-slice, tomographic concept. MRI does not use ionizing radiation, thereby making it less hazardous than CT. For example, there are no proven long-term adverse effects of exposure to strong static fields and because of this the patient can be scanned as many times as necessary, in contrast with X-ray and CT. Moreover, there are some health-related risk related with tissues getting heated due to exposure to the RF field and also to various external devices that are implanted in the body such as pace maker. Those risks are reduced through the design of the machine and the scanning methodology used. Computer Tomography and Magnetic Resonance Imaging are sensitive to various anatomical characteristics of the body, the images captured by the two process differ significantly. In Computer Tomography, X-rays should be obstructed by any type of high density tissue to generate an image, hence the quality of images while examining of soft tissues might be not of good quality. A Magnetic Resonance Image will visualize hydrogen-based objects, like bones, that contain calcium, will not be visible in the image, and hence will not obstruct the soft tissue. Thus, it is suitable for investigating joints or the brain.

MRI, or NMR imaging as it was originally known, has been in use only since the early 1980s. Effects from longterm or repeated exposure to the intense static magnetic.

11.2.5 Positron emission tomography (PET)

Positron emission tomography (PET) is used to diagnose various cerebral anomalies. Using nuclear medicine, a short-lived isotope, such as 18F, is transfused into the body through glucose that gets assimilated through blood into the tumor to be investigated. PET scans are mostly done in addition to CT scans, that are done using the same machine and hence do not require movement of the patient. Hence, tumors found by the PET scan can be analyzed parallel with the images obtained using a CT scan. PET uses nuclear medicine that might have an adverse effect on body issues in the long run.

11.3 Review of Different Existing Segmentation Techniques and Their Analysis

There has been large, multifaceted, and fast growth in the field on image processing in the last decade. In today's time, capturing, storage, and analysis of biomedical images is digitized [4]. With the latest state-of-the-art technologies, complete understanding of biomedical images is a challenging task because of time and accuracy constraints. The challenge for radiologists is high especially in anatomical abnormalities with different color and shape that need to be identified for further studies. The main objective in developing image processing and computer vision applications is to maintain the highest standards of accuracy in segmenting medical images. Image segmentation is the process of partitioning several distinct regions of the image based on various criteria.

Medical image segmentation is necessary in planning surgery, investigating growth after surgery, abnormal growth or tumor detection, and various other medical applications. Although there are many automatic and semi-automatic algorithms of segmenting images, they fail in many instances mostly due to unknown or non-regular noises, homogeneity, low contrast, or weaker boundaries that are very common to biomedical images. MR images or various other biomedical images consist of several complicated and minute anatomical variations which require very precise and exact segmentation for doing diagnosis clinically [5].

Segmentation of the brain from an MR image is very critical and challenging, but extremely precise and accurate segmentation is required for various clinical diagnoses like detection, analysis, and classifying various tumor categories such as edema, hemorrhage detection, and necrotic tissues. Unlike a CT scan, MR image acquisition parameters are greatly adjustable for creating high-contrast images having distinct gray levels for different cases of neuropathology [6]. Hence, segmenting MR images is the most recent research focus in the biomedical image processing domain.

In neuroscience, segmenting of the MR image is required in the diagnosis of neurodegenerative and also various psychiatric disorders [7].

11.3.1 Existing denoising methods

Although there are many state-of-the-art algorithms for denoising, accurate removal of noise from an MR image remains a challenge. There are various technologies, such as using standard filters or advanced filters, nonlinear filtering methods, an isotropic nonlinear diffusion filtering, Markov random field (MRF) models, wavelet models, non-local methods (NL-methods) models, and analytically correction schemes. Computational cost, denoising, quality of denoising, and boundary-preserving capabilities in those methods are similar. Hence, denoising still remains a research problem and requires innovation and improvement. Linear filters mostly reduce noise by the updation of pixel value by calculating the weighted average of neighborhood pixels but degrade the image quality. On the contrary, nonlinear filters preserve edges but degrade fine structures.

11.3.1.1 Markov random field method

In MRF, spatial correlation information is used to retain various fine details [8], i.e., the estimated noise is regularized in the spatial domain. In this method, the value of a pixel is updated by the iteration of conditional modes and simulated annealing having a function that maximizes a posterior estimate.

11.3.1.2 Wavelet-based methods

In the frequency domain, wavelet-based methods are implemented for removing noise and preserving the signal. The use of wavelets on MR images biases the wavelet and scaling coefficient. To remove the issue, the MR image is squared by the noncentral chi-square distribution method. Hence the scaling coefficient becomes independent of the signal and is eliminated. Although, in this case, having low signal-to-noise-ratio images, linear details are not preserved [9].

11.3.1.3 Analytical correction method

The analytical correction method works by estimating noise and thereby generating a noise-free signal from the initial image. It uses maximum likelihood estimation (MLE) [10] for the estimation of noise in the image and then generates images devoid of noise. Neighborhood smoothing is essential to calculate noise-free images by taking the signal in a small region to be constant. Image boundaries are quite degraded.

11.3.1.4 Non-local (NL) methods

NL methods uses the repetitive information of images [11]. The image pixel values are substituted by calculating the weighted average of the neighborhood. MR images have non-redundant details because of noise, complicated structures, blur in acquisition, and the partial volume effect generating because of low sensor resolution, which gets removed by the NL method.

11.3.2 Image segmentation methods

Thresholding, region growing, statistical models, active control models, and clustering have been implemented for segmenting images. As the intensity distribution in biomedical images is complex, thresholding becomes a critical task and fails in most cases [12].

Fuzzy clustering means (fuzzy C-means, or FCM) is a widely used technology for biomedical image segmentation but it considers only the image intensity thereby giving unsatisfactory output in noisy images. A set of algorithms is proposed to make FCM robust against noise and in homogeneity, but it's still far from perfect.

In probabilistic classification, a very accurate estimation of the probability density function (PDF) is necessary. The nonparametric approach does not make any assumption in getting the parameters of PDF; hence, it is precise but costly. In the parametric approach, a function is considered to be a PDF function. It is relatively simple to implement but in certain cases it lacks in preciseness and does not correlate with real data distribution.

11.3.2.1 FCM

Fuzzy C-means technology uses the initial cluster center positions from the SOM (self-organizing map) clustering algorithm. If the summation of the variances of the weight vector when divided by the size of the weight vector is less than the element of the weight vector, then the weight vector is expanded [13].

11.3.2.2 Learning vector quantization

It is the supervised competitive learning algorithm which calculates the decision boundaries of the input domain depending on the training dataset. It creates prototypes of class boundaries, based on a nearest-neighbor rule and uses a winner-takes-all paradigm. It consists of input, competitive, and output layers. The initial data gets classified in the competitive layer which are then mapped onto the target class of the output layer. The neuron weight gets updated in the learning phase depending on training datasets. The winner neuron is selected using Euclidean distance, then its weight is adjusted [6]. There are different methods to learn learning vector quantization networks.

11.3.2.3 Self-organizing maps

An SOM is an unsupervised clustering network which maps inputs that can be multidimensional to a 1D or 2D discrete lattice of neuron units [6]. Input is organized into several patterns based on a similarity factor, such as Euclidean distance, and every pattern assigns to a neuron. Every neuron had a weight which depends on the pattern that is assigned to the neuron [6]. The input is classified as per their grouping in the input space and neighboring neuron. SOM also learns distribution and topology of input [6]. SOP have two layers: first is the input layer where the number of neurons in the layer is equal to the dimension of the input, and second is the competitive layer where each neuron corresponds to one class or pattern. The number of neurons in this layer relates to the number of clusters and gets arranged in regular geometric mesh structure. Every connection from the input layer to a neuron in the competitive layer is assigned a weight vector. The SOM functions in two steps [6]: first, by finding the winning neuron where the maximum similar neuron is input by a similarity factor like Euclidean distance, and second, by updating the weight of the winning neuron and its neighbor pixels based on input.

11.3.2.4 Watershed

Watershed is an image gradient-based segmentation algorithm where distinct gradient values are taken as distinct height. A void is created in every local minimum and submersed in water. Hence, the water level will increase until local maximums. When any two bodies of water connect, a dam is erected between those two water bodies. The water level increases slowly until all points (heights) of the map are submersed. The image gets divided by the dam boundaries. Those dams are said to be watersheds and the segmented zones are known as catchments basins [14,15]. The over-segmentation problem still remains in this algorithm.

11.3.2.5 Active control method

The active control method is a model for delineating the outline of an object from the noisy image which is dependent on a curve, $X(s) = [x(s), y(s)]$, where s in range of [0,1] is length of arc. It changes such that it reduces the energy function. The tension and rigidity of the deforming curve is controlled by the internal energy whereas external energy is used to modify the deforming curve toward the target [16]. The active control method uses Gaussian gradient force to calculate external force. The advantages of this algorithm are insensitiveness to initialization of image contour, boundary concavities, reduced computational time, and high preciseness.

11.3.2.6 Markov random field method

The Markov network or undirected graph models is a set of random variables that have Markov properties that is defined by an undirected graph. The MRF model is the statistical model that is implemented to model spatial relations which exist in the neighbor of pixels [17]. Image segmentation techniques use it to gain the advantage of neighborhood information in the segmentation process; such as, in biomedical images most neighborhood pixels have the same class and hence by using neighborhood information, effect of noise in segmentation is reduced.

11.3.2.7 Graph cut-based

In this approach, the image segmentation is treated as a graph partitioning problem and global optimization criteria which calculates both total dissimilarity within various groups and also total similarity within them is used. An efficient algorithm uses generalized eigenvalue approach for optimizing the criteria of image segmentation [18].

11.3.2.8 Segmentation for brain with anatomical deviations

The main problem that exists in segmentation of brain images is varying anatomical deviation or abnormalities such as tumors having various shapes, dimensions, locations, and intensities. A deviation not only makes variations in the part of brain where the tumor exists but also affects shape and intensities of other anatomical variations of the brain [19].

11.3.2.9 FFT-based segmentation for brain

Different noises that exists in biomedical images are multiplicative in nature and reductions of those noises is challenging. The anatomical variations must not be compromised because of denoising from the clinical perspective. In the case of the brain, the tumor comprises a dead necrotic part and an edema, which is the active part in the adjacent brain. A radix-4 fast Fourier transform (FFT) recursively partitions a discrete Fourier transform (DFT) into four quarter-length DFTs of groups for every fourth time sample. The cost of computation gets minimized by shorter FFT outputs that can be reutilized for output computation.

11.4 Spatial Domain Segmentation of Brain Tumor Using Multiple Images of Brain MR

Segmentation and accurate detection of a brain tumor are critical tasks in MRI. MRI is an imaging technology used in radiology to picturize detailed internal anatomical structures in different body parts. An MR image is used to differentiate between pathologic tissue, like a brain tumor, and other tissue, like white matter and gray matter of the brain [20]. MRI applies powerful fields of magnets and non-ionizing emissions in the RF range, unlike CT scans and X-rays, which use different radiation that is ionizing in nature. MRI generates a high-contrast image, showing regular and abnormal tissues, that is used to differentiate between overlapping segments.

The objective of image segmentation is to interpret or modify the representation of the image such that it is more impressive, has extra information, and is easier to investigate. Segmentation of image is required to identify objects and their borders [21]. Segmentation is the technique to segregate an image into different discrete segments, i.e., sets of pixels, commonly known as super pixels. It is the process of giving a label to every pixel of the image so that the pixels having the same label share the same visual characteristics. Each pixel in a segment is similar in some computed property or characteristic, like color, intensity, or texture. Adjoining regions are considerably dissimilar in comparison to the similar visual characteristics. The output of segmentation is the set of discrete sections that communally wrap the whole image, or a particular set of contours removed from the image. It is important when working with biomedical images, where diagnosis before and after surgery is essential for the administration of starting and speeding up the improvement process. Labor-intensive segmentation of the abnormal tissue cannot be measured up to with today's high-speed computers that allow us to visually observe the volume and position of unwanted tissues. Computer-assisted identification of unwanted growth of tissues is greatly defined by the necessity of obtaining maximum possible truthfulness. The main objective of the opening of computer systems is accuracy, simplicity, ease of use, and reliability. A biomedical image is taken in biomedical imaging, digitized and progressed for the purpose of segmenting the images and getting necessary information [22]. Hence, there must be well-established algorithms that can precisely classify the border of brain tissues and also minimize the end-user interaction with the system, thereby reducing human error. Moreover, the manual segmentation procedure needs approximately 3 hours to generate output. The traditional algorithms to calculate tumor volume are not well established and is also error prone.

The steps in the proposed algorithm are discussed in the following subsections.

11.4.1 Enhancement of MR image–preprocessing

Here the image is enhanced by executing a 3×3 "unsharp" filter which enhances contrast. It is defined by calculating the negative of a Laplacian filter having parameter α, where α determines the curve of the Laplacian; α will have a value between 0.0 and 1.0. The default value of α is 0.2. A blurrier copy of the image is substracted from the original image in this filtering technique. The double precision floating point is used for computing every element in the output image.

11.4.2 Multiple images registration

Multiple MR images from various adjacent layers are used in this approach. In the preprocessing section, two operations are done. Initially, various MR images get registered in to a base image. Hence, multiple image registration is necessary to implement and then the registered image has to be fused. Because of image diversity that is to be registered and also because of different degradation, it is extremely difficult to develop a generalized algorithm that can be applied for all types of registration. The algorithm must take into consideration not only the type of geometric deformations among the images but also various radiometric deformations and noise corruption effects that require the precision of registration process and data characteristics which are dependent on applications. Most registration algorithms consist of the following steps:

- *Feature detection*: Distinct and salient objects such as contours, boundary regions, line intersections, edge, and corner, etc. that are automatically found by analysis of image histogram and filtering by a spatial mask of 9 $\times$ 9 matrix across the ROI. Then, the feature is designated by the point representative; here, center of gravity, known as control point (CP). A minimum of three control points are designated for this case.

- *Feature matching*: Here, correlation between the feature found in the sensed image and the feature observed in the referenced image are established.
- *Transform model estimation*: The sensed image is aligned with the reference image based on the mapping function parameters that are calculated by the establishing feature correspondence. Here, rigid geometric transformation is implemented. Rotation and translation of the image are performed for aligning with the base image.
- *Image transformation and re-sampling*: Image values in non-integer coordinates. The values are calculated by interpolation. Here, transformation is done using nearest neighbor interpolation.

11.4.3 Registered image fusion

The critical aspect of fusioning images is determining the process of combining sensor images. The naive image fusion technique is to simply take the average pixel gray level value of the source image. The algorithm fuses two or more pre-registered images in one high-quality image. Gray and color image are supported. Here, α factor is changed to alter the degree of fusion of every constituent image. Having $\alpha = 0.5$, both images are fused in equal proportion. Having $\alpha < 0.5$, the background image will have greater impact. Having $\alpha > 0.5$, the foreground image will have higher impact.

11.4.4 Stripping of skull

The skull bone is removed by generating a skull-mask from the MRI. Otsu's algorithm is implemented for image histogram shape-based thresholding [23,24], i.e., reducing the gray level image to the binary image. The Otsu algorithm for thresholding defines that images that need thresholding have two classes of pixel value, i.e., foreground and background. After that, the estimation of the optimum threshold separating these two classes is one such that the intra-class variance is minimal; $\{(weightofforeground * varianceofforeground) + (weightofbackground * varianceofbackground)\}$ is calculated for the image and hence the minimum value is considered for thresholding. After doing complete thresholding, the skull is extracted by the subtraction of the binary image of the skull from the original image.

11.4.5 Segmentation

A three-step refining segmentation algorithm is discussed here. The steps are as follows:

- Initial segmentation using the K-means algorithm
- Grid-based coarse grain localization using local standard deviation
- Grid-based fine grain localization using local standard deviation

It is the clustering problem in which image gray levels is clustered in a pre-defined number of gray levels that are clustered. The most used unsupervised learning methodologies which explains well-known clustering problem is the K-means [25]. The algorithm uses a simple and efficient method to classify a set of data using a finite number of clusters, assuming k clusters, that is predefined. Those centroids should be placed randomly and separated from one another. The challenge lies in labeling k centroids for each cluster. Next for each point the location is calculated for a given dataset and correlated with the adjacent centroid. After this step, k new centroids are recalculated for those cluster resulting from previously. After calculating those k new centroids, a new binding is done among those equivalent data set points and their neighboring new centroids. As an effect of iteration, it is concluded that those k centroids change their position at each iteration until no further changes are made in the subsequent iterations, i.e., centroids do not shift any more [26–28].

Lastly, the algorithm tends to minimize the objective function, using squared error function in which the square is calculated as the distance between a data point and cluster center.

The algorithm executes through these steps:

1. Place k distinct points in original image which are characterized by objects which are getting clustered. Initial group centroids are signified by those points.

2. Every object is assigned to a group by the closest centroid.

3. After the assignment of every object, recalculation of the position of k centroids is executed.

4. Repeat Steps 2 and 3 until the centroids stop movement.

Though the algorithm always converges, the K-means algorithm does not always detect the best case assignment, which is equal to the global minimization objective. The technique is quite effective to the primary randomly selected cluster centers.

Hence, to achieve optimal segmentation cluster configuration we have proposed a grid-based coarse grain localization using local standard deviation, that will be executed after having done segmentation using K-means. The local standard deviation of output of the K-means segmented image is calculated. This image is projected onto a big grid; ideally such a grid is 8×8 pixels. Local standard deviation of every pixel is generated depending on the pixel values of these 64 pixels of the grid. After that, a histogram of every grid is generated and depending on this local standard deviation and histogram in every grid, the boundary of segmentation in those grids is recalculated to give a better result. Selecting a larger grid dimension helps to reduce noise in segmentation. But, on the contrary, the larger grid dimension removes the fine anatomic variations like sharp curves in the boundary region of the tumor or the overlapping area of the gray and white matter of the brain.

Hence, to achieve segmentation with optimal results, we reprocess the output segmented image using the method of grid-based fine-grain localization using local standard deviation. A grid size of 3×3 pixels is chosen for this purpose. Hence, selecting a smaller grid helps to focus on fine anatomical variations of the MR image which requires proper preservation. It restores the minute details of the tumor border and fine analysis of the overlapping region of gray and white matter.

11.4.6 Extracting GM, WM, and tumor–postprocessing

Upon successful execution of the above steps of image segmentation, a histogram is generated of the segmented image. The analysis of this generated histogram depicts distinct peaks in three different image gray levels, corresponding to gray matter (GM), white matter, and the tumor, respectively. Based on the histogram, the subsequent segments of gray matter, white matter (WM), and the tumor are extracted.

11.4.7 Analysis of tumor dimension–postprocessing

Line scan algorithm is used with the derived output of the detected tumor after the image segmentation and the maximum length and width of the tumor along the two axes, i.e., the y-axis and x-axis, are calculated (Figures 11.1 through 11.6).

The output of the test case is that the MR image is in coronal view.

Similar outcomes are available in all the imaging views of MR images. Test cases are executed in over 100 MR images where successful tumor detection was observed in nearly 97% of cases.

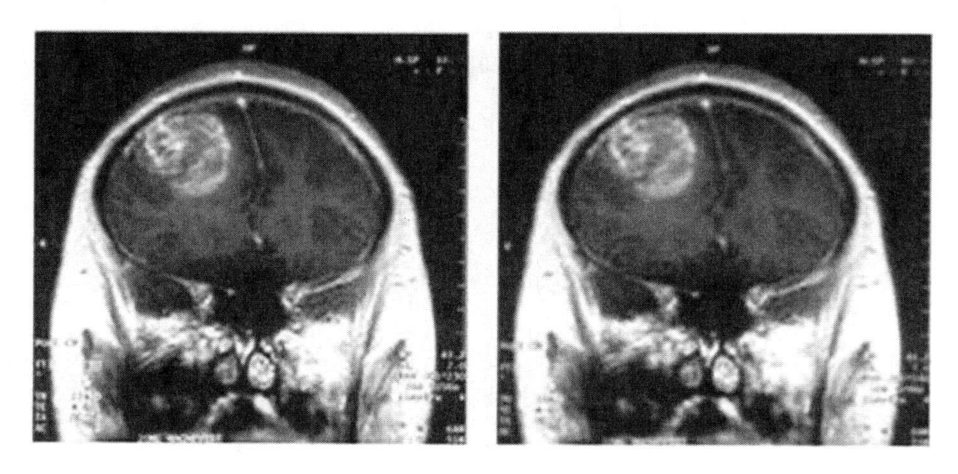

FIGURE 11.1: Input and enhanced image.

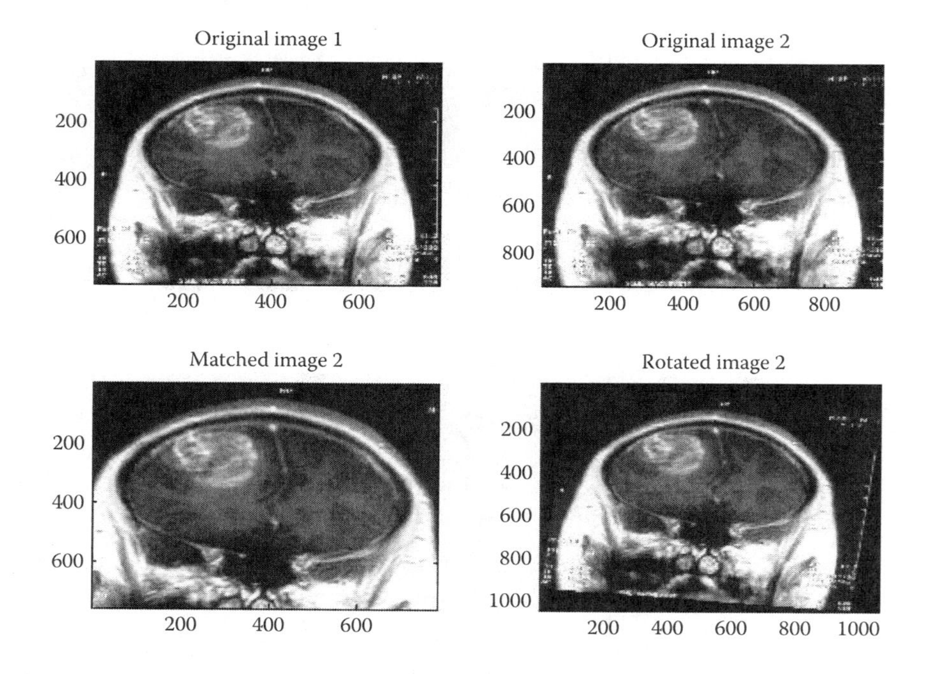

FIGURE 11.2: Registered image.

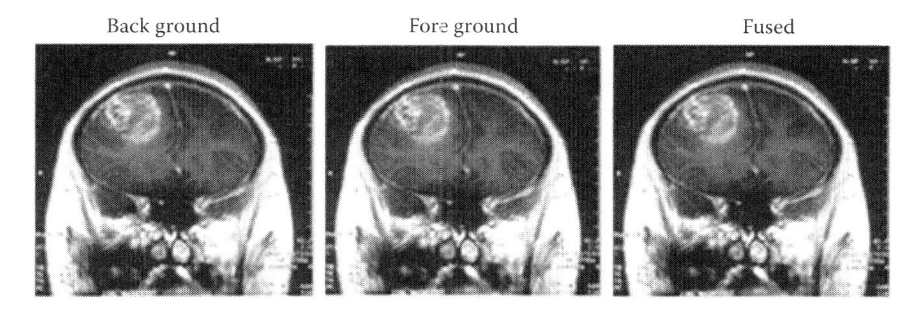

FIGURE 11.3: Fused image.

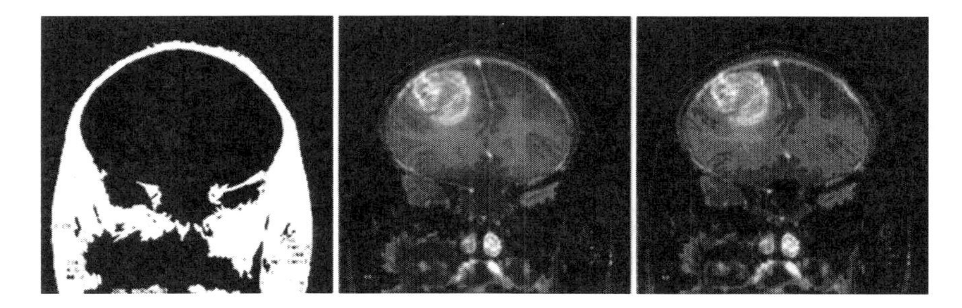

FIGURE 11.4: Skull mask, skull removed, and K-means segmented image.

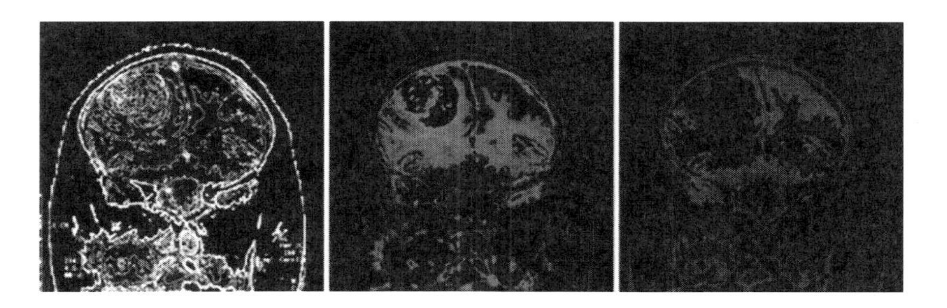

FIGURE 11.5: Local standard deviation, gray matter, white matter.

FIGURE 11.6: Tumor.

This proposed segmentation algorithm is restricted by the constraint that the MRI needs to be of adjacent layers of imaging. Moreover, transformation of the rigid geometric type is used. But in this restricted environment, the registration process generates a satisfactory result. However, this registration could be made more accurate and precise by applying affine transformations. This fusion algorithm gives satisfactory results in fusing multiple MR images. But in certain specific cases it causes loss of intensity. This drawback can be overcome by executing fusion in the frequency domain.

The algorithm for segmentation of the brain proposed here removes the drawbacks of the conventional K-means segmentation method and generates a satisfactory outcome in terms of quantitative and qualitative analysis. The outcomes of the brain segmentation, as shown in Figures 11.1 through 11.6, are in accordance with the latest medical standard. Moreover, the success ratio of segmentation in the brain MR images taken from all the three angles of imaging is quite good and satisfactory. The machine simulation execution time in the test cases, as shown, is less than 12 seconds in most cases, which proves it to be as good as the current industry standard and also much less than the manual process. Hence, it is a noble segmentation algorithm in the family of unsupervised clustering.

11.5 Wavelet-Based Brain Tumor Segmentation Using Fuzzy K-Means in Frequency-Domain

Segmentation of an image may be done using various techniques, such as the edge-detection technique, thresholding technique, region-based technique, clustering technique, and neural network technique. The method of segmentation, which involves classification of objects into certain groups or clusters, depending on some of the specific properties of the objects, is known as clustering. In this technique, initially an attempt is made to extract a vector from local areas of the image. The standard way of clustering is to assign each pixel to the nearest cluster mean. We can classify the clustering algorithm into K-means clustering, hard clustering, fuzzy clustering, etc. [29].

An example of a hard-clustering method (HCM) algorithm is K-means. This process needs a value of either one or zero to every patterned data [30]. Then an initial hard c partition is applied to it c that is, the c center is evaluated and each object is connected to the center which is nearest to it to reduce the within-cluster variance. After each iteration a test is performed to compare the current partition with the former, and if it is found that the difference is lower than a predefined threshold, the iteration stops. The K-means algorithm is a statistical clustering algorithm. The K-means algorithm depends on the index of similarity or dissimilarity among various pairs of data components. This algorithm is quite popular for its simple nature and ease of implementation and it is being widely used to group pixels in the image.

On the other hand, the fuzzy clustering method is a type of clustering algorithm which depicts the relationship between the input data pattern and clusters more naturally. One of the popular clustering algorithm is fuzzy C-means, which is highly effective in the case of spherical clusters. Both these techniques have their own disadvantages as well as advantages. So, we have combined the merits of both the algorithms to create a more powerful algorithm which is known as the wavelet-based fuzzy K-means algorithm. Figure 11.7 depicts the flow diagram of our algorithm [31].

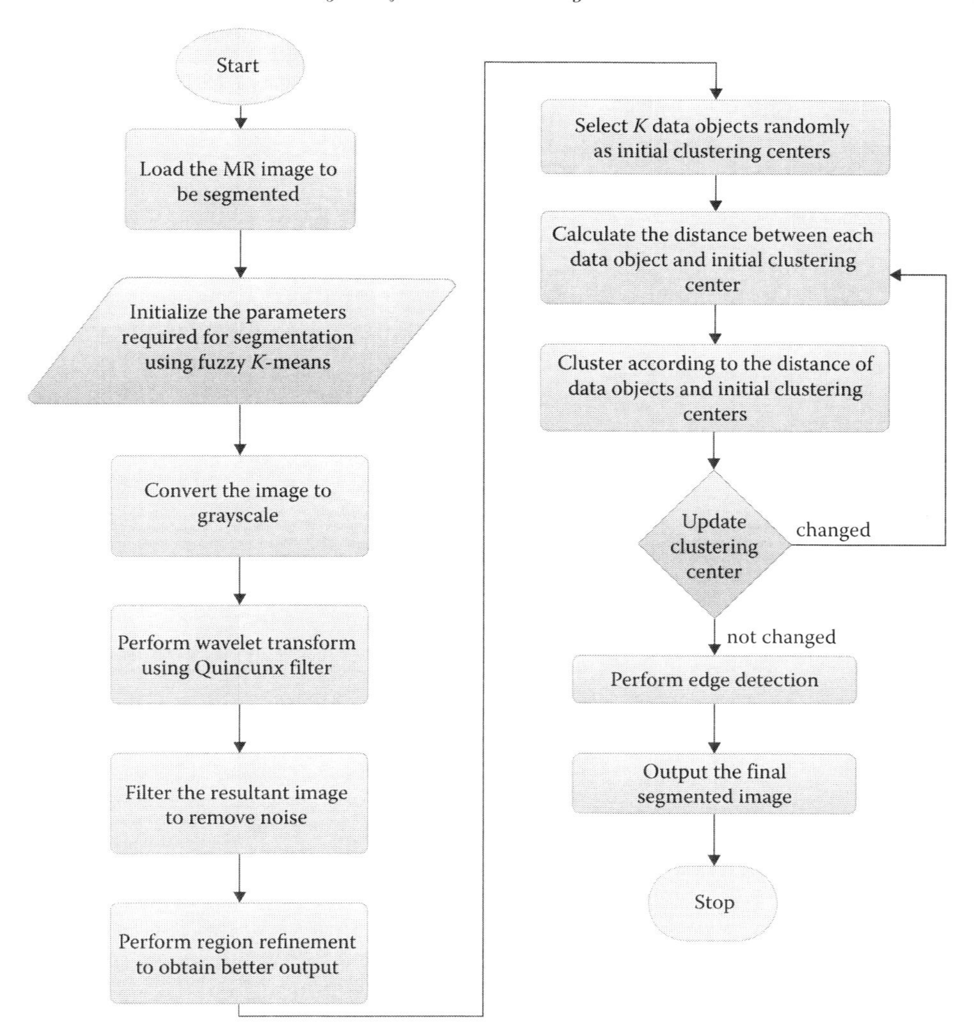

FIGURE 11.7: Flowchart of wavelet-based image segmentation using fuzzy K-means.

11.5.1 Preprocessing

11.5.1.1 Image loading and initialization of different variables and parameters

These are primary variables. The initialization of these variables is very necessary as proper tuning of these variables gives accurate results. That is, if more of the variables are properly tuned, then the tumor will be more efficiently segmented from the image.

A parameter is introduced to check if an image is a 2D image or the image contains rhizome-like objects. Another parameter is initialized to one if the tumor in the MR image is darker than the background; otherwise the value of the parameter is initialized to zero.

To indicate the number of classes we set a variable. Here we use three classes to perform the task. In order to implement the fuzzy logic, we have to set the fuzzy exponent to a value that must be greater than one. Now, to make this algorithm run for a finite number of times it demands a value that would highlight the highest number of iterations of the proposed algorithm. As the algorithm mainly deals with the medical images, it is necessary for it to deal with high-precision value. Hence, we have introduced a parameter or variable whose main job is to control the limit of error or to specify the maximum error value that can be accepted. Here we have evaluated the Mahalanobis distance wherever necessary. The Mahalanobis distance is the multidimensional generalization of the idea of measuring the amount of standard deviation of P from the mean of D. Zero distance denotes that P is the mean of D. As the distance increases, the deviation also increases along with the principal component axis. Rescaling the axes to have unit variance makes the Mahalanobis distance correspond to the standard Euclidean distance. In geometry, measurements of variables are done in the same unit of length. In modeling problems, variables have different scales. The Mahalanobis distance uses the covariance among variables to calculate distances. Hence, the problem of scale and correlation of the Euclidean distance has been overcome. While using the Euclidean distance, all equidistant points from a particular location form a sphere. The Mahalanobis distance corrects the corresponding scales of the variables and accounts for them. In Euclidean geometry, the data is considered to be isotropic Gaussian, treating each of them equally. Whereas, the Mahalanobis distance measures the correlation between the variables, assuming anisotropic Gaussian distribution. If the priorities are known among the features, then using the Mahalanobis distance is a better option. The Z-score feature scaling can overcome the usefulness of choosing a Mahalanobis distance over Euclidean distance. Lastly a variable is used to control the number of cases to find a perfect solution. We have introduced a parameter which determines the nature of the input image and the quantity of noise that is present in it. To mark a low-contrast object another variable or parameter is needed. This parameter is tuned to a low value. In order to control noise we utilize a parameter whose value is one for noisy image, or else it can be initialized to zero.

11.5.1.2 Wavelet transform

After loading the parameters, wavelet transform is performed. Many algorithms are available to convert a signal in discrete wavelet transform (DWT). The oldest algorithm is the Mallat, or pyramidal, algorithm. This algorithm uses two filters: a smoothing filter and a non-smoothing filter. These are manufactured from the wavelet coefficients. The two filters are repeatedly used to get information for every scale. Here, the length of the output array is the same as that of the input array. The outcomes are then generally sorted from large scale to small scale within the output array. Within Gwyddion, the Mallat algorithm is performed to compute the DWT. DWT in 2D can be achieved by using the DWT module. The shape of an object is solely dependent on edges and boundaries. The most important feature of an image is the intersection between surfaces and textures which is necessary for image segmentation. An edge inside an image may be defined as a very well-defined as well as finite collection of contiguous pixels which have shown a rapid change in the intensity with respect to other neighborhood pixels in a single direction. It is a local property of an image. For a nth degree polynomial, it is assumed that the wavelet transform is the same as that of the wavelet transform of a constant value, provided that the Fourier transform of the wavelet has a zero of the order (n+1) at f=0. Hence, the wavelet transform is a powerful tool to identify the singularity of a function and edge detection. Image compression is another application of wavelet transform. Transformation coding is a powerful weapon to remove the spatial redundancy to compress an image so that the transformation coefficients are decorrelated. The multiresolution wavelet disintegration is a mapping onto subspaces spanned by the wavelet basis and scaling function basis. Approximating the signal is the result of the extension on the scaling-function basis. The extension on a wavelet basis gives the disparity between two contiguous resolution levels. Thus, signals are decorrelated and hence compressed. DWT is used to de-noise a noisy signal easily. Within Gwyddion,

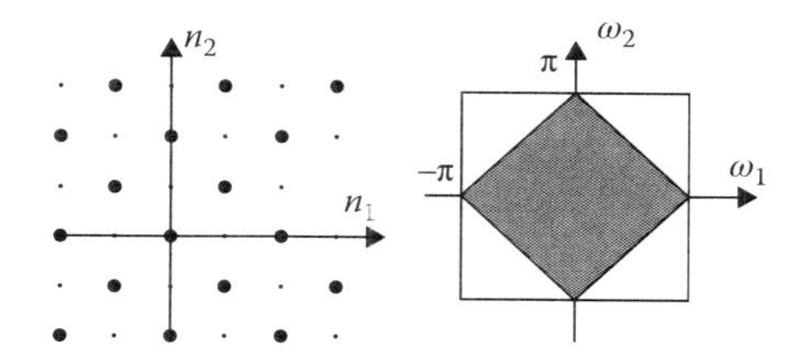

FIGURE 11.8: Lattice structure of Quincunx sampling.

scale adaptive thresholding and the universal thresholding is implemented. The DWT is less complex to compute: it takes O(N) time as compared to FFT that takes O(NlogN) time. For discrete wavelet transform, the lesser time complexity is not for its inherent feature; rather, it indicates the proper selection of logarithmic frequency division, unlike the equispaced frequency divisions of the DFT that need the same basis functions as FFT.

- **Quincux wavelet**

The lattice structure of Quincunx sampling based on some basic outputs is shown in Figure 11.8 [32]:

The initial grid with $\vec{k} = [k_1, k_2]\epsilon Z^2$ represents the discrete signal. The 2D Z-transform of $X[\vec{k}]$ is given by Equation 11.1.

$$X(\vec{z}) = \sum X[\vec{k}]\vec{Z}^{-\vec{k}} \tag{11.1}$$

where, $\vec{Z}^K = Z_1^{K_1} Z_2^{K_2}$

Its continuous time Fourier transform is:

$$X(e^{j\vec{\omega}}) = \sum X[\vec{K}]e^{-j(\vec{\omega},\vec{K})} \tag{11.2}$$

where, $\vec{\omega} = [W_1, W_2]$. The discrete time 2D Fourier transform of $X[\vec{K}]$ on a NxN grid $(k_1, k_2 = 0, 1, 2, 3, 4, \ldots N)$ is:

$$X[\vec{n}] = \sum X[\vec{k}]e^{-j2\pi\frac{(\vec{n},\vec{k})}{N}} \tag{11.3}$$

The Quincunx sampling version of $X[\vec{k}]$ is:

$$[X]_D[\vec{K}] = X[D\vec{K}] \tag{11.4}$$

The link between D and the identity matrix is $D^2 = 2I$. Representation of Equation 11.4 in Fourier domain is:

$$[X]_D[\vec{K}] \implies \frac{1}{2}[X(e^{jD-\tau\vec{\omega})]} + X(e^{j(D^{-\tau}\vec{\omega}+\vec{\pi})} \tag{11.5}$$

where $\vec{\pi} = (\pi, \pi)$.

The definition of up-sampling is given by:

$$[X]_D[\vec{K}] = \begin{cases} X[\vec{D}\ \vec{K}], & \text{where } K_1 + K_2 \text{ is even} \\ 0, & \text{otherwise} \end{cases} \tag{11.6}$$

Its Fourier domain representation is given by:

$$[X]_D[\vec{K}] \implies e^{(jD\tau\vec{\omega})} \tag{11.7}$$

The up-sampling operator and the down-sampling operator on chaining together produce:

$$[X]_D[\vec{K}] = \begin{cases} X[\vec{D}\ \vec{K}], & \text{where } K_1 + K_2 \text{ is even} \\ 0, & \text{otherwise} \end{cases} \tag{11.8}$$

implies

$$\frac{1}{2}[X(e^{j\vec{\omega}} + X(e^{j(\vec{\omega}+\vec{\pi}}))] \tag{11.9}$$

From this equation, the conditions for the perfect reconstruction are:

$$\begin{cases} \tilde{H}(\vec{z})H(\vec{z}) + \tilde{G}(\vec{z})G(\vec{z}) = 2 \\ \tilde{H}(-\vec{z})H(\vec{z}) + \tilde{G}(-\vec{z})G(\vec{z}) = 0 \end{cases} \tag{11.10}$$

The wavelet filter $\tilde{G}$ is an inflect version of the low-pass filter $\tilde{H}$.

The Qunicunx filter is:

$$h_\lambda(e^{j\vec{\omega}}) = \frac{\sqrt{2}(2 + cos\omega_1 + cos\omega_2)^{\frac{\lambda}{2}}}{\sqrt{(2 + cos\omega_1 + cos\omega_2)^\lambda + (2 - cos\omega_1 - cos\omega_2)^\lambda}} \tag{11.11}$$

It is an orthogonal filter for which the λth order zero is at $\omega = \pi$, mapped at $\vec{\omega} = (\pi, \pi)$. This is the absolute condition for 2D wavelet transform of order λ. The orthogonal wavelet filter (obtained by some modulation) is given by:

$$G_\lambda = Z_1 H_\lambda(-\vec{Z}^{-1}) \tag{11.12}$$

The orthogonal scaling function $\psi(x)$ is achieved implicitly as a result of the Quincunx two-scale relation:

$$\psi_\lambda(\vec{X}) = \sqrt{2}\sum h_\lambda[\vec{k}]\psi_\lambda(D_x - \vec{k}) \tag{11.13}$$

The refinement filter is orthogonal with respect to the Quincunx lattice and follows the relation $\vec{\phi}_\lambda(\vec{X})\epsilon L_2(R^2)$. It is quadratic for its integer translates. The partition of the unity condition is fulfilled by it for all $\lambda > 0$, implying the vanishing of the filter at $(\omega_1, \omega_2) = (\pi, \pi)$. Hence the orthogonal Quincunx wavelet is given by:

$$\psi_\lambda(\vec{X}) = \sqrt{2}\sum g_\lambda[\vec{K}]\phi_\lambda(Dx - \vec{K}) \tag{11.14}$$

The wavelet is acting as a λth order differentiator at low frequencies as $\vec{\phi}_\lambda(\vec{x})\alpha \mid \vec{W} \mid^\lambda$. The vanishing moment property, in 2D case, is given by:

$$\int X_1^{n_1} X_2^{n_2} \phi_\lambda(\vec{x})dx \qquad for\, n_1 + n_2 <= [\lambda - 1] \tag{11.15}$$

The center of the wavelet is at $(\frac{1}{2}, \frac{1}{2})$ and as the value of the λ increases the wavelet gets smoother.

11.5.1.3 Filtering

After wavelet transform the resulting image is filtered. During the process of acquisition or transmission MR images are subjected to noise. Noise may also be a result of an imperfect instrument used during processing and compression. Noise in an image may be defined as a random variation of brightness or the color information of the image which is produced by the sensor or the circuitry of the scanner. Noise in MRI mostly follows Rician distribution. It has a non-zero mean which depends only on the local intensity in the image. Also, the Rician noise is signal dependent as well as very problematic in high resolution and low

SNR regime where it creates random fluctuation. It blurs the images and hence reduces the high frequency content of the image such as edge and contours and at the same time alters the intensity values of pixels. Rician noise affects the image in both a quantitative and qualitative manner and thus it provides obstruction to image analysis and interpretation as well as feature detection. In this segmentation technique, the filter which has been used is a 2D Gaussian filter. The purpose of filtering is to remove noise and details [34].

The impulse response of Gaussian filter is given as follows:

$$g(x, y) = \frac{1}{\sqrt{2\pi}} e^{-\frac{1}{2}(\frac{x^2}{\sigma_x{}^2} + \frac{y^2}{\sigma_y{}^2})} \tag{11.16}$$

For a 2D Gaussian filter the frequency response is as follows:

$$H(u, v) = 2\pi\sigma_x\sigma_y e^{-2\pi^2[u^2\sigma_x{}^2 + v^2\sigma_y{}^2]} \tag{11.17}$$

11.5.1.4 Region refinement

The next step is region refinement. During this stage three basic operations are performed:

1. If any holes are detected after the completion of filtering, then they are filled so that the plateau regions get merged into its surrounding object regions.
2. The morphological disks are used for filtering so as to discard any noisy objects.
3. In this process filtering is performed with erosion and then it is followed by dilation, i.e., opening.

This last step is included so as to enhance the quality of the output.

11.5.2 Regions of interest (ROI) detection

11.5.2.1 Clustering using fuzzy K-means algoritm

Clustering is performed using fuzzy K-means logic. The process involved in fuzzy K-means is to primarily partition the data points in such a way such that a given set of vectors gets represented in an improved way. Initially a set of cluster centers is considered randomly and then the above mentioned mapping process is repeated until it satisfies a given condition. When the condition is satisfied the iteration stops. No two clusters can be described by the same cluster center. In order to avoid such a situation, a cluster center must be changed randomly to neglect this type of coincidence in the continual process [35]. Consider that $d_{ij} < \eta$, (where η can be any positive number which is very small in magnitude), then $u_{ij} = 1$ and $u_{i1} = 0$ for $1 \neq j$.

The fuzzy K-means clustering algorithm can now be represented in the following manner:

Step 1: Input a set of random initial clusters centers, say $TC_0 = c_j(0)$, and the value of ϵ, acceptable error limit. Initialize p to 1.

Step 2: With the given cluster center set TC_p, evaluate d_ij for i=1 to N and j= 1 to k. Modify the membership u_{ij} with the help of the following expression:

$$u_{ij} = ((d_{ij})^{\frac{1}{m-1}} \sum_{l=1}^{k} (\frac{1}{d_{ij}})^{\frac{1}{m-1}}) - 1 \tag{11.18}$$

Step 3: Evaluate the new center for each cluster using Equation 11.19, which gives a new set of cluster centers, TC_{p+1}:

$$C_j P = \frac{\sum_{i=1}^{n} u_{ij}^m X_i}{\sum_{i=1}^{n} u_{ij}^m} \tag{11.19}$$

Step 4: If $\mid c_j(p+1) - c_j(p) \mid < \epsilon$ for all values of j then terminate the loop else set $p+1 \implies p$ and go to step 2 [12].

11.5.2.2 Edge detection

The final step of the algorithm is edge detection. Edge detection is the technique to determine and place sharp discontinuities, which is the sharp change in pixel concentration. They determine the object boundaries. First and/or second derivative measurements are used to determine edges. After determining the edge strength, which is the magnitude of the gradient, threshold is applied to determine the presence of edges in the image. Edge detection is generally the first stage of an image interpretation process and is hence a very important step. It locates the pixels in an image that relate to the edges of the object in the image.

Edge detection is performed with the help of the Roberts operator. In this process, the input image is first convolved with the help of the following two kernels:

$$\begin{bmatrix} +1 & 0 \\ 0 & -1 \end{bmatrix}$$

and

$$\begin{bmatrix} 0 & +1 \\ -1 & 0 \end{bmatrix}$$

Consider I(x,y) to be a point in the input image and another point Gx(x,y) in an image which is obtained by convolving the initial image with the first kernel, and Gy(x,y) to be another point in the image which is formed by convolving the original image with the second kernel.

Then we can define the gradient as follows:

$$I(x,y) = G(x,y) = \sqrt{G_x^2 + G_y^2} \tag{11.20}$$

The gradient direction is given by:
$\theta(x,y) = \arctan$

$$\left[\frac{G_y(x,y)}{G_x(x,y)}\right]$$

The Roberts edge filter helps in detecting the edges with the help of a horizontal and vertical filter. These two filters are applied to the input signal and the results are summed up to form the output. The two filters are of the form:
HORIZONTAL FILTER

$$\begin{bmatrix} +1 & 0 \\ 0 & -1 \end{bmatrix}$$

and VERTICAL FILTER

$$\begin{bmatrix} 0 & +1 \\ -1 & 0 \end{bmatrix}$$

For example, if a 2 × 2 window is used as such

$$\begin{bmatrix} p_1 & p_2 \\ p_3 & p_4 \end{bmatrix}$$

where the filter is centered on p_1 with p_2 being pixel[y][x+1] and p_3 being pixel[y+1][x], etc. then the formula to calculate the resulting new p_1 pixel is

$$pixel = abs(p_4 - p_1) + abs(p_3 - p_2) \tag{11.21}$$

which is then converted to the 0–255 range.

The Roberts edge detector is fast in its operation as the filter is small but at the same time it is also subjected to noise which interferes with the output. If the edges of the source image are not very distinct or sharp the filter fails to detect the edge [36].

11.5.3 Final output

After the above process, the final output is obtained, which is much better and more accurate than the tradition methods of segmentation.

Some of the segmented images, which are the outputs of our algorithm, are shown in Figures 11.9 through 11.12. The segmented part is bounded by the fine dark line.

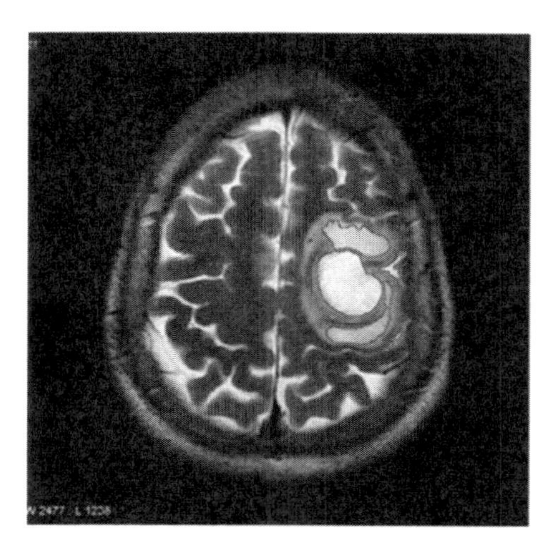

FIGURE 11.9: Output–Case 1.

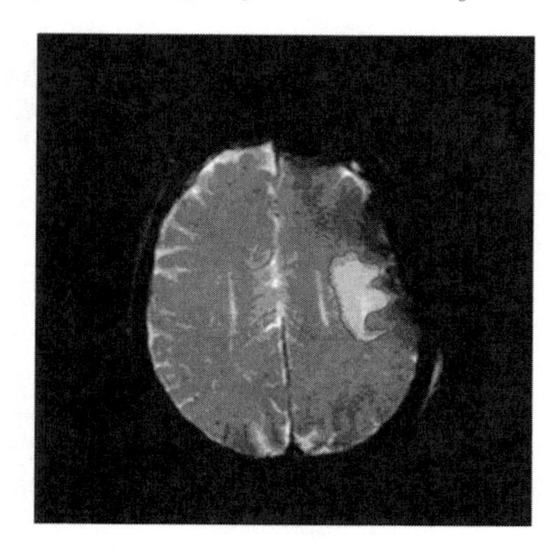

FIGURE 11.10: Output–Case 2.

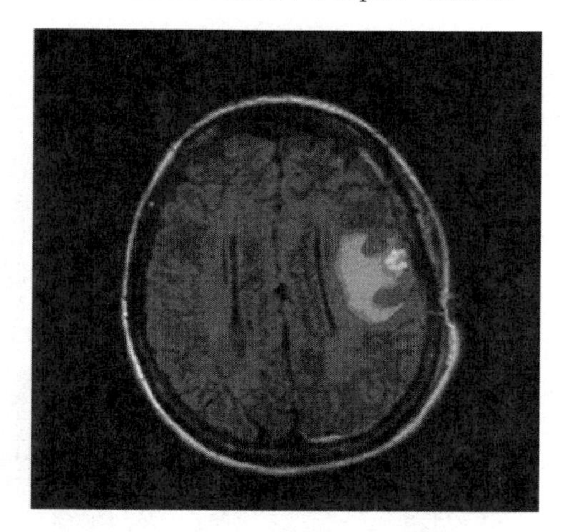

FIGURE 11.11: Output–Case 3.

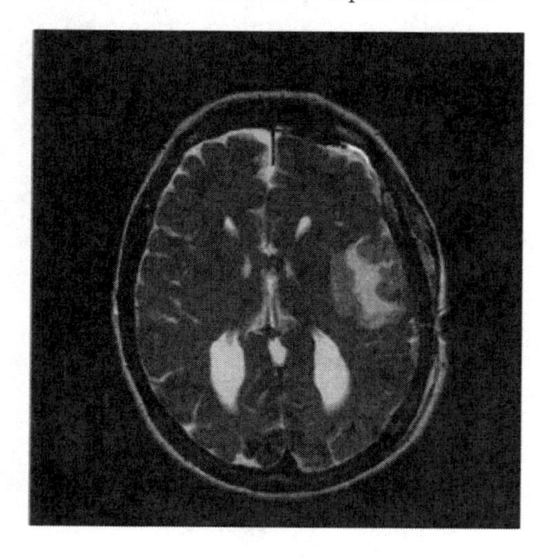

FIGURE 11.12: Output–Case 4.

11.6 Comprehensive Study (Tables 11.1 through 11.3)

11.6.1 Graphical comparison of various parameters of the three algorithms

11.6.1.1 Parameter 1: True positive

The true positive parameter is defined as follows:

$$True\ Positive = \frac{No.\ of\ resulting\ images\ having\ brain\ tumor}{Total\ no.\ of\ images} \tag{11.22}$$

TABLE 11.1: Comparison of outputs of traditional segmentation algorithms with our wavelet-based fuzzy K-means algorithm

Outputs	Fuzzy C-means	K-means	Wavelet-based fuzzy K-means
Output 1	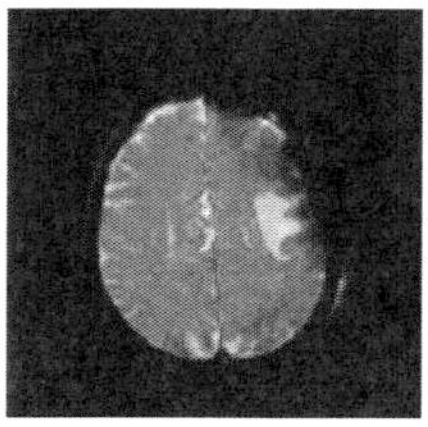	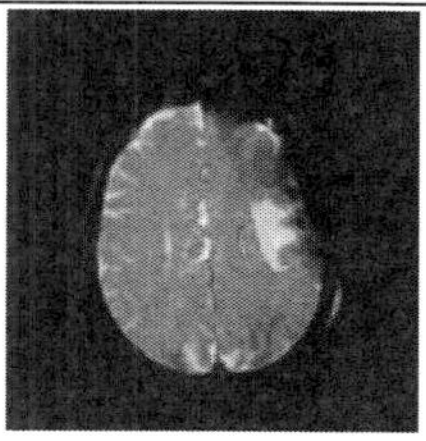	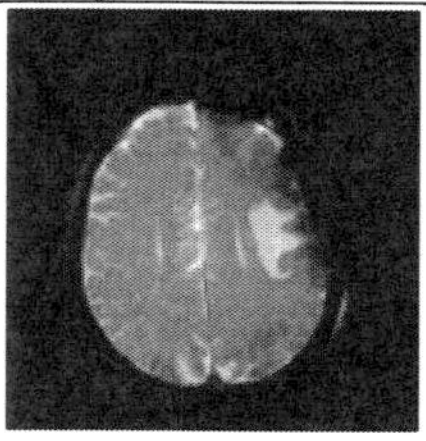
Time Taken	4.891 sec	5.662 sec	6.823 sec
Output 2	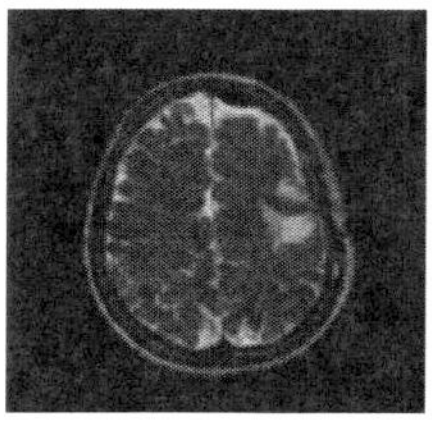	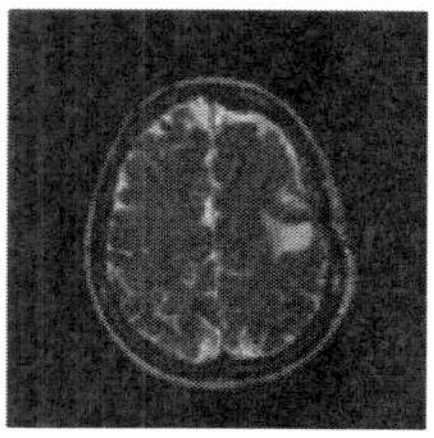	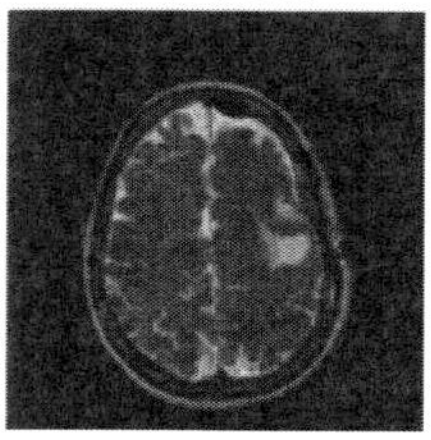
Time Taken	5.141 sec	5.679 sec	6.596 sec
Output 3	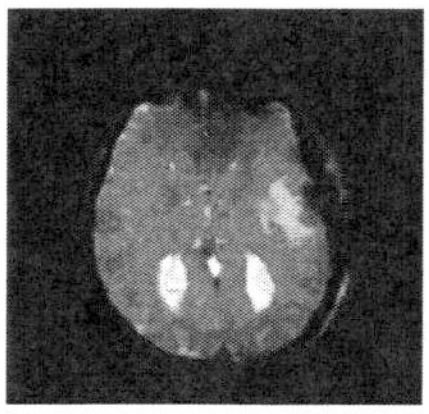	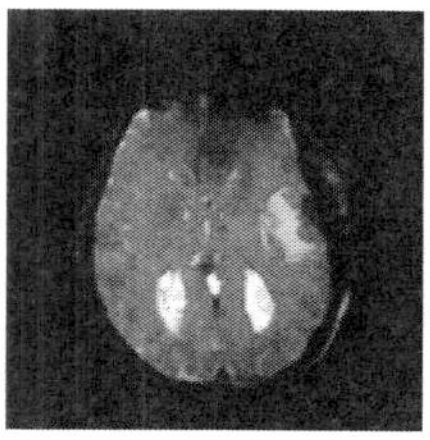	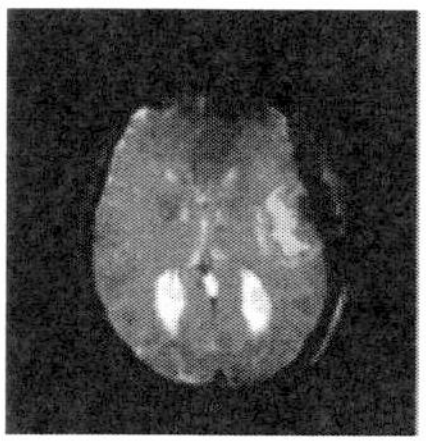
Time Taken	4.797 sec	5.312 sec	6.231 sec
Output 4	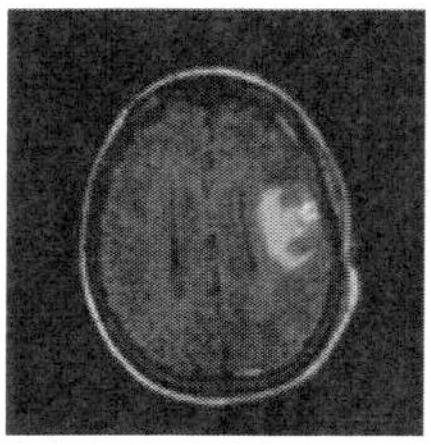	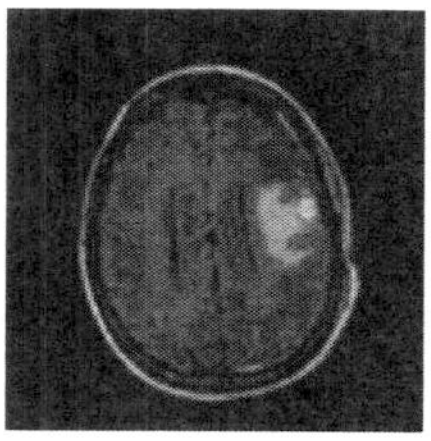	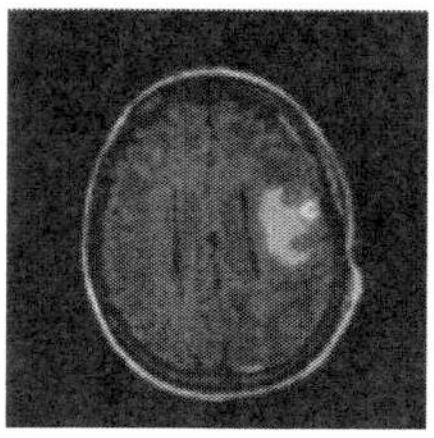
Time Taken	4.912 sec	5.892 sec	6.927 sec

TABLE 11.2: Comparison table for a fixed dataset (40 images)

Parameters	Wavelet-based fuzzy K-means	Advanced K-means*	Fuzzy C-means
True positive (TP)	0.850	0.820	0.830
True negative (TN)	0.900	0.880	0.870
False positive (FP)	0.100	0.120	0.130
False negative (FN)	0.150	0.180	0.170
Precision	0.895	0.872	0.865
Recall	0.850	0.820	0.830
Accuracy	0.875	0.850	0.843

Note: *The proposed modified K-means algorithm

TABLE 11.3: Comparison table for varying dataset

No. of Dataset	Wavelet-based fuzzy K-means				Fuzzy C-means				Advanced K-means			
	TP	TN	FP	FN	TP	TN	FP	FN	TP	TN	FP	FN
40	0.85	0.9	0.1	0.15	0.83	0.89	0.11	0.17	0.82	0.87	0.13	0.18
100	0.92	0.93	0.07	0.08	0.86	0.91	0.09	0.14	0.84	0.9	0.1	0.16
150	0.96	0.95	0.05	0.04	0.9	0.94	0.06	0.1	0.87	0.91	0.09	0.13
200	0.97	0.98	0.02	0.03	0.92	0.96	0.04	0.08	0.91	0.93	0.07	0.09

Source: Aly A. et al., *IEEE Transactions on Medical Imaging*, 2002; Vijay Kumar B. et al. MRI brain image segmentation using modified fuzzy C-means clustering algorithm. In *International Conference on Communication Systems and Network Technologies*, 2011.

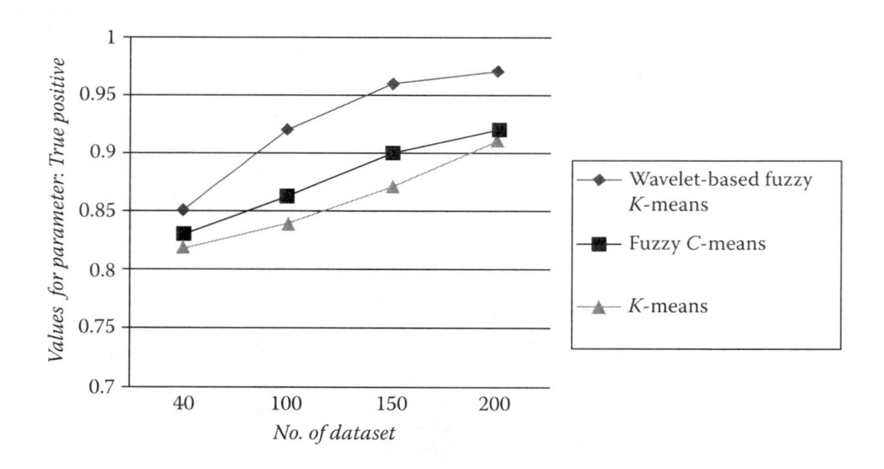

FIGURE 11.13: Comparison–True positive.

In Figure 11.13, we have plotted the various obtained values for the parameter "true positive" (for the three algorithms) for increasing datasets. As the number of datasets are increased the parameter value increases, thereby suggesting that the algorithms work much more efficiently with a wide range of images. However, from the graph it can be seen that the slope is much steeper in the case of our proposed algorithm, i.e., wavelet-based fuzzy K-means, which clearly depicts that the algorithm is superior to the other two.

11.6.1.2 Parameter 2: True negative

The true negative parameter is defined as follows:

$$True\ negative = \frac{No.\ of\ resulting\ images\ without\ brain\ tumor}{Total\ no.\ of\ images} \tag{11.23}$$

In Figure 11.14, we have plotted the various obtained values for the parameter "true negative" (for the three algorithms) for increasing datasets. As the number of datasets increases the parameter value increases, which suggests that the algorithms work much more efficiently with a wide range of images. However, from the graph it can be seen that the slope is much steeper in the case of our proposed algorithm, i.e., wavelet-based fuzzy K-means, which clearly depicts that the algorithm is superior to the other two. It is more successful in differentiating between an image with a brain tumor and an image without it.

11.6.1.3 Parameter 3: False positive

This parameter is defined as follows:

$$False\ Positive = \frac{No.\ of\ resulting\ images\ without\ brain\ tumor\ and\ detected\ positive}{Total\ no.\ of\ images} \tag{11.24}$$

In Figure 11.15, we have plotted the various obtained values for the parameter "false positive" (for the three algorithms) for increasing datasets. As the datasets are increased the value of the parameters decrease, which implies the effectiveness of the algorithm with a wide range of images. The decrement of slope implies the fact that a wide range of datasets reduces the probability of wrong detection, and the definition of false positive suggests the decrement of its value with a wide range of images. However, from the graph it can be seen that the slope is much steeper in the case of our proposed algorithm, i.e., wavelet-based fuzzy K-means, which clearly depicts that the algorithm is much superior to the other two. It is more successful in differentiating an image with brain tumor from an image without it.

11.6.1.4 Parameter 4: False negative

This parameter is defined as follows:

$$False\ Negative = \frac{No.\ of\ resulting\ images\ have\ brain\ tumor\ and\ detected\ negative}{Total\ no.\ of\ images} \tag{11.25}$$

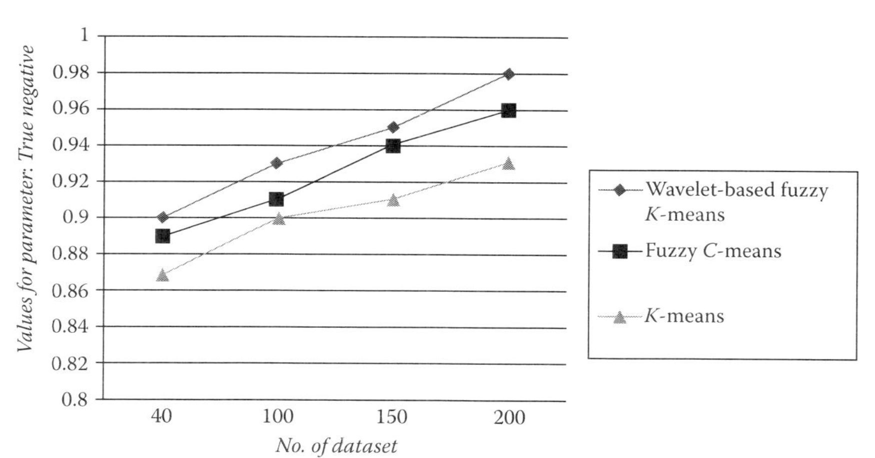

FIGURE 11.14: Comparison–True negative.

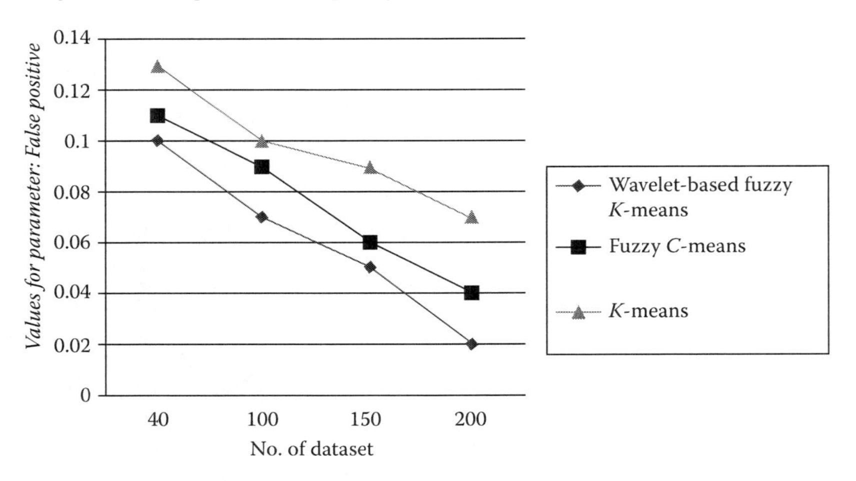

FIGURE 11.15: Comparison–False positive.

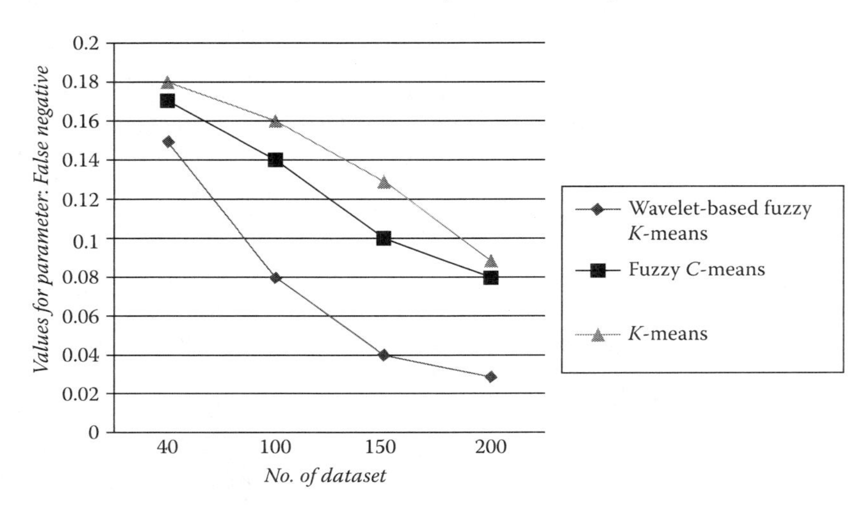

FIGURE 11.16: Comparison–False negative.

In Figure 11.16, we have plotted the various obtained values for the parameter 'false negative' (for the three algorithms) for increasing datasets. As the datasets are increased the value of the parameters decrease, since categorizing a MRI image with tumor or blob-like object from an image without tumor is more readily done with increased dataset. However from the graph it can be seen that the slope is much steeper in the case of our proposed algorithm, i.e., wavelet-based fuzzy K-means, which clearly depicts that the algorithm is superior to the other two.

11.7 Conclusion

In this chapter, we have discussed various imaging modalities of the brain. The main challenge in biomedical image processing of the brain lies in segmenting the images. Many different methods of segmentation have been proposed in recent years but the challenge still remains to increase the efficiency and precision of segmentation. Since the brain is the most complex and sensitive organ of the human body, it requires extremely high precision in segmentation. Two different approaches are discussed here: the first one in spatial domain and the second one in frequency domain. A comparative analysis of the algorithm is also given that concludes that the frequency domain approach is more precise and gives more positive results. There is still room for improvement in segmentation techniques by incorporating more parameters in analyzing.

Glossary

FFT: Fast Fourier transform algorithm computes the discrete Fourier transform (DFT) of a sequence, or its inverse.

Frequency: Frequency is the number of occurrences of a repeating event per unit time.

Image: An image (from Latin: *imago*) is an artifact that depicts visual perception, for example a two-dimensional picture, that has a similar appearance to some subject—usually a physical object or a person, thus providing a depiction of it.

MRI: Magnetic resonance imaging is a non-invasive pathological test that uses a magnetic field and pulses of radio-wave energy to make pictures of organs and structures inside the body.

Segmentation: Image segmentation is the process of partitioning a digital image into multiple segments (sets of pixels, also known as super-pixels).

Wavelet: A wavelet is a mathematical function useful in digital signal processing and image compression.

References

1. Woods R. E. and Gonzalez R. C. *Image Processing*. India: PHI.

2. Bandhyopadhyay S. K. and Paul T. U. Segmentation of brain MRI image – A review. *International Journal of Advanced Research in Computer Science and Software Engineering*, 2, 409–413, 2012.

3. Bandhyopadhyay S. K. and Paul T. U. Automatic segmentation of brain tumour from multiple images of brain MRI. *International Journal of Application or Innovation in Engineering and Management*, 2, 240–248, 2013.

4. Chang P. L. and Teng W. G. Exploiting the self-organizing map for medical image segmentation. In *Twentieth IEEE International Symposium on Computer-Based Medical Systems*, Maribor, Slovenia, pp. 281–288. IEEE, 2007.

5. Hall L. O. et al. A comparison of neural network and fuzzy clustering techniques in segmenting magnetic resonance images of the brain. *IEEE Transactions on Neural Network*, 3, 672–682, 1992.

6. Tian D. and Fan L. A brain MR images segmentation method based on SOM neural network. In *The 1st International Conference on Bioinformatics and Biomedical Engineering*, Wuhan, China, pp. 686–689, 2007.

7. Balafar M. A. et al. New multi-scale medical image segmentation based on fuzzy c-mean (FCM). In *IEEE Conference on Innovative Technologies in Intelligent Systems and Industrial Applications*, Selangor, Malaysia, pp. 66–70, 2008.

8. An S. and An D. Stochastic relaxation, Gibbs distributions, and the Bayesian restoration of images. *IEEE Transactions on Pattern Analysis and Machine Intelligence*, PAMI-6, 672–682, 1984.

9. Tisdall D. and Atkins M. S. MRI denoising via phase error estimation. In *Proceedings of SPIE*, pp. 646–654. SPIE, 2005.

10. Sijbers J. et al. Estimation of the noise in magnitude MR images. *Magnetic Resonance Imaging*, 16, 87–90, 1998.

11. Buades A., Coll B., and Morel J. A non-local algorithm for image denoising. In *IEEE Computer Society Conference on Computer Vision and Pattern Recognition*, San Diego, CA, pp. 60–65, 2005.

12. Lions P. L., Morel J. M., and Coll T. Image selective smoothing and edge detection by nonlinear diffusion. *SIAM Journal of Numerical Analysis*, 29, 182–193, 1992.

13. Pohle R. and Toennies K. D. Segmentation of medical images using adaptive region growing. In *Proceedings of SPIE Medical Imaging*, pp. 1337–1346. SPIE, 2001.

14. Li N., Liu M., and Li Y. Image segmentation algorithm using watershed transform and level set method. In *IEEE International Conference on Acoustics, Speech and Signal Processing*, Honolulu, Hawaii, pp. 613–616. IEEE, 2007.

15. Vincent L. and Soille P. Watersheds in digital spaces: An efficient algorithm based on immersion simulations. *IEEE Transactions on Pattern Analysis and Machine Intelligence*, 13, 583–598, 1991.

16. Yoon S. W. et al. Medical endoscopic image segmentation using snakes. *IEICE Transactions on Information and Systems*, E87-D, 785–789, 2004.

17. Li S. Z. Lecture notes in computer science. In Stan Z. Li (ed.), *Markov Random Field Models in Computer Vision*, pp. 361–370, Tokyo, Japan: Springer 1994.

18. Rajeswari R. and Anandhakumar P. Segmentation and identification of brain tumor MRI image with Radix4 FFT techniques. *European Journal of Scientific Research*, 52, 100–109, 2011.

19. Clark M. C. et al. Automatic tumor segmentation using knowledge-based techniques. *IEEE Transaction on Medical Imaging*, 17, 187–201, 1998.

20. Squire L. F. and Novelline R. A. *Squires Fundamentals of Radiology* (5th ed.). Cambridge, MA: Harvard University Press, 1997.

21. Linda G. S. and George C. S. *Computer Vision*. Upper Saddle River, NJ: Prentice-Hall, 2001.

22. Wareld S. et al. Automatic identication of grey matter structures from MRI to improve the segmentation of white matter lesions. *Journal of Image Guided Surgery*, 1, 326–338, 1995.

23. Sezgin M. and Sankur B. Survey over image thresholding techniques and quantitative performance evaluation. *Journal of Electronic Imaging*, 13, 146–165, 2004.

24. Otsu N. A threshold selection method from gray-level histograms. *IEEE Transactions on Systems, Man, and Cybernetics: Systems*, 9, 62–66, 1979.

25. MacQueen J. B. Some methods for classification and analysis of multivariate observations. In *Proceedings of 5th Berkeley Symposium on Mathematical Statistics and Probability*, pp. 281–297. Berkeley, CA: University of California Press, 1967.

26. Kaur A., Singh G., and Sharma P. Different techniques of edge detection in digital image processing. *International Journal of Engineering Research and Applications*, 39, 458–461, 2013.

27. Aly A. et al. A modified fuzzy C-means algorithm for bias field estimation and segmentation of MRI data. *IEEE Transactions on Medical Imaging*, 89, 193–199, 2002.

28. Vijay Kumar B. et al. MRI brain image segmentation using modified fuzzy C-means clustering algorithm. In *International Conference on Communication Systems and Network Technologies*, 2011.

29. Piatko C. et al. An efficient K-means clustering algorithm: Analysis and implementation. In *IEEE Conference on Computer Vision and Pattern Recognition*, Quebec City, Canada, pp. 881–892. IEEE, 2002.

30. Pakhira M. K. A modified K-means algorithm to avoid empty clusters. *International Journal of Recent Trends in Engineering*, 1, 220–226, 2009.

31. Abdul Nazeer K. A. and Sebastian M. P. Improving the accuracy and efficiency of the K-means clustering algorithm. In *Proceedings of the World Congress on Engineering*, 1, 2009.

32. Van De Ville D. and Feilner M. An orthogonal family of quincunx wavelet with continuously adjustable order. *IEEE Transactions on Image Processing*, 14, 499–510, 2005.

33. Bessaid A. et al. Medical image compression using Quincunx wavelets and SPIHT coding. *Journal of Electrical Engineering and Technology*, 7, 264–272, 2012.

34. Chandel. Image filtering algorithms and techniques: A review. *International Journal of Advanced Research in Computer Science and Software Engineering*, 3, 198–202, 2013.

35. Arthur D. and Vassilvitski S. K-mean++: The advantage of careful seeding. In *Symposium of Discrete Algorithms*, pp. 1027–1035. Philadelphia, PA: Society for Industrial and Applied Mathematics, 2007.

36. Chhabra A. and Chinu. Overview and comparative analysis of edge detection techniques in digital image processing. *International Journal of Information and Computation Technology*, 4, 973–980, 2014.

Chapter 12

Haralick's Texture Descriptors for Classification of Renal Ultrasound Images

Komal Sharma and Jitendra Virmani

12.1 Introduction

The kidney performs many important roles for normal functioning of the human body, such as (a) removing excess water; (b) retaining it when the body needs more; (c) adjusting the levels of other minerals and removing waste products, such as urea and creatinine; (d) making important chemicals called hormones which regulate blood pressure, red blood cell production, and calcium balance in the body. These functions mainly involve the parenchyma region of the kidney. Thus, early diagnosis of renal diseases that hamper the normal functioning of parenchyma is clinically significant to avoid further complications such as renal cancer [1–13]. The real-time, inexpensive, nonradioactive, feasible, and noninvasive nature of ultrasound (US) imaging modality are some of its significant advantages and why it is considered the primary examination for the imaging of soft tissues. However, there are certain associated disadvantages, such as (a) variations associated with interpretation of ultrasound image, (b) motion artifacts, and (c) limitations associated with equipment [14–20]. Addressing these limitations is important for development of an efficient computer-assisted coding (CAC) system based on US imaging modality. Although medical resonance imaging (MRI) and computed tomography (CT) scans offer high sensitivity for diagnosis of renal disease, they are costly and, in CT, the patient is also exposed to ionizing radiation and therefore it is necessary to address the limitations which are posed by the US imaging modality [21–24]. Hence, it is highly desired to design an efficient CAC system for characterization of renal diseases using US images. The need for the benchmark database for renal US images is the requirement of the hour so as to evaluate the soft-computing algorithms for developing computed tomography CAD systems for diagnosis of renal diseases. However, the studies carried out to date, have used the data collected from the hospitals by the efforts of individual research groups.

The brief description of sonographic appearance of these image classes follows:

(a) Normal: A normal bean-shaped kidney has a parenchyma region which is composed of the renal cortex and renal medulla. The outer portion of the kidney forms the cortex, between the renal capsule and renal medullary pyramids. The centrally located hyperechoic region is termed the renal sinus which is mainly composed of renal fat, renal vessels, calices, nerve tissue, and lymphatic channels. The differentiation between the renal sinus and the parenchyma region can be made easily in case of normal US image class as renal sinus region exhibits more echogenicity in comparison to parenchyma region and cortico-medullary differentiation can also be made [4,16,17,20] as shown in Figure 12.1a.

(b) MRD (multicystic renal dysplasia): Any acquired kidney disease may cause renal infection leading to destruction of kidney tissues which may cause renal failure. Renal disorders like polyuria, pyuria, proteinuria, and hematuria may involve diseased glomerulus, nephrons and blood vessels of the parenchyma region. Such disorders that occur in renal parenchyma cannot be diagnosed easily; they are referred to as MRD [1–13]. In MRD US image class, the texture of renal parenchyma region becomes hyperechoic and thus, the differentiation between renal sinus and renal parenchyma cannot be made along with the loss of cortico-medullary differentiation as shown in Figure 12.1b.

(c) Cyst: Renal cysts are regions in the kidney that are filled with fluid, and are frequently observed on US. Renal cysts can be associated with serious disorders that may impair kidney function. They are thin-walled, anechoic fluid-filled regions in the kidney which exhibit posterior acoustic enhancement as shown in Figure 12.1c. A cyst found in kidney can be of any type like simple renal cyst, complex cyst, or polycyst [25,26].

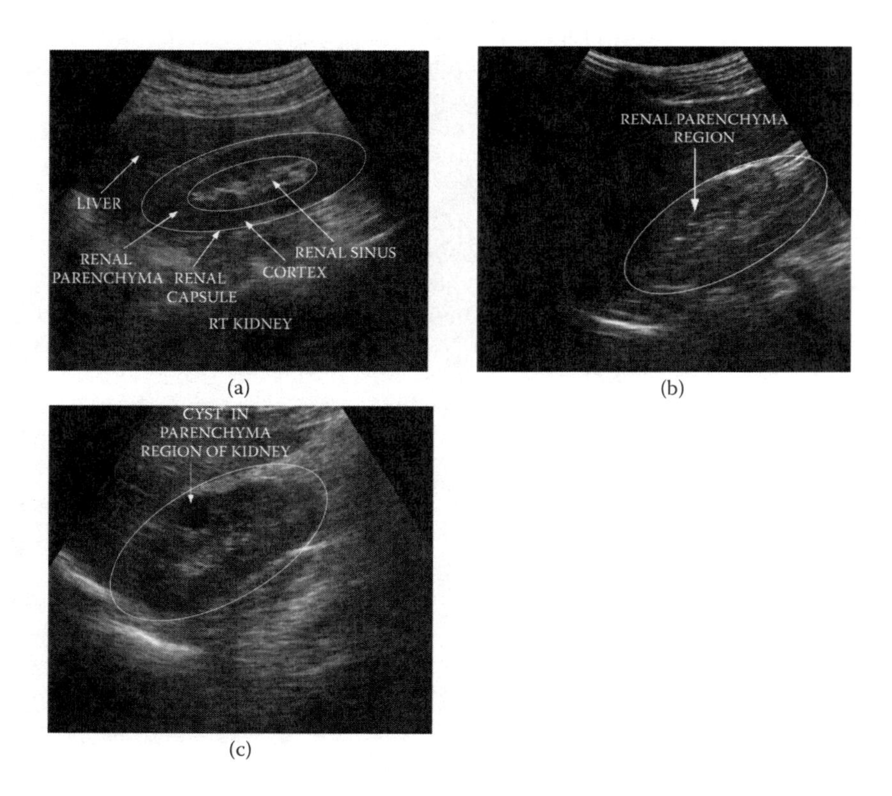

FIGURE 12.1: (a) Sample US image of normal class. *Note*: RT kidney: Right kidney. *Note*: Renal sinus region exhibits more echogenicity in comparison to parenchyma region. (b) Sample US image of MRD class. *Note*: Parenchyma and renal sinus appear to be isoechoic with respect to each other. (c) Sample US image of renal cyst. *Note*: Renal cyst is visible in parenchyma region.

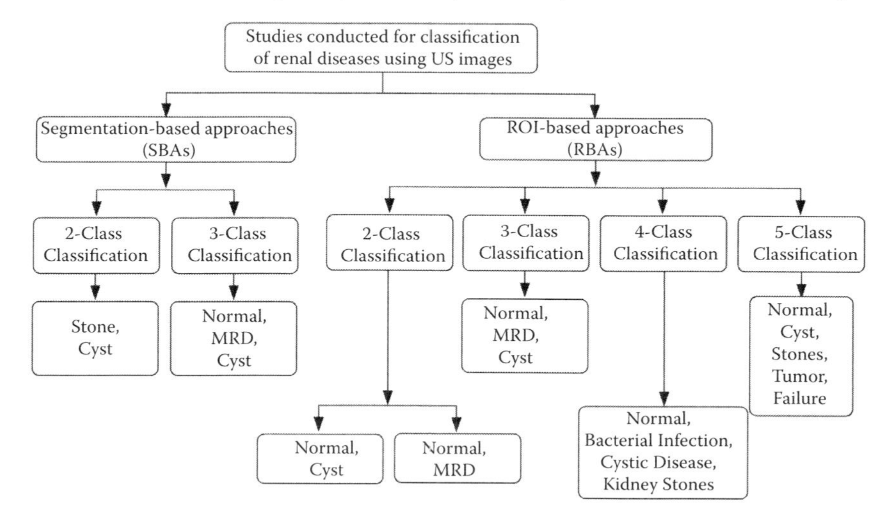

FIGURE 12.2: Brief description of studies conducted for classification of renal diseases on US images.

The relevant studies in the literature for classification of renal US images are few and can be classified into two categories, viz. (a) segmentation-based approaches (SBAs) [27–35] and (b) ROI-based approaches (RBAs) [36–40].

The brief overview of studies conducted for classification of renal diseases on US images is given in Figure 12.2.

(a) **CAC system designs based on segmentation-based approaches (SBAs):** Various researchers in the past have developed different CAC systems for classifying the renal US images into various classes using segmentation-based approaches (SBAs). In SBA, the kidney region is first segmented and the texture features are applied to the segmented kidney region. A brief description of these related studies is given in Table 12.1.

B. K. Raja et al. [27] reported classification of three kidney classes, namely, normal, MRD and cyst, using power spectral features. The results revealed the high content descriptive features that provide a discrete range of values for each kidney class. Such isolated feature values enabled the identification of the kidney categories objectively.

B. K. Raja et al. [28] proposed a set of kidney area independent features for classification of renal US images into normal, MRD, and cyst classes. Six different features, viz. first-order statistics (FOS), multi-scale differential features (MSDF), spatial gray-level dependence matrix (SGLDM), features extracted using fast fourier transform (FFT), and Gabor wavelet transform, were used. The study revealed that the performance of the fuzzy-neural system was better as compared to the multi-layer back propagation network (MBPN). The individual class accuracy achieved by the fuzzy-neural system is 84.6%, 80.7% and 84.6% for normal, MRD, and cyst classes respectively.

B. K. Raja et al. [29] reported classification of normal, MRD, and cyst with statistical texture analysis methods such as first-order statistics (FOS), multi-scale, power spectral features for classification of kidney disorders with US scan. In this study 28 features are extracted. Artificial neural network (ANN) classifier is used for the classification task.

TABLE 12.1: Summary of studies carried out for renal US image classification using SBAs

Author	No. of Classes	No. of Images	Features	Classifiers	Acc.
B. K. Raja et al. [27]	3 (Normal, MRD, Cyst)	150	Power spectral features	–	–
B. K. Raja et al. [28]	3 (Normal, MRD, Cyst)	150	FOS, SGLDM, MI, MSDF, FFT, Gabor	MBPN and HFNN	83.3
B. K. Raja et al. [29]	3 (Normal, MRD, Cyst)	150	FOS, 2nd-order gray-level statistics, multi-scale, power spectral features	ANN and MBPN	87.5
B. K. Raja et al. [30]	3 (Normal, MRD, Cyst)	150	FOS, MI features	–	–
B. K. Raja et al. [31]	3 (Normal, MRD, Cyst)	150	FOS, 2nd-order statistical features, MI, MSDF, power spectral, Gabor features	MBPN and HFNN	94.8
B. K. Raja et al. [32]	3 (Normal, MRD, Cyst)	150	Gabor wavelet features	*k*NN	82.1
K. Dhanalakshmi et al. [33]	3 (Normal, MRD, Cyst)	150	GLCM features	ARCKi	93.0
J. S. Jose, et al. [34]	3 (Normal, MRD, Cyst)	–	Fractal- and histogram-based features	Bayesian	–
K. D. Krishna et. al. [35]	2 (Stone, Cyst)	200	Haralick's- and histogram-based features	SVM with MLP	86.0

Note: SBAs: segmentation-based approaches; Acc.: accuracy; MRD: medical renal disease; FOS: first order statistics; ANN: artificial neural network; MBPN: multi-layer back propagation network; MI: moment invariant; MSDF: multi-scale differential features; FFT: fast fourier transform; HFNN: hybrid fuzzy-neural network; *k*NN: *k*-nearest neighbor; ARCKi: generation of association rules with high confidence for kidney images; SGLDM: special gray-level dependence matrix; SVM: support vector machine; MLP: multi-layer perceptron. The value of accuracy is expressed in percentage.

B. K. Raja et al. [30] reported the analysis of US kidney images for evaluating the tissue characteristic for implementation of an unbiased diagnosis procedure and to classification of important renal disorders. In this study, three kidney classes, namely, normal, MRD, and cyst are considered for the analysis of statistical texture analysis methods and MI (movement invariant) features. In this study 15 features are extracted out of which six features are highly significant.

B. K. Raja et al. [31] proposed an approach for automated diagnosis and classification of US kidney images into normal, MRD, and cyst classes using statistical first- and second-order features, algebraic MI features, multi-scale differential features (MSDF), spectral features estimated by the FFT and dominant Gabor wavelet features. Higher-order spline interpolated contour was used for segmentation of kidney region. The classification efficiency obtained from the fuzzy-neural system is 96.1%, 92.3%, and 96.1% for normal, MRD, and cyst US image classes respectively. Also, it was found that ranking the features enhances the classification accuracy.

B. K. Raja et al. [32] reported classification of three classes, namely, normal, MRD, and cyst for the analysis of dominant Gabor wavelet features by using k-nearest neighbor (kNN) classifier, which provides the classification efficiency of 86.6% for normal, 76.6% for MRD, and 83.3% for cyst.

K. Dhanalakshmi et al. [33] classified US renal images into three classes namely normal, MRD, and cyst by second-order statistical features and feature selection by using a preprocessing solution for association rule generation (PreSAGe) algorithm which performs feature selection and discretization. Association rules with high confidence for kidney images (ARCKi) is a new associative classifier and proves to produce good classification results.

J. S. Jose et al. [34] presented an automated classification of renal US images into normal, MRD, and cyst classes using fractal features; histogram-based features like mean, variance, kurtosis, skewness, energy, and entropy; FOS features like mean, dispersion, variance, average energy, skewness, kurtosis, median and mode; and spatial gray-level dependence matrix (SGLDM) features and Bayesian classifier.

K. D. Krishna et al. [35] proposed a field-programmable gate array (FPGA)–based CAC system for normal and abnormal classes of renal US images. An intensity histogram and Harlick's texture features were used for feature extraction from the segmented region of the kidney. Standard deviation of each metric was calculated for all images. The classification was further extended to classify renal abnormal images into stone and cyst classes using SVM classifier with multilayer perceptron (MLP) kernel that resulted in an accuracy of 86.0%.

(b) **CAC system designs using ROI-based approaches (RBAs):** Different CAC system designs have been proposed in the recent past for classifying the renal US images into different classes using an ROI-based approach. In this approach, the acquired images are manually cropped to extract regions of interest (ROIs) from the kidney.

A brief description of CAC system designs for classification of renal diseases using RBA is given in Table 12.2.

Prema T. Akkasaligar et al. [36] proposed an approach that aimed at classifying the renal US images into two classes, namely, normal and cyst, using three different filters, viz. Gaussian low pass, median filter, and Weiner filter, for preprocessing. Features from gray-level co-occurrence matrix (GLCM) and run length matrix (RLM) were extracted and kNN classifier was used for the classification task.

TABLE 12.2: Summary of studies carried out for renal US image classification using RBAs

Author	No. of Classes	No. of Images	Features	Classifiers	Acc.
Prema T. Akkasaligar et al. [36]	2 (Normal, cyst)	52 [–]	GLCM, RLM	*k*NN	84.0
M. B. Subramanya et al. [37]	2 (Normal, MRD)	35 [32 × 32]	FOS, MI, gradient, RLM, GLCM, Laws' features	SVM	85.8
M. B. Subramanya et al. [38]	3 (Normal, MRD, cyst)	35 [32 × 32]	FOS, gradient, MI, GLCM, RLM features	SVM	86.3
W. M. Hafizah et al. [39]	4 (Normal, Bacterial infection, Cystic disease, Kidney stones)	– [–]	Intensity histogram features, GLCM	–	–
M. W. Attia et. al. [40]	5 (Normal, cyst, Stone, Tumor, Failure)	– [256 × 256]	DWT, GLCM	Neural network classifier	97.0

Note: RBAs: ROI-based approaches; Acc.: accuracy; MRD: medical renal disease; GLCM: gray level co-occurrence matrix; RLM: run length matrix; FOS: first order statistics; MI: moment invariant; SVM: support vector machine; *k*NN: *k*-nearest neighbor; DWT: discrete wavelet transform. The value of accuracy is expressed in percentage.

M. B. Subramanya et al. [37] presented a CAC system design for the classification of normal and MRD classes using B-mode US images and observed that an average classification accuracy of 85.8% has been obtained using gradient and GLCM features together with the SVM classifier.

M. B. Subramanya et al. [38] reported classification of US images into normal, MRD, and cyst classes using texture analysis methods such as statistical features like first-order statistics (FOS), gradient-based features, MI, GLCM, RLM, and Laws' mask features, which have been computed by using SVM classifier.

W. M. Hafizah et al. [39] reported the classification of four classes of renal US images, namely, normal, bacterial infection, cystic disease, and kidney stones, using intensity histogram features and GLCM matrix features. The ROIs were cropped from images in order to remove complicated background. After ROI cropping, contour detection was performed. Results of the study indicated that five texture features including kurtosis, mean, skewness, cluster shade, and cluster prominence dominate over other texture features for classification task.

M. W. Attia et al. [40] proposed an automated method for classification of renal US images into five classes, namely, normal, cyst, stone, tumor, and failure by using a set of statistical and multi-scale wavelet based features.

The present study belongs to a second category, i.e., RBAs, as we have acquired fixed-size ROIs and there are certain limitations of following SBAs, as a separate algorithm is required to segment the entire kidney region. This additional step has been eliminated in the CAC system design that uses RBAs and therefore our study can directly be compared with these studies. Moreover, the present work does not require any preprocessing (i.e., de-speckling or enhancement of US images) prior to feature computation.

The present study is carried out using a fixed-sized ROIs-based approach utilizing Haralick's texture descriptors and SVM classifier, and can be directly compared with the study in [38] as both have been conducted using a fixed-sized ROI-based approach for classification between normal, MRD, and cyst renal US images. It is different than the other fixed-size RBAs, as (a) the ROIs have been extracted carefully from the renal parenchyma, excluding the centrally placed renal sinus region, and the renal sinus portion has been excluded, keeping in view that the renal sinus region bears the same echogenicity in the case of normal as well as MRD US renal images [2,4,16]; (b) the size of ROIs used in the present study is 25×25; and (c) the potential of Haralick's texture descriptors for classification between normal, MRD, and cyst renal US image class has been exhaustively tested using SVM classifier.

12.2 Methodology

In the present work, a CAC system for renal US images is proposed to evaluate the potential of Haralick's texture descriptors exhaustively. The work has been carried out on 35 B-mode renal US images that consists of 11 normal, 8 MRD, and 16 cyst images. Potential of signal processing-based and transform domain Gabor wavelet features is also investigated for comparison purpose. The workflow diagram proposed in the present work is depicted in Figure 12.3.

12.2.1 Image database

The image database consists of 35 B-mode renal US images, including (a) 11 normal, (b) 8 MRD, and (c) 16 cyst images, collected from Biomedical Instrumentation Laboratory,

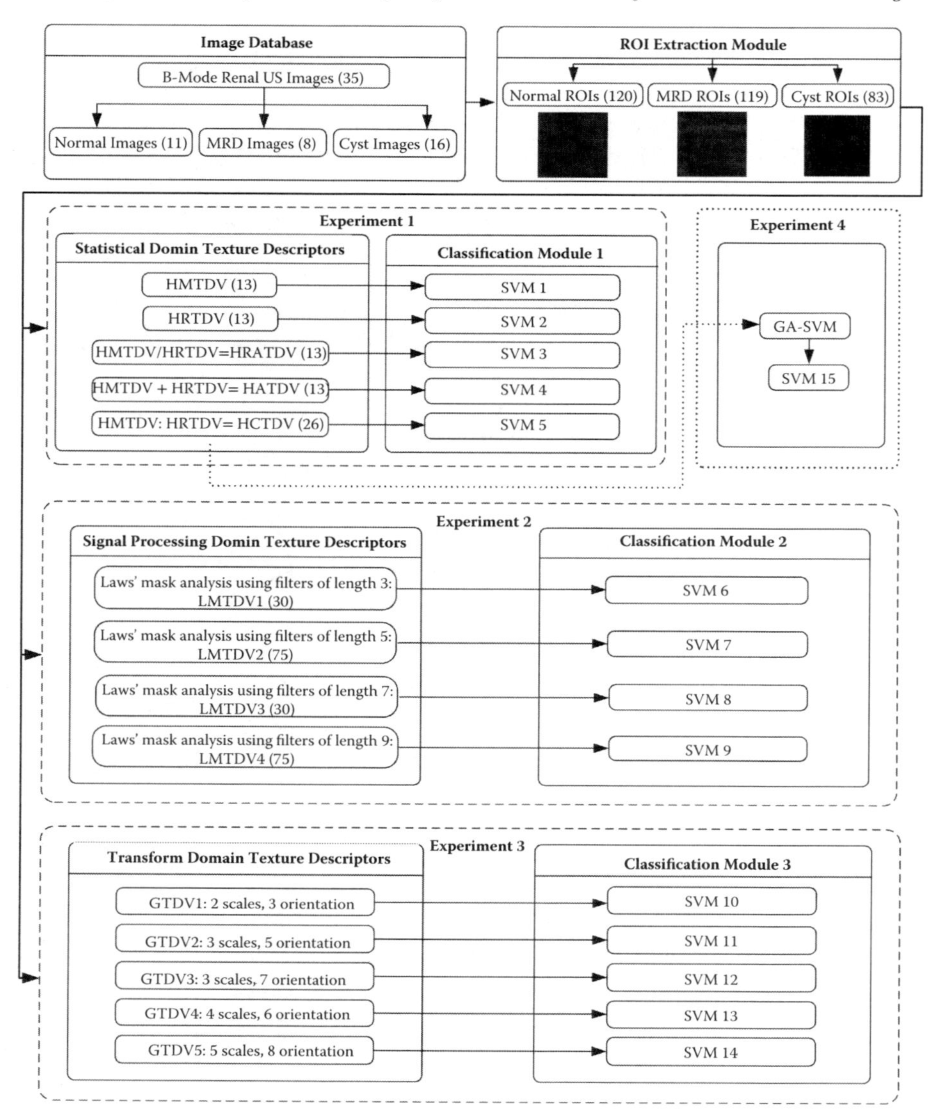

FIGURE 12.3: Experimental workflow. *Note:* HMTDV: Haralick's mean texture descriptor vector; HRTDV: Haralick's range texture descriptor vector; HRATDV: Haralick's ratio texture descriptor vector; HATDV: Haralick's additive texture descriptor vector; HCTDV: Haralick's combined texture descriptor vector; LMTDV: Laws' mask texture descriptor vector; GTDV: Gabor texture descriptor vector; SVM: support vector machine; GA-SVM: genetic algorithm support vector machine; MRD: medical renal disease.

Indian Institute of Technology–Roorkee, Uttarakhand. The protocols followed for data collection are (a) the images are acquired in the longitudinal plane as the longitudinal cross section of kidney includes renal sinus, renal medulla and renal cortex region; (b) the cases of hydro-nephrotic kidney are excluded; and (c) the images of the patients with only MRD and no other renal disorders are included.

These images with 256 gray tones and resolution of 96 dpi have been acquired from 35 patients. A total of 322 ROIs are extracted from 35 renal US (120 normal ROIs from 11 normal kidney images + 119 MRD ROIs from 8 MRD images + 83 cyst ROIs from 16 cyst

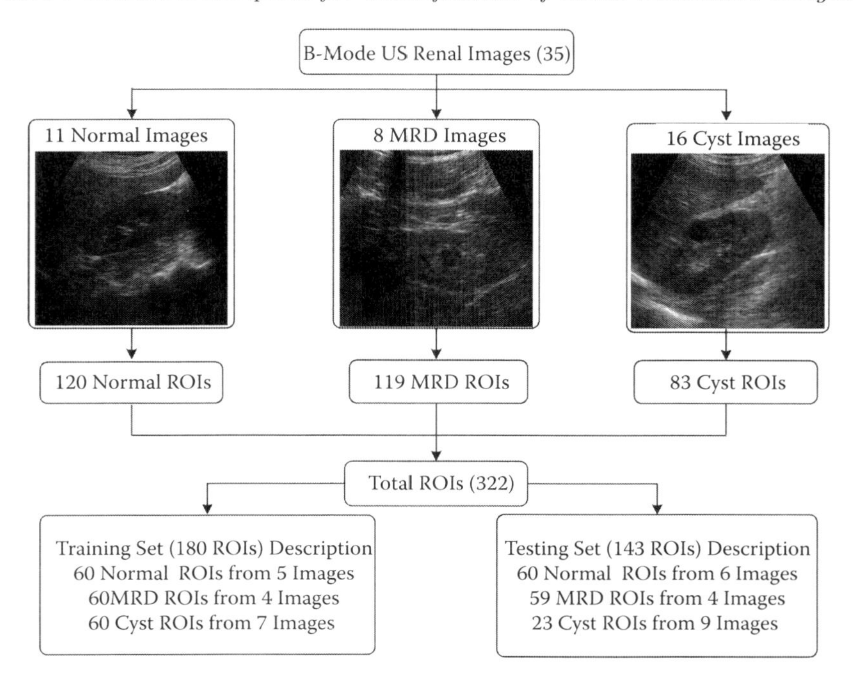

FIGURE 12.4: Dataset description and bifurcation into training and testing sets.

images). Distribution of image database amongst various renal US image classes and its further division into training and testing data set is shown in Figure 12.4.

12.2.2 ROI extraction module

In the ROI extraction module, 322 non-overlapping ROIs of size 25 × 25 are extracted from 35 B-mode renal US images. The protocols followed for ROI extraction are (a) maximum number of non-overlapping ROIs are extracted carefully from the parenchymal region of kidney in case of normal and MRD; (b) the centrally located hyperechoic renal sinus region is excluded; and (c) in the case of a renal cyst, the ROIs are extracted from region inside lesions.

The ROIs are manually extracted from B-mode renal US images. Sample images of ROIs extraction from normal, MRD, and cyst renal US image respectively are shown in Figure 12.5a through c respectively. The dataset consisting of 120 normal ROIs from 11 normal images, 119 MRD ROIs from eight MRD images and 83 cyst ROIs from 16 cyst images, is stored in a PC (Intel Core i5-2450M CPU, 2.5 GHz with 4 GB RAM).

12.2.3 Feature extraction module

In the present study, classification of renal ultrasound images have been carried out by using the following:

(a) statistical texture description using five different Haralick's texture descriptor vectors that include (i) Haralick's mean texture descriptor vector (HMTDV); (ii) Haralick's range texture descriptor vector (HRTDV); (iii) Haralick's ratio texture descriptor vector (HRATDV), formed by taking the ratio of mean and range; (iv) Haralick's additive texture descriptor vector (HATDV), formed by adding

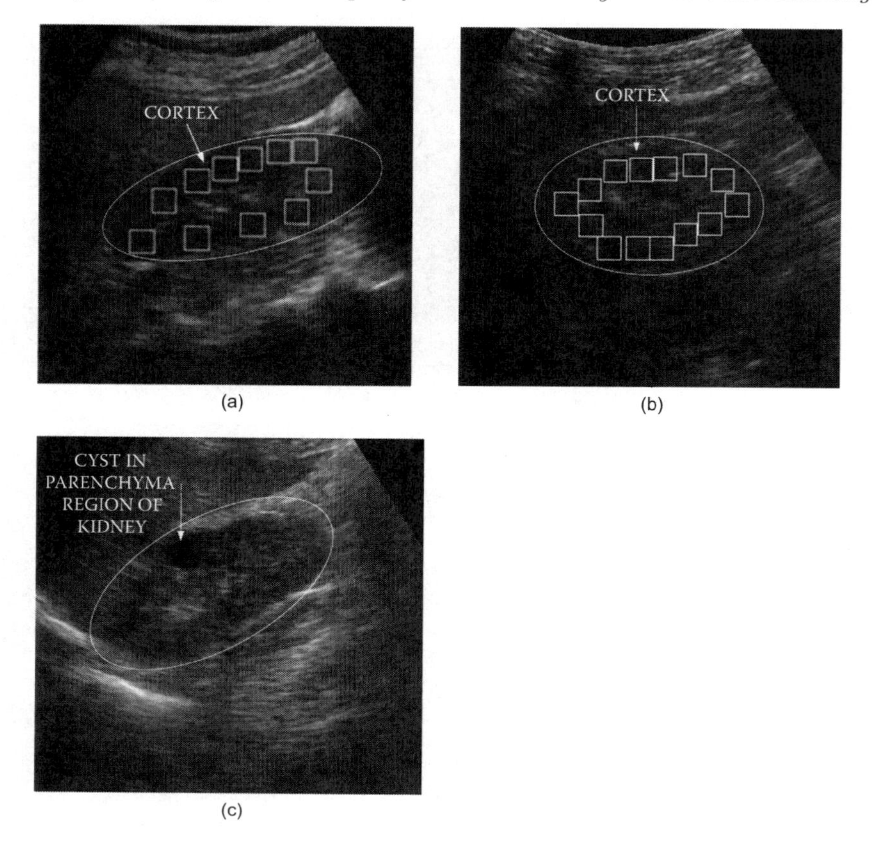

FIGURE 12.5: Sample US image of (a) normal, (b) MRD, and (c) cyst with ROIs marked. *Note*: ROI are extracted from renal parenchyma region in case of normal and MRD class and region inside lesion for cyst class.

mean and range; and (v) Haralick's combined texture descriptor vector (HCTDV), formed by combining mean and range.

(b) Texture description in signal processing-based methods using four different Laws' mask texture descriptor vectors derived from Laws' mask of different resolution, i.e., (i) texture descriptor vector derived from Laws' mask of length 3 (LMTDV1), (ii) texture descriptor vector derived from Laws' mask of length 5 (LMTDV2), (iii) texture descriptor vector derived from Laws' mask of length 7 (LMTDV3), and (iv) texture descriptor vector derived from Laws' mask of length 9 (LMTDV4).

(c) Texture description in transform domain using five different Gabor texture descriptor vectors evaluated at different scales and orientations, i.e., (i) Gabor texture descriptor vectors evaluated at two scales and three orientations (GTDV1), (ii) three scales and five orientations (GTDV2), (iii) three scales and seven orientations (GTDV3), (iv) four scales and six orientations (GTDV4), and (v) five scales and eight orientations (GTDV5).

12.2.3.1 Statistical texture description

Haralick's texture descriptors are derived by considering the spatial relationship that exists between two pixels. The occurrence of different combinations of pixel pairs of a specific gray level in an image is tabulated. It is formed for each ROI image for interpixel distance $d \in 1, 2, 3, \ldots, 10$ and direction $\theta° \in 0°, 45°, 90°, 135°$. The Haralick's texture descriptors (i.e., angular second moment (ASM), contrast, correlation, inverse difference moment, variance, sum average, sum variance, difference variance, entropy, sum entropy, difference entropy, information measures of correlation-1 and information measures of correlation-2) are computed in four directions for each interpixel distance $d \in 1, 2, 3, \ldots 10$.

From the computed four directional Haralick's texture descriptors, the Haralick's mean texture descriptor vector (HMTDV) and Haralick's range texture descriptor vector (HRTDV) are derived for different interpixel distances $d = i \in 1, 2, 3 \ldots, 10$ [39,41–54]. For example, the value of ASM_{mean} and ASM_{range} is computed by using Equations 12.1 through 12.4:

$$ASM_{mean(d=i)} = \frac{ASM_{0°,d=i} + ASM_{45°,d=i} + ASM_{90°,d=i} + ASM_{135°,d=i}}{4} \tag{12.1}$$

$$ASM_{max(d=i)} = max\,(ASM_{0°,d=i}, ASM_{45°,d=i}, ASM_{90°,d=i}, ASM_{135°,d=i}) \tag{12.2}$$

$$ASM_{min(d=i)} = min\,(ASM_{0°,d=i}, ASM_{45°,d=i}, ASM_{90°,d=i}, ASM_{135°,d=i}) \tag{12.3}$$

$$ASM_{range(d=i)} = ASM_{max(d=i)} - ASM_{min(d=i)} \tag{12.4}$$

Here, $ASM_{0°}$, $ASM_{45°}$, $ASM_{90°}$, $ASM_{135°}$ are angular second moments values, calculated at $0°, 45°, 90°$ and 135° respectively. ASM_{max} and ASM_{min} denotes the maximum and minimum values among all the four directions.

The HMTDV and HRTDV are combined in various ways to form three other texture descriptor vectors, namely, (i) Haralick's ratio texture descriptor vector (HRATDV), formed by taking the ratio of mean and range; (ii) Haralick's additive texture descriptor vector (HATDV), formed by adding mean and range; and (iii) Haralick's combined texture descriptor vector (HCTDV), formed by combining mean and range.

12.2.3.2 Texture description in signal processing–based methods

The Laws' texture descriptors are extracted from total of 322 ROIs. In this method, the 2D Laws' masks computed by using 1D filters are convolved with these filters in order to enhance the underlying texture characteristics. These filters perform averaging, edge detection, spot detection, wave detection, and ripple detection to determine the properties of the texture.

Laws' masks analysis using filters of lengths 3, 5, 7 and 9 are used to compute different features forming texture descriptor vectors as (a) LMTDV1 (Laws' mask texture descriptor vector computed using 1D filter of length 3), (b) LMTDV2 (Laws' mask texture descriptor vector computed using 1D filter of length 5), (c) LMTDV3 (Laws' mask texture descriptor vector computed using 1D filter of length 7) and (d) LMTDV4 (Laws' mask texture descriptor vector computed using 1D filter of length 9). Description of Laws' masks features computed using 1D filter of different lengths is given in Table 12.3.

The ROIs are convolved with each of the above 25 (for Laws' 5 and 9) 2D Laws' masks (L5S5 for example) to form texture image (TI_{L5S5}) which is then normalized. These normalized TIs (TI_N) are passed through a 15 × 15 square window to derive 25 texture energy images (TEMs). Out of 25 TEMs, 15 rotationally invariant texture energy images (TR_{L5S5}) are obtained by averaging. Five statistical parameters, i.e., mean, standard deviation, skewness, kurtosis, and entropy, are computed from the derived TRs, thus computing 75 Laws' texture features (15 TRs × 5 statistical parameters) for each ROI [55–60]. An algorithm to carry out Laws' mask analysis using special 1D filters of length 5 is shown in Figure 12.6.

TABLE 12.3: Description of Laws' mask of different lengths

Length of 1D filter	1D filter coefficients	No. of 2D Laws' mask	No. of TR images
3	L3= [1, 2, 1] E3= [-1, 0, 1] S3= [-1, 2, -1]	9	6
5	L5= [1, 4, 6, 4, 1] E5= [−1, −2, 0, 2, 1] S5= [−1, 0, 2, 0, −1] W5= [−1, 2, 0, −2 1] R5= [1, −4, 6, −4, 1]	25	15
7	L7= [1, 6, 15, 20, 15, 6, 1] E7= [−1 −4, −5, 0, 5, 4, 1] S7= [−1, −2, 1, 4, 1, −2, −1]	9	6
9	L9= [1, 8, 28, 56, 70, 56, 28, 8, 1] E9= [1, 4, 4, −4, −10, −4, 4, 4, 1] S9= [1, 0, −4, 0, 6, 0, −4, 0, 1] W9= [1, −4, 4, −4, −10, 4, 4, −4, 1]	25	15
9	R9= [1, −8, 28, −56, 70, −56, 28, −8, 1]		

Note: TR: rotation invariant texture images.

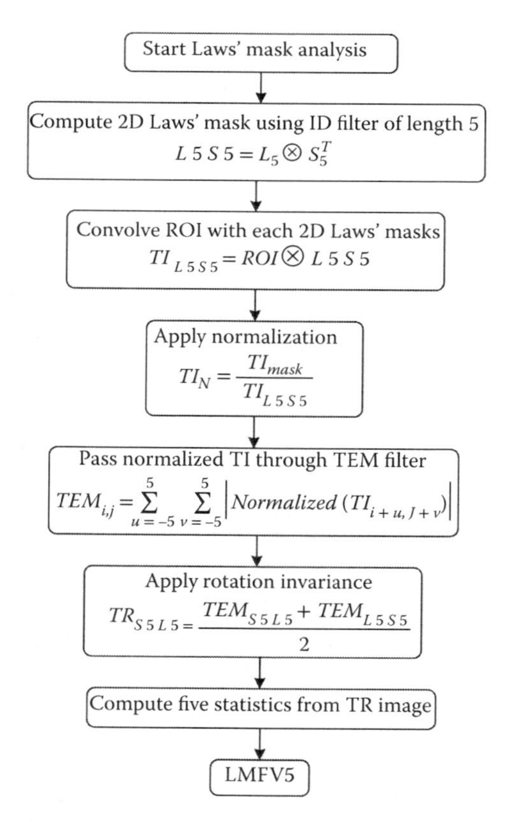

FIGURE 12.6: Laws' mask analysis algorithm using special 1D filters of length 5.

12.2.3.3 Texture description in transform domain

Two-dimensional Gabor Wavelet Transform: A set of frequency and orientation selective filters (i.e., capturing energy at a specific frequency and orientation), are applied to an image that results in 2D Gabor wavelet transform. For instance, if three scales (0, 1 and 2) and seven angles ($22.5°, 45°, 67.5°, 90°, 112.5°, 135°$ and $157.5°$) are considered, then it results in the formation of 21 (7×3) wavelets. Hence, 21 feature images are obtained when these 21 wavelets are convolved with an ROI image. Each of these image represents image information at a specific scale and orientation [32,61–70]. Mean and standard deviation are computed as texture descriptors from these 21 feature images, forming a texture descriptor vector of length 42 (21×2). The real part of 21 wavelets resulting from a 13×13 convolution mask with three scales and seven orientations are shown in Figure 12.7.

Five different Gabor texture descriptor vectors (GTDVs) are evaluated at different scales and orientations, i.e., (a) Gabor texture descriptor vector evaluated at two scales and three orientations (GTDV1), (b) Gabor texture descriptor vector evaluated at three scales and five orientations (GTDV2), (c) Gabor texture descriptor vector evaluated at three scales and seven orientations (GTDV3), (d) Gabor texture descriptor vector evaluated at four scales and six orientations (GTDV4), and (e) Gabor texture descriptor vector evaluated at five scales and eight orientations (GTDV5). A brief description of scales and orientations of GTDVs is given in Table 12.4.

12.2.4 Classification module

SVM Classifier: The goal of the SVM classifier is to design a hyper plane that classifies all training vectors in two classes with the maximum margin. The margin is the distance between the hyper plane and the closest data points from this hyper plane. The SVM classifier is implemented using LibSVM library. In the present work, the Guassian radial basis function kernel has been used for mapping of data points from input space to higher dimensional feature space [36,44,57,71–74]. The optimal values for (a) kernel parameter γ (controls the curvature of the decision boundary) and (b) soft margin constant C (increase the margin with minimum possible error) are required to build an efficient training model for the SVM classifier. In the present work, the potential of different Haralick's texture

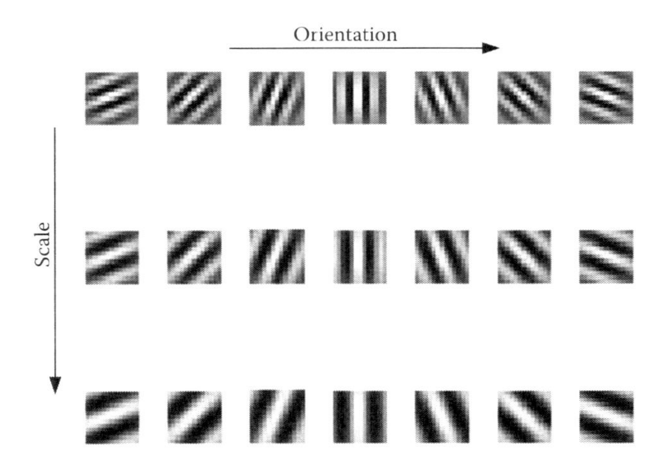

FIGURE 12.7: Real part of Gabor filter family of 21 wavelets evaluated at three scales and seven orientations. *Note*: Orientation: 22.5°, 45°, 67.5°, 90°, 112.5°, 135°, and 157.5° from left to right. Scale: 0, 1, and 2 from top to bottom.

TABLE 12.4: Brief description of scales and orientations used in respective GTDVs

FV	Scale	Orientation
GTDV1	0, 1	45°, 90°, 135°
GTDV2	0, 1, 2	30°, 60°, 90°, 120°, 150°
GTDV3	0, 1, 2	22.5°, 45°, 67.5°, 90°, 112.5°, 135°, 157.5°
GTDV4	0, 1, 2, 3	25.71°, 51.4°, 77.13°, 102.84°, 128.55°, 154.26°
GTDV5	0, 1, 2, 3, 4	20°, 40°, 60°, 80°, 100°, 120°, 140°, 160°

Note: FV: feature vector; GTDV1: Gabor texture descriptor vector evaluated at two scales and three orientations; GTDV2: Gabor texture descriptor vector evaluated at three scales and five orientations; GTDV3: Gabor texture descriptor vector evaluated at three scales and seven orientations; GTDV4: Gabor texture descriptor vector evaluated at four scales and six orientations; GTDV5: Gabor texture descriptor vector evaluated at five scales and eight orientations.

descriptor vectors for different interpixel distances $d \in 1, 2, 3, \ldots, 10$ has been evaluated using SVM classifiers. The optimal values of C and γ for each SVM classifier has been obtained using grid search procedure such that $\gamma \in 2^{-12}, 2^{-11}, \ldots, 2^{5}$ and $C \in 2^{-4}, 2^{-3}, \ldots, 2^{15}$ using 10-fold cross-validation on training data i.e. for $\gamma = 2^{-12}$ the value of C is varied from $2^{-4}, 2^{-3}, \ldots, 2^{15}$ then for $\gamma = 2^{-11}$ the value of C is varied from $2^{-4}, 2^{-3}, \ldots, 2^{15}$. In this way for each value of C and γ in the specified grid, 10-fold cross-validation training accuracy is obtained. The values of C and γ yielding the max training accuracy are used to freeze the training model. The unseen testing data is then projected on the trained model.

12.3 Results and Discussions

Exhaustive experiments are conducted to evaluate the potential of Haralick's texture descriptors. The potential of signal processing-based Laws' mask texture features and transform domain 2D Gabor wavelet features is also evaluated. The results and discussion of all the experiments are given below:

12.3.1 Experiment 1: The classification performance of Haralick's texture descriptors is evaluated in this experiment. The HMTDV, HRTDV, HRATDV, HATDV and HCTDV are subjected to SVM classifier for the classification task

12.3.1.1 Experiment 1(a): Evaluating the classification performance of Haralick's mean texture descriptor vector (HMTDV) using SVM classifier

In this experiment, the classification performance of HMTDV obtained by varying interpixel distance d from 1 to 10 is tested by using the SVM classifier. The brief description of the results obtained is given in Table 12.5.

From Table 12.5, it is observed that a maximum overall classification accuracy (OCA) of 84.5% has been obtained from HMTDV computed at interpixel distance $d = 1$ along with individual class accuracy (ICA) values of 83.3%, 83.0%, and 91.3% for normal, MRD, and cyst class respectively.

TABLE 12.5: Classification performance of HMTDV for d varying from 1 to 10

FV (l)	d	OCA	ICA$_{NOR}$	ICA$_{MRD}$	ICA$_{Cyst}$
HMTDV (13)	1	84.5	83.3	83.0	91.3
	2	83.1	81.6	79.6	95.6
	3	73.9	75.0	64.4	95.6
	4	78.1	75.0	74.5	95.6
	5	81.6	50.8	67.6	95.6
	6	66.9	70.0	52.5	95.6
	7	66.9	56.6	66.1	95.6
	8	76.0	73.3	71.1	95.6
	9	69.7	60.0	69.4	95.6
	10	70.4	61.6	69.4	95.6

Note: FV: feature vector; l: length of feature vector; HMTDV: Haralick's mean texture descriptor vector; d: interpixel distance; OCA: overall classification accuracy; ICA: individual class accuracy; ICA$_{NOR}$: ICA value for NOR class; ICA$_{MRD}$: ICA value for MRD class; ICA$_{Cyst}$: ICA value for cyst class. The OCA and ICA values are expressed in percentage.

TABLE 12.6: Classification performance of HRTDV for d varying from 1 to 10

FV (l)	d	OCA	ICA$_{NOR}$	ICA$_{MRD}$	ICA$_{Cyst}$
HRTDV (13)	1	83.8	88.3	76.2	91.3
	2	78.1	80.0	72.8	86.9
	3	76.7	76.6	71.1	91.3
	4	66.2	63.3	61.0	86.9
	5	66.2	68.3	55.9	86.9
	6	69.0	76.6	54.2	86.9
	7	62.6	75.0	40.6	86.9
	8	64.0	73.3	47.4	82.6
	9	64.0	70.0	52.5	78.2
	10	67.1	83.3	45.7	82.6

Note: FV: feature vector; l: length of feature vector; HRTDV: Haralick's range texture descriptor vector; d: interpixel distance; OCA: overall classification accuracy; ICA: individual class accuracy; ICA$_{NOR}$: ICA value for NOR class; ICA$_{MRD}$: ICA value for MRD class; ICA$_{Cyst}$: ICA value for Cyst class. The OCA and ICA values are expressed in percentage.

12.3.1.2 Experiment 1(b): Evaluating the classification performance of Haralick's range texture descriptor vector (HRTDV) using SVM classifier

In this experiment, the classification performance of HRTDV obtained by varying the interpixel distance d from 1 to 10 is tested by using SVM classifier. The brief description of the results obtained is given in Table 12.6.

From Table 12.6, it is observed that maximum OCA of 83.8% has been obtained from HRTDV computed at interpixel distance $d = 1$ along with ICA values of 88.3%, 76.2%, and 91.3% for normal, MRD, and cyst class respectively.

12.3.1.3 Experiment 1(c): Evaluating the classification performance of Haralick's ratio texture descriptor vector (HRATDV) using SVM classifier

In this experiment, the classification performance of HRATDV obtained by varying interpixel distance d from 1 to 10 is tested by using SVM classifier. The brief description of the results obtained is given in Table 12.7.

TABLE 12.7: Classification performance of HRATDV for d varying from 1 to 10

FV (l)	d	OCA	ICA_{NOR}	ICA_{MRD}	ICA_{Cyst}
HRATDV (13)	1	78.8	73.3	81.3	86.9
	2	78.1	76.6	77.9	82.6
	3	72.5	61.6	77.9	86.9
	4	71.8	70.0	67.7	86.9
	5	66.2	56.6	69.4	82.6
	6	64.7	60	62.7	82.6
	7	58.4	60.0	49.1	78.2
	8	56.3	65.0	38.9	78.2
	9	54.2	45.0	52.2	78.2
	10	63.3	75.0	44.0	82.6

Note: FV: feature vector; l: length of feature vector; HRATDV: Haralick's ratio texture descriptor vector; d: interpixel distance; OCA: overall classification accuracy; ICA: individual class accuracy; ICA_{NOR}: ICA value for NOR class; ICA_{MRD}: ICA value for MRD class; ICA_{Cyst}; ICA value for Cyst class. The OCA and ICA values are expressed in percentage.

From Table 12.7, it is observed that maximum OCA of 78.8% has been obtained from HRATDV computed at interpixel distance $d = 1$ along with ICA values of 73.3%, 81.3%, and 86.9% for normal, MRD, and cyst class respectively.

12.3.1.4 Experiment 1(d): Evaluating the classification performance of Haralick's additive texture descriptor vector (HATDV) using SVM classifier

In this experiment, the classification performance of HATDV obtained by varying the interpixel distance d from 1 to 10 is tested by using SVM classifier. A brief description of the results obtained is given in Table 12.8.

From Table 12.8, it is observed that maximum OCA of 80.2% has been obtained from HATDV computed at interpixel distance $d = 1$ along with ICA values of 81.6%, 74.5%, and 91.3% for normal, MRD, and cyst class respectively.

12.3.1.5 Experiment 1(e): Evaluating the classification performance of Haralick's combined texture descriptor vector (HCTDV) using SVM classifier

In this experiment, the classification performance of HCTDV obtained by varying the interpixel distance d from 1 to 10 is tested by using SVM classifier. The brief description of the results obtained is given in Table 12.9.

From Table 12.9, it is observed that maximum OCA of 85.9% has been obtained from HCTDV computed at interpixel distance $d = 1$ along with ICA values of 86.6%, 83.0%, and 91.3% for normal, MRD, and cyst class respectively.

The result of the Experiment 1 indicates that among the five statistical domain Haralick's texture descriptor vectors, the vector consisting of combination of mean and range texture descriptors (HCTDV) yields the highest overall classification accuracy of 85.9% at $d = 1$. Hence, the detailed result of HCTDV is given in Table 12.10.

TABLE 12.8: Classification performance of HATDV for d varying from 1 to 10

FV (l)	d	OCA	ICA$_{NOR}$	ICA$_{MRD}$	ICA$_{Cyst}$
HATDV (13)	1	80.2	81.6	74.5	91.3
	2	80.2	78.3	76.2	95.6
	3	74.6	65.0	77.9	91.3
	4	71.8	70.0	64.4	95.6
	5	71.8	66.7	67.7	95.6
	6	68.3	63.3	64.4	91.3
	7	72.5	61.6	74.5	95.6
	8	69.7	56.6	74.5	91.3
	9	64.7	53.3	64.4	95.6
	10	61.2	50.0	61.0	91.3

Note: FV: feature vector; l: length of feature vector; HATDV: Haralick's additive texture descriptor vector; d: interpixel distance, OCA: overall classification accuracy; ICA: individual class accuracy; ICA$_{NOR}$: ICA value for NOR class; ICA$_{MRD}$: ICA value for MRD class; ICA$_{Cyst}$: ICA value for Cyst class. The OCA and ICA values are expressed in percentage.

TABLE 12.9: Classification performance of HCTDV for d varying from 1 to 10

FV (l)	d	OCA	ICA$_{NOR}$	ICA$_{MRD}$	ICA$_{Cyst}$
HCTDV (26)	1	85.9	86.6	83.0	91.3
	2	81.6	83.3	89.7	95.6
	3	79.5	78.3	74.5	95.6
	4	70.4	70.0	61.0	95.6
	5	73.2	65.0	72.8	95.6
	6	69.7	63.3	67.7	91.3
	7	65.4	68.3	52.5	91.3
	8	69.7	63.3	67.7	91.3
	9	68.3	58.3	69.4	91.3
	10	59.8	43.3	66.1	86.9

Note: FV: feature vector; l: length of feature vector; HCTDV: Haralick's combined texture descriptor vector; d: interpixel distance; OCA: overall classification accuracy; ICA: individual class accuracy; ICA$_{NOR}$: ICA value for NOR class; ICA$_{MRD}$: ICA value for MRD class; ICA$_{Cyst}$: ICA value for Cyst class. The OCA and ICA values are expressed in percentage.

TABLE 12.10: Classification performance of HCTDV using SVM classifier for three classes of renal US images at $d = 1$

FV (l)		CM			OCA	ICA
		NOR	MRD	CYST		
HCTDV (26)	NOR	52	7	1	85.9	86.6
	MRD	10	49	0		83.0
	CYST	2	0	21		91.3

Note: CM: confusion matrix; FV: feature vector; l: length of feature vector; HCTDV: Haralick's combined texture descriptor vector. The OCA and ICA values are expressed in percentage.

12.3.2 Experiment 2: The classification performance of signal processing based Laws' mask texture descriptors computed using 1D filters of lengths 3, 5, 7 and 9 is evaluated in this experiment. The LMTDV1, LMTDV2, LMTDV3 and LMTDV4 are subjected to SVM classifier for the classification task

12.3.2.1 Experiment 2(a): Evaluating the classification performance of Laws' mask texture descriptor vector computed using 1D filter of length 3 (LMTDV1) using SVM classifier

In this experiment, the classification performance of LMTDV1 is tested by using the SVM classifier. The brief description of the results obtained is given in Table 12.11.

From Table 12.11, it is observed that an OCA of 76.6% has been obtained with 30 Laws' mask texture descriptors obtained using 1D filter of length 3, along with ICA values of 71.6%, 74.5%, and 91.3% for normal, MRD, and cyst renal US image classes respectively.

12.3.2.2 Experiment 2(b): Evaluating the classification performance of Laws' mask texture descriptor vector computed using 1D filter of length 5 (LMTDV2) using SVM classifier

In this experiment, the classification performance of LMTDV2 is tested by using SVM classifier. The brief description of the results obtained is given in Table 12.12.

From Table 12.12, it is observed that an OCA of 79.5% has been obtained with 75 Laws' mask texture descriptors computed using 1D filter of length 5, along with ICA values of 80.0%, 74.5%, and 91.3% for normal, MRD, and cyst renal US image classes respectively.

TABLE 12.11: Classification performance of LMTDV1 using SVM classifier.

FV (l)		CM			OCA	ICA
		NOR	MRD	CYST		
LMTDV1 (30)	NOR	43	17	0	76.6	71.6
	MRD	15	44	0		74.5
	CYST	2	0	21		91.3

Note: CM: confusion matrix; FV: feature vector; *l*: length of feature vector; LMTDV1: Laws' mask texture descriptor vector computed using 1D filter of length 3. The OCA and ICA values are expressed in percentage.

TABLE 12.12: Classification performance of texture features obtained by LMTDV2 using SVM classifier

FV (l)		CM			OCA	ICA
		NOR	MRD	CYST		
LMTDV2 (75)	NOR	48	11	1	79.5	80.0
	MRD	15	44	0		74.5
	CYST	2	0	21		91.3

Note: CM: confusion matrix; FV: feature vector; *l*: length of feature vector; LMTDV2: Laws' mask texture descriptor vector computed using 1D filter of length 5. The OCA and ICA values are expressed in percentage.

12.3.2.3 Experiment 2(c): Evaluating the classification performance of Laws' mask texture descriptor vector computed using 1D filter of length 7 (LMTDV3) using SVM classifier

In this experiment, the classification performance of LMTDV3 is tested by using SVM classifier. The brief description of the results obtained is given in Table 12.13.

From Table 12.13, it is observed that OCA of 71.8% has been obtained with 30 Laws'mask texture descriptors computed using 1D filter of length 7, along with ICA values of 60.0%, 76.2%, and 91.3% for normal, MRD, and cyst renal US image classes respectively.

12.3.2.4 Experiment 2(d): Evaluating the classification performance of Laws' mask texture descriptor vector computed using 1D filter of length 9 (LMTDV4) using SVM classifier

In this experiment, the classification performance of LMTDV4 is tested by using SVM classifier. The brief description of the results obtained is given in Table 12.14.

From Table 12.14, it is observed that an OCA of 71.8% has been obtained with 75 Laws' mask texture descriptors computed using 1D filter of length 9, along with ICA values of 85.0%, 52.5%, and 86.9% for normal, MRD, and cyst renal US image classes respectively.

From Experiment 2, it is observed that, in signal processing-based texture description, among four Laws'mask texture descriptor vectors, the vector computed using Laws'1D filter of length 5 (LMTDV2) yields maximum overall classification accuracy of 79.5% along with individual class accuracy values of 80.0%, 74.5%, and 91.3% for normal, MRD, and cyst image classes respectively.

TABLE 12.13: Classification performance of texture features obtained by LMTDV3 using SVM classifier

FV (l)		**CM**			**OCA**	**ICA**
		NOR	MRD	CYST		
LMTDV3 (30)	NOR	36	24	0		60.0
	MRD	14	45	0	71.8	76.2
	CYST	2	0	21		91.3

Note: CM: confusion matrix; FV: feature vector; *l*: length of feature vector; LMTDV3: Laws' mask texture descriptor vector computed using 1D filter of length 7. The OCA and ICA values are expressed in percentage.

TABLE 12.14: Classification performance of texture features obtained by LMTDV4 using SVM classifier

FV (l)		**CM**			**OCA**	**ICA**
		NOR	MRD	CYST		
LMTDV4 (75)	NOR	51	6	3		85.0
	MRD	28	31	0	71.8	52.5
	CYST	3	0	20		86.9

Note: CM: confusion matrix; FV: feature vector; *l*: length of feature vector; LMTDV4: Laws' mask texture descriptor vector computed using 1D filter of length 9. The OCA and ICA values are expressed in percentage.

12.3.3 Experiment 3: The classification performance of five different Gabor texture descriptor vectors evaluated at different scales and orientations (S, θ), i.e., (2, 3), (3, 5), (3, 7), (4, 6), and (5, 8) is evaluated in this experiment.The GTDV1, GTDV2, GTDV3, GTDV4 and GTDV5 are subjected to SVM classifier for the classification task

12.3.3.1 Experiment 3(a): Evaluating the classification performance of Gabor texture descriptor vector evaluated at two scales and three orientations (GTDV1) using SVM classifier

In this experiment, the classification performance of GTDV1 is tested by using SVM classifier. A brief description of the results obtained is given in Table 12.15.

From Table 12.15, it is observed that an OCA of 69.0% has been obtained with 12 different 2D Gabor features, along with ICA values of 55.0%, 72.8%, and 95.6% for normal, MRD, and cyst renal US image classes respectively.

12.3.3.2 Experiment 3(b): Evaluating the classification performance of Gabor texture descriptor vector evaluated at three scales and five orientations (GTDV2) using SVM classifier

In this experiment, the classification performance of GTDV2 is tested by using SVM classifier. A brief description of the results obtained is given in Table 12.16.

From Table 12.16, it is observed that an OCA of 68.3% has been obtained with 30 different 2D Gabor features, along with ICA values of 53.3%, 72.8%, and 95.6% for normal, MRD, and cyst renal US image classes respectively.

TABLE 12.15: Classification performance of GTDV1 using SVM classifier

FV (l)		**CM**			**OCA**	**ICA**
		NOR	MRD	CYST		
GTDV1 (12)	NOR	33	27	0		55.0
	MRD	16	43	0	69.0	72.8
	CYST	1	0	22		95.6

Note: CM: confusion matrix; FV: feature vector; *l*: length of feature vector; GTDV1: Gabor texture descriptor vector evaluated at two scales and three orientations. The OCA and ICA values are expressed in percentage.

TABLE 12.16: Classification performance of GTDV2 using SVM classifier

FV (l)		**CM**			**OCA**	**ICA**
		NOR	MRD	CYST		
GTDV2 (30)	NOR	32	28	0		53.3
	MRD	16	43	0	68.3	72.8
	CYST	1	0	22		95.6

Note: CM: confusion matrix; FV: feature vector; *l*: length of feature vector; GTDV2: Gabor texture descriptor vector evaluated at three scales and five orientations. The OCA and ICA values are expressed in percentage.

12.3.3.3 Experiment 3(c): Evaluating the classification performance of Gabor texture descriptor vector evaluated at three scales and seven orientations (GTDV3) using SVM classifier

In this experiment, the classification performance of GTDV3 is led by using SVM classifier. A brief description of the results obtained is given in Table 12.17.

From Table 12.17, it is observed that an OCA of 80.2% has been obtained with 42 different 2D Gabor features, along with ICA values of 66.6%, 88.1%, and 95.6% for normal, MRD, and cyst renal US image classes respectively.

12.3.3.4 Experiment 3(d): Evaluating the classification performance of Gabor texture descriptor vector evaluated at four scales and six orientations (GTDV4) using SVM classifier

In this experiment, the classification performance of GTDV4 is tested by using SVM classifier. A brief description of the results obtained is given in Table 12.18.

From Table 12.18, it is observed that an OCA of 65.4% has been obtained with 48 different 2D Gabor features, along with ICA values of 43.3%, 76.2%, and 95.6% for normal, MRD, and cyst renal US image classes respectively.

12.3.3.5 Experiment 3(e): Evaluating the classification performance of Gabor texture descriptor vector evaluated at five scales and eight orientations (GTDV5) using SVM classifier

In this experiment, the classification performance of GTDV5 is tested by using SVM classifier. A brief description of the results obtained is given in Table 12.19.

From Table 12.19, it is observed that an OCA of 73.9% has been obtained with 80 different 2D Gabor features, along with ICA values of 60.0%, 79.6%, and 95.6% for normal, MRD, and cyst renal US image classes respectively.

TABLE 12.17: Classification performance of GTDV3 using SVM classifier

FV (l)		**CM**			**OCA**	**ICA**
		NOR	MRD	CYST		
GTDV3 (42)	NOR	40	20	0		66.6
	MRD	7	52	0	80.2	88.1
	CYST	1	0	22		95.6

Note: CM: confusion matrix; FV: feature vector; *l*: length of feature vector; GTDV3: Gabor texture descriptor vector evaluated at three scales and seven orientations. The OCA and ICA values are expressed in percentage.

TABLE 12.18: Classification performance of GTDV4 using SVM classifier

FV (l)		**CM**			**OCA**	**ICA**
		NOR	MRD	CYST		
GTDV4 (48)	NOR	26	34	0		43.3
	MRD	14	45	0	65.4	76.2
	CYST	1	0	22		95.6

Note: CM: confusion matrix; FV: feature vector; *l*: length of feature vector; GTDV4: Gabor texture descriptor vector evaluated at four scales and six orientations. The OCA and ICA values are expressed in percentage.

TABLE 12.19: Classification performance of GTDV5 using SVM classifier

FV (l)		CM			OCA	ICA
		NOR	MRD	CYST		
GTDV5 (80)	NOR	36	24	0	73.9	60.0
	MRD	12	47	0		79.6
	CYST	1	0	22		95.6

Note: CM: confusion matrix; FV: feature vector; l: length of feature vector; GTDV5: Gabor texture descriptor vector evaluated at five scales and eight orientations. The OCA and ICA values are expressed in percentage.

TABLE 12.20: Comparative analysis of Haralick's, Laws' and Gabor wavelet texture descriptors

Experiments	FV (l)	OCA	ICA_{NOR}	ICA_{MRD}	ICA_{Cyst}
Exp. 1: Haralick's texture descriptors	HCTDV (26)	85.9	86.6	83.0	91.3
Exp. 2: Laws' mask texture features	LMTDV2 (75)	79.5	80.0	74.5	91.3
Exp. 3: 2D Gabor wavelet features	GTDV3 (42)	80.2	66.6	88.1	95.6

Note: FV: feature vector; l: length of feature vector; HCTDV: Haralick's combined texture descriptor vector; LMTDV2: Laws' mask texture descriptor vector computed using 1D filter of length 5; GTDV3: Gabor texture descriptor vector evaluated at three scales and seven orientations; OCA: overall classification accuracy; ICA: individual class accuracy; ICA_{NOR}: ICA value for NOR class; ICA_{MRD}: ICA value for MRD class; ICA_{Cyst}: ICA value for Cyst class. The OCA and ICA values are expressed in percentage.

From Experiment 3, it is observed that out of five different Gabor texture descriptor vectors evaluated at different scales and orientations, Gabor texture descriptor vector evaluated at three scales and seven orientations (GTDV3) yields maximum OCA of 80.2% along with ICA values of 66.6%, 88.1%, and 95.6% for normal, MRD, and cyst renal US image classes respectively.

On conducting exhaustive experiments by using statistical methods based Haralick's texture descriptors, signal processing-based Laws' mask texture features and transform domain 2D Gabor wavelet texture descriptors for feature extraction, a comparison can be drawn between the highest OCAs achieved by the texture features in respective methods. A brief description of the comparative analysis is given in Table 12.20.

From Table 12.20, it is observed that the HCTDV consisting of combination of Haralick's mean and range texture descriptors is most efficient for classification of renal US images into normal, MRD and cyst classes. This HCTDV is further subjected to wrapper-based Genetic algorithm SVM (GA-SVM) method for selection of optimal features.

Considering the above result, one more experiment is conducted.

12.3.4 Experiment 4: Evaluating the classification performance of Haralick's combined texture descriptor vector (HCTDV) using GA-SVM

The classification performance of the prominent features selected from HCTDV is evaluated in this experiment. The HCTDV is subjected to wrapper based GA-SVM technique for feature selection. Classification is performed using SVM classifier.

GA-SVM: GA-SVM is a hybrid algorithm that utilizes genetic-algorithm simultaneously in order to optimize the parameters of SVM and select relevant features with classification. The difference between the GA-SVM method and normal GA lies in the ways to design the chromosome.

For application of genetic algorithm, two steps are required, (a) adequate representation, (b) appropriate fitness function. For the purpose of feature selection, binary representation is used to extract the appropriate features. SVM classifier gives the training accuracy which is applied as fitness function [50, 65].

It involves seven main steps:

(i) Random creation of the initial population of possible candidate solution.

(ii) A 48-bit binary mask represents each chromosome where each bit corresponds to a single feature. A zero at any location indicates that the corresponding feature is irrelevant.

(iii) The performance of each individual is measured by the appropriate fitness function.

(iv) Selection of individuals (roulette wheel selection) which are believed to be fit and have high probability to enter into the mating pool.

(v) Recombination of the selected individuals in the mating pool using crossover operator for the generation of next-generation offspring.

(vi) Application of mutation operator to the offspring with low probability to ensure variations in the pool of solutions.

(vii) Evaluation of offspring using fitness function, and those with the higher fitness value are considered to form a new population.

A total of 16 out of 26 Haralick's combined texture descriptors are selected using GA-SVM method. In this experiment, classification performance of these 16 prominent features from HCTDV of length 26 is evaluated. A brief description of the results obtained using GA-SVM is given in Table 12.21.

From Table 12.21, it is observed that on subjecting HCTDV to wrapper-based GA-SVM technique for selection of optimal features, 16 optimal texture features consisting of seven Haralick's mean texture descriptors (ASM, contrast, variance, sum average, sum variance, difference variance, information measures of correlation-1 from mean) and nine Haralick's range texture descriptors (ASM, correlation, variance, inverse difference moment, sum average, sum variance, entropy, difference variance, information measures of correlation-1 from

TABLE 12.21: Classification performance of prominant features from HCTDV using SVM classifier

FV (l)		**CM**			**OCA**	**ICA**
		NOR	MRD	CYST		
HCTDV (16)	NOR	52	7	1		86.6
	MRD	13	46	0	83.8	77.9
	CYST	2	0	21		91.3

Note: CM: Confusion Matrix, FV: feature vector; l: length of feature vector; HCTDV: Haralick's combined texture descriptor vector. The OCA and ICA values are expressed in percentage.

range) yield an OCA of 83.8% with ICA values of 86.6%, 77.9%, and 91.3% for normal, MRD, and cyst classes respectively.

A brief description of all the experiments conducted for classification of renal diseases using Haralick's texture descriptors, signal processing-based Laws' texture descriptors and transform domain 2D Gabor wavelet texture descriptors along with their OCA values are given in Figure 12.8.

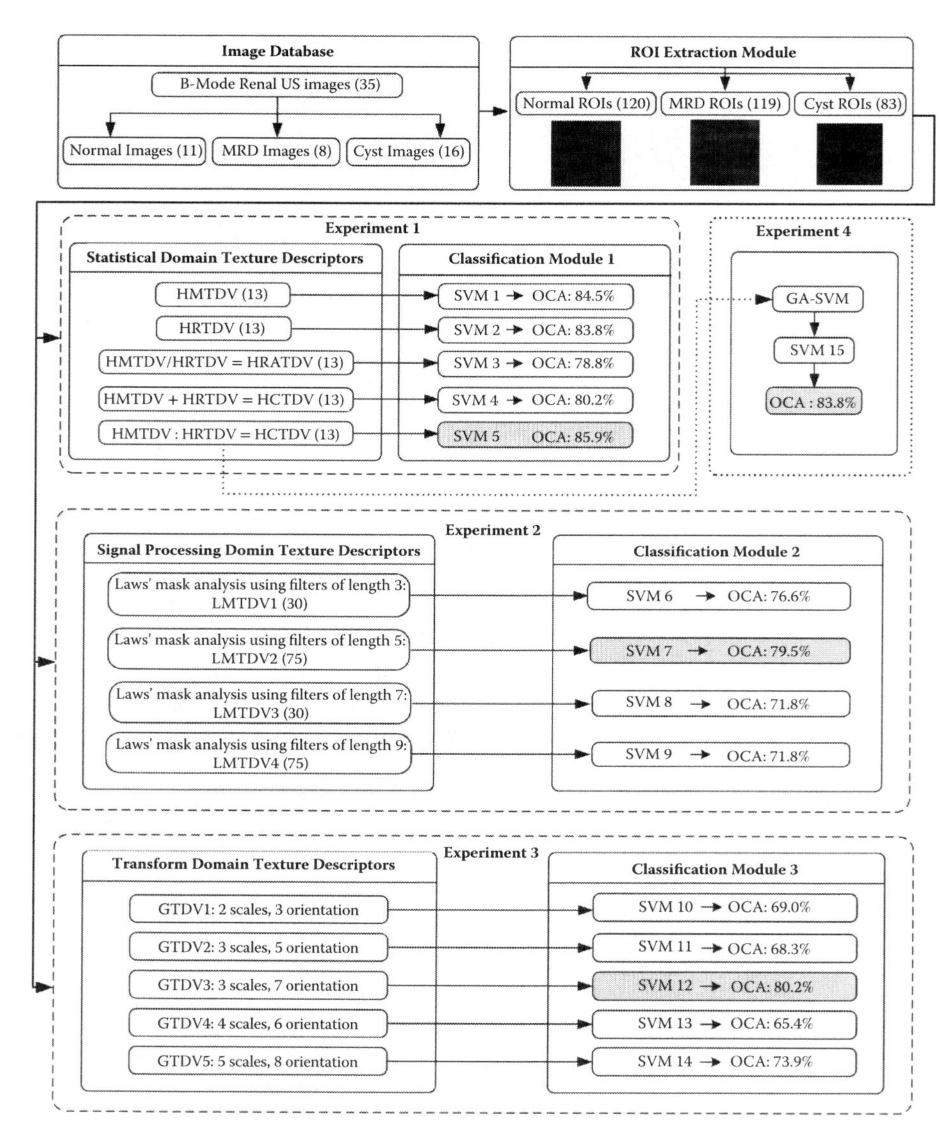

FIGURE 12.8: Experimental workflow with results. *Note:* HMTDV: Haralick's mean texture descriptor vector; HRTDV: Haralick's range texture descriptor vector; HRATDV: Haralick's ratio texture descriptor vector; HATDV: Haralick's additive texture descriptor vector; HCTDV: Haralick's combined texture descriptor vector; LMTDV: Laws' mask texture descriptor vector; GTDV: Gabor texture descriptor vector; SVM: support vector machine; GA-SVM: genetic algorithm support vector machine; MRD: medical renal disease.

12.4 Conclusion

The result of the study indicates that Haralick's combined texture descriptor vector (HCTDV) yields the highest OCA of 85.9% at interpixel distance $d = 1$ with ICA values of 86.6%, 83.0%, and 91.3% for normal, MRD, and cyst classes respectively. Hence, it is observed that the HCTDV is most efficient for classification of renal ultrasound images into normal, MRD, and cyst classes.

It can be concluded that out of 26 Haralick's mean and range texture descriptors, 16 texture descriptors consisting of seven Haralick' s mean texture descriptors (angular second moment (ASM), contrast, variance, sum average, sum variance, difference variance, information measures of correlation-1 from mean) and nine Haralick's range texture descriptors (ASM, correlation, variance, inverse difference moment, sum average, sum variance, entropy, difference variance, information measures of correlation-1 from range) yield an OCA of 83.8% with ICA values of 86.6%, 77.9%, and 91.3% for normal, MRD, and cyst classes respectively. Hence, both Haralick's mean and range texture descriptors are prominent for classification of normal, MRD, and cyst renal US images.

The promising results obtained from the study indicate that the proposed CAC system design using selected Haralick's mean and range texture descriptors can be routinely used in clinical environment for classification of renal diseases. The diagram of the proposed GA-SVM-based CAC system using Haralick's mean and range texture descriptors for diagnosis of renal disease is shown in Figure 12.9.

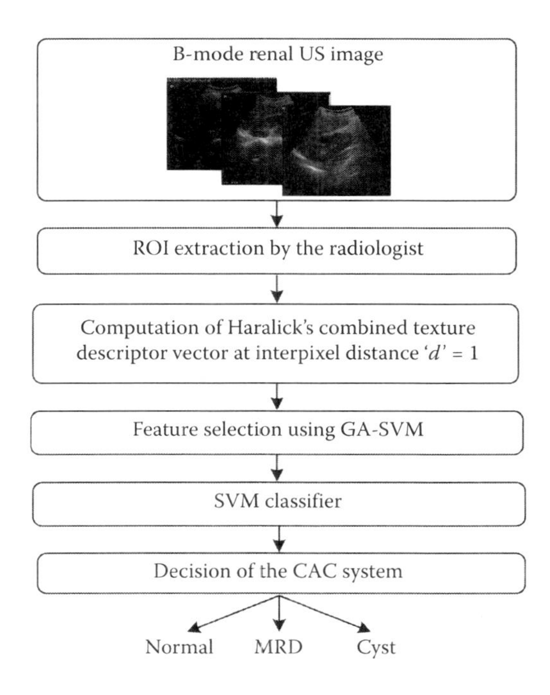

FIGURE 12.9: Proposed GA-SVM-based CAC system for diagnosis of renal diseases. *Note:* GA-SVM: Genetic algorithm support vector machine.

Acknowledgment

The authors are thankful to M. Subramanya, biomedical instrumentation laboratory, IIT-Roorkee, Uttarakhand, for providing the labeled renal ultrasound images. The author Komal Sharma is also thankful to Thapar University–Patiala, Punjab, for providing constant patronage and support.

References

1. MR Ghalib, S Bhatnagar, S Jayapoorani, and U Pande. Artificial neural network based detection of renal tumors using CT scan image processing. *International Journal of Engineering and Technology* 6(1):28–35, 2014.

2. E Quaia and M Bertolotto. Renal parenchymal diseases: Is characterization feasible with ultrasound? *European Radiology*, 12(8):2006–2020, 2002.

3. H Zarnow and LR Bigongiari. Parenchymal diseases of the kidneys. In Lang, EK (ed.), *Radiology of the Upper Urinary Tract*, pp. 71–101. London, UK: Springer, 1991.

4. DK Huntington, SC Hill, and MC Hill. Sonographic manifestations of medical renal disease. *Seminars in Ultrasound, CT, and MR*, 12:290–307, 1991.

5. H Rayner, M Thomas, and D Milford. *Understanding Kidney Diseases*. Springer, 2016.

6. JE Page, SH Morgan, JB Eastwood, SA Smith, DJ Webb, SA Dilly, J Chow, A Pottier, and AEA Joseph. Ultrasound findings in renal parenchymal disease: Comparison with histological appearances. *Clinical Radiology*, 49(12):867–870, 1994.

7. D Soldo, B Brkljacic, V Bozikov, I Drinkovic, and M Hauser. Diabetic nephropathy: Comparison of conventional and duplex doppler ultrasonographic findings. *Acta Radiologica*, 38(2):296–302, 1997.

8. PJ Kenney. Imaging of chronic renal infections. *American Journal of Roentgenology*, 155(3):485–494, 1990.

9. AT Rosenfield and NJ Siegel. Renal parenchymal disease: Histopathologic-sonographic correlation. *American Journal of Roentgenology*, 137(4):793–798, 1981.

10. X Chen, RM Summers, and J Yao. Kidney tumor growth prediction by coupling reaction–diffusion and biomechanical model. *IEEE Transactions on Biomedical Engineering*, 60(1):169–173, 2013.

11. K-C Hung. Echocardiographic characteristics of chronic kidney disease: The Taiwanese experience. *Journal of Medical Ultrasound*, 23(1):14–16, 2015.

12. JK Siddappa, S Singla, SC Mohammed Al Ameen, and N Kumar. Correlation of ultrasonographic parameters with serum creatinine in chronic kidney disease. *Journal of Clinical Imaging Science*, 3:28, 2013.

13. X Chen, R Summers, and J Yao. Fem-based 3-d tumor growth prediction for kidney tumor. *IEEE Transactions on Biomedical Engineering*, 58(3):463–467, 2011.

14. S Strauss, E Gavish, P Gottlieb, and L Katsnelson. Interobserver and intraobserver variability in the sonographic assessment of fatty liver. *American Journal of Roentgenology*, 189(6):W320–W323, 2007.

15. S İçer, A Coşkun, and T İkizceli. Quantitative grading using grey relational analysis on ultrasonographic images of a fatty liver. *Journal of Medical Systems*, 36(4):2521–2528, 2012.

16. R Momenan, RF Wagner, MH Loew, MF Insana, and BS Garra. Characterization of tissue from ultrasound images. *IEEE Control Systems Magazine*, 8(3):49–53, 1988.

17. F Fiorini and L Barozzi. The role of ultrasonography in the study of medical nephropathy. *Journal of Ultrasound*, 10(4):161–167, 2007.

18. K Viswanath and R Gunasundari. Design and analysis performance of kidney stone detection from ultrasound image by level set segmentation and ANN classification. In *ICACCI, 2014 International Conference on Advances in Computing, Communications and Informatics*, Delhi, India, pp. 407–414. IEEE, 2014.

19. NA Patel and PP Suthar. Ultrasound appearance of congenital renal disease: Pictorial review. *The Egyptian Journal of Radiology and Nuclear Medicine*, 45(4):1255–1264, 2014.

20. A Kadioglu. Renal measurements, including length, parenchymal thickness, and medullary pyramid thickness, in healthy children: What are the normative ultrasound values? *American Journal of Roentgenology*, 194(2):509–515, 2010.

21. HC Thoeny, FD Keyzer, RH Oyen, and RR Peeters. Diffusion-weighted MR imaging of kidneys in healthy volunteers and patients with parenchymal diseases: Initial experience 1. *Radiology*, 235(3):911–917, 2005.

22. AW Leung, GM Bydder, RE Steiner, DJ Bryant, and IR Young. Magnetic resonance imaging of the kidneys. *American Journal of Roentgenology*, 143(6):1215–1227, 1984.

23. MG Linguraru, S Wang, F Shah, R Gautam, J Peterson, WM Linehan, and RM Summers. Computer-aided renal cancer quantification and classification from contrast-enhanced CT via histograms of curvature-related features. In *2009 Annual International Conference of the IEEE Engineering in Medicine and Biology Society*, Minneapolis, MN, pp. 6679–6682. IEEE, 2009.

24. SP Raman, Y Chen, JL Schroeder, P Huang, and EK Fishman. CT texture analysis of renal masses: Pilot study using random forest classification for prediction of pathology. *Academic Radiology*, 21(12):1587–1596, 2014.

25. KW Fong, MR Rahmani, TH Rose, MB Skidmore, and TP Connor. Fetal renal cystic disease: Sonographic-pathologic correlation. *American Journal of Roentgenology*, 146(4):767–773, 1986.

26. R Hoffman and J Riley. The diagnostic approach to the parenchymal renal mass. *American Journal of Roentgenology*, 100(3):698–708, 1967.

27. K Bommanna Raja, M Madheswaran, and K Thyagarajah. Ultrasound kidney image analysis for computerized disorder identification and classification using content descriptive power spectral features. *Journal of Medical Systems*, 31(5):307–317, 2007.

28. K Bommanna Raja and M Madheswaran. Determination of kidney area independent unconstrained features for automated diagnosis and classification. In *International Conference on Intelligent and Advanced Systems (ICIAS 2007)*, Kuala Lumpur, Malaysia, pp. 724–729. IEEE, 2007.

29. K Bommanna Raja, M Madheswaran, and K Thyagarajah. Analysis of ultrasound kidney images using content descriptive multiple features for disorder identification and ANN-based classification. In *International Conference on Computing: Theory and Applications (ICCTA'07)*, Kolkata, India, pp. 382–388. IEEE, 2007.

30. K Bommanna Raja, M Madheswaran, and K Thyagarajah. Evaluation of tissue characteristics of kidney for diagnosis and classification using first order statistics and RTS invariants. In *2007 International Conference on Signal Processing, Communications and Networking*, Dubai, United Arab Emirates, pp. 483–487. IEEE, 2007.

31. K Bommanna Raja, M Madheswaran, and K Thyagarajah. A hybrid fuzzy-neural system for computer-aided diagnosis of ultrasound kidney images using prominent features. *Journal of Medical Systems*, 32(1):65–83, 2008.

32. K Bommanna Raja, M Madheswaran, and K Thyagarajah. Texture pattern analysis of kidney tissues for disorder identification and classification using dominant Gabor wavelet. *Machine Vision and Applications*, 21(3):287–300, 2010.

33. K Dhanalakshmi and V Rajamani. An efficient association rule-based method for diagnosing ultrasound kidney images. In *2010 IEEE International Conference on Computational Intelligence and Computing Research (ICCIC)*, Coimbatore, India, pp. 1–5. IEEE, 2010.

34. JS Jose, R Sivakami, N Uma Maheswari, and R Venkatesh. An efficient diagnosis of kidney images using association rules. *International Journal of Computer Technology and Electronics Engineering*, 12(2):14–20, 2012.

35. K Divya Krishna, V Akkala, R Bharath, P Rajalakshmi, AM Mohammed, SN Merchant, and UB Desai. Computer aided abnormality detection for kidney on FPGA based IoT enabled portable ultrasound imaging system. *IRBM*, 37(4):189–197, 2016.

36. PT Akkasaligar and S Biradar. Classification of medical ultrasound images of kidney. In *IJCA Proceedings on International Conference on Information and Communication Technologies (ICICT)*, vol. 3, pp. 24–28, 2014.

37. MB Subramanya, V Kumar, S Mukherjee, and M Saini. Classification of normal and medical renal disease using B-mode ultrasound images. In *2015 2nd International Conference on Computing for Sustainable Global Development (INDIACom)*, BVICAM, New Delhi, India, pp. 1914–1918. IEEE, 2015.

38. MB Subramanya, V Kumar, S Mukherjee, and M Saini. SVM-based CAC system for B-mode kidney ultrasound images. *Journal of Digital Imaging*, 28(4):448–458, 2015.

39. WM Hafizah, E Supriyanto, and J Yunus. Feature extraction of kidney ultrasound images based on intensity histogram and gray level co-occurrence matrix. In *2012 Sixth Asia Modelling Symposium*, Bali, Indonesia, pp. 115–120. IEEE, 2012.

40. MW Attia, FEZ Abou-Chadi, H El-Din Moustafa, and N Mekky. Classification of ultrasound kidney images using PCA and neural networks. *International Journal of Advanced Computer Science and Applications*, 6(4):53–57, 2015.

41. M Bevk and I Kononenko. A statistical approach to texture description of medical images: A preliminary study. In *Proceedings of the 15th IEEE Symposium on Computer-Based Medical Systems, 2002 (CBMS 2002)*, Maribor, Slovenia, pp. 239–244. IEEE, 2002.

42. J Virmani, V Kumar, N Kalra, and N Khandelwal. Neural network ensemble based CAD system for focal liver lesions from B-mode ultrasound. *Journal of Digital Imaging*, 27(4):520–537, 2014.

43. RM Haralick, K Shanmugam, and IHDinstein. Textural features for image classification. *IEEE Transactions on Systems, Man, and Cybernetics*, (6):610–621, 1973.

44. J Virmani, V Kumar, N Kalra, and N Khandelwal. SVM-based characterisation of liver cirrhosis by singular value decomposition of GLCM matrix. *International Journal of Artificial Intelligence and Soft Computing*, 3(3):276–296, 2013.

45. M Amadasun and R King. Textural features corresponding to textural properties. *IEEE Transactions on Systems, Man, and Cybernetics*, 19(5):1264–1274, 1989.

46. CC Gotlieb and HE Kreyszig. Texture descriptors based on co-occurrence matrices. *Computer Vision, Graphics, and Image Processing*, 51(1):70–86, 1990.

47. A Baraldi and F Parmiggiani. An investigation of the textural characteristics associated with gray level cooccurrence matrix statistical parameters. *IEEE Transactions on Geoscience and Remote Sensing*, 33(2):293–304, 1995.

48. J Virmani, V Kumar, N Kalra, and N Khandelwal. Prediction of cirrhosis based on singular value decomposition of gray level co-occurence marix and anneural network classifier. In *Developments in E-systems Engineering (DeSE), 2011*, Dubai, United Arab Emirates, pp. 146–151. IEEE, 2011.

49. C-M Wu, Y-C Chen, and K-S Hsieh. Texture features for classification of ultrasonic liver images. *IEEE Transactions on Medical Imaging*, 11(2):141–152, 1992.

50. J Virmani, V Kumar, N Kalra, and N Khandelwal. Characterization of primary and secondary malignant liver lesions from B-mode ultrasound. *Journal of Digital Imaging*, 26(6):1058–1070, 2013.

51. D Balasubramanian, P Srinivasan, and R Gurupatham. Automatic classification of focal lesions in ultrasound liver images using principal component analysis and neural networks. In *2007 29th Annual International Conference of the IEEE Engineering in Medicine and Biology Society*, Lyon, France, pp. 2134–2137. IEEE, 2007.

52. S Nawaz and AH Dar. Hepatic lesions classification by ensemble of svms using statistical features based on co-occurrence matrix. In *4th International Conference on Emerging Technologies, 2008 (ICET 2008)*, Rawalpindi, Pakistan, pp. 21–26. IEEE, 2008.

53. M Partio, B Cramariuc, M Gabbouj, and A Visa. Rock texture retrieval using gray level co-occurrence matrix. In *Proceedings of 5th Nordic Signal Processing Symposium*, Hurtigruten, Norway, vol. 75. Citeseer, 2002.

54. JR Carr and FP De Miranda. The semivariogram in comparison to the co-occurrence matrix for classification of image texture. *IEEE Transactions on Geoscience and Remote Sensing*, 36(6):1945–1952, 1998.

55. R Shenbagavalli and K Ramar. Classification of soil textures based on laws features extracted from preprocessing images on sequential and random windows. *Bonfring International Journal of Advances in Image Processing*, 1:15, 2011.

56. Kriti and J Virmani. Breast density classification using Laws' mask texture features. *International Journal of Biomedical Engineering and Technology*, 19(3):279–302, 2015.

57. Kriti, J Virmani, N Dey, and V Kumar. PCA-PNN and PCA-SVM based CAD systems for breast density classification. In Hassanien, A-E, Grosan, C, Tolba, MF (eds.), *Applications of Intelligent Optimization in Biology and Medicine*, pp. 159–180. Springer, 2016.

58. HA Elnemr. Statistical analysis of law's mask texture features for cancer and water lung detection. *International Journal of Computer Science Issues*, 10(6):196–202, 2013.

59. AS Setiawan, Elysia, J Wesley, Y Purnama. Mammogram classification using Law's texture energy measure and neural networks. *Procedia Computer Science*, 59:92–97, 2015.

60. OF Ertugrul. Adaptive texture energy measure method. *International Journal of Intelligent Information Systems*, 3(2):13–18, 2014.

61. R Yoshikawa, A Teramoto, T Matsubara, and H Fujita. Automated detection of architectural distortion using improved adaptive Gabor filter. In *International Workshop on Digital Mammography*, Gifu City, Japan pp. 606–611. Springer, 2014.

62. S Banik and RM Rangayyan. Detection of architectural distortion in prior mammograms using Gabor filters, phase portraits, fractal dimension, and texture analysis. *International Journal of Computer Assisted Radiology and Surgery*, 5(4):421–423, 2010.

63. V Van Raad. Design of gabor wavelets for analysis of texture features in cervical images. In *Proceedings of the 25th Annual International Conference of the IEEE Engineering in Medicine and Biology Society, 2003*, Cancun, Mexico, vol. 1, pp. 806–809. IEEE, 2003.

64. RJ Ferrari, RM Rangayyan, JE Leo Desautels, and AF Frere. Analysis of asymmetry in mammograms via directional filtering with Gabor wavelets. *IEEE Transactions on Medical Imaging*, 20(9):953–964, 2001.

65. RM Rangayyan and FJ Ayres. Gabor filters and phase portraits for the detection of architectural distortion in mammograms. *Medical and Biological Engineering and Computing*, 44(10):883–894, 2006.

66. S Lahmiri and M Boukadoum. Hybrid discrete wavelet transform and Gabor filter banks processing for features extraction from biomedical images. *Journal of Medical Engineering*, (2013):1–14, 2013.

67. M Hussain, S Khan, G Muhammad, M Berbar, and G Bebis. Mass detection in digital mammograms using Gabor filter bank. In *IET Conference on Image Processing (IPR 2012)*, London, UK, pp. 1–5. IET, 2012.

68. J Virmani, V Kumar, N Kalra, and N Khandelwal. Prediction of liver cirrhosis based on multiresolution texture descriptors from B-mode ultrasound. *International Journal of Convergence Computing*, 1(1):19–37, 2013.

69. D Dunn, WE Higgins, and J Wakeley. Texture segmentation using 2-D Gabor elementary functions. *IEEE Transactions on Pattern Analysis and Machine Intelligence*, 16(2):130–149, 1994.

70. C-C Lee and S-H Chen. Gabor wavelets and SVM classifier for liver diseases classification from CT images. In *2006 IEEE International Conference on Systems, Man and Cybernetics*, Taipei, Taiwan, vol. 1, pp. 548–552. IEEE, 2006.

71. A Ben-Hur and J Weston. A user's guide to support vector machines. In Carugo, O, Eisenhaber, F (eds.) *Data Mining Techniques for the Life Sciences*, pp. 223–239, 2010.

72. TS Furey, N Cristianini, N Duffy, DW Bednarski, M Schummer, and D Haussler. Support vector machine classification and validation of cancer tissue samples using microarray expression data. *Bioinformatics*, 16(10):906–914, 2000.

73. S Tong and E Chang. Support vector machine active learning for image retrieval. In *Proceedings of the Ninth ACM International Conference on Multimedia*, Ottawa, Canada, pp. 107–118. ACM, 2001.

74. C-C Chang and C-J Lin. Libsvm: A library for support vector machines. *ACM Transactions on Intelligent Systems and Technology (TIST)*, 2(3):27, 2011.

Chapter 13

A Self-Organizing Map-Based Spectral Clustering on Shortest Path of a Graph

Parthajit Roy, Swati Adhikari, and J. K. Mandal

13.1 Introduction

Cluster analysis is a kind of multivariate data analysis which currently plays an important role. Clustering is a sub-branch of pattern recognition [1]. Clustering is done when any labeled data is not available. The aim of cluster analysis is to assign the related data into the same group and the dissimilar ones into different groups.

The need of cluster analysis is increasing in recent years. Clustering has a wide range of applications in different fields of studies, from medical sciences [2] to image processing [3]. The challenges of cluster analysis lie on the data with huge sample size and also on the data with higher dimension (attributes). When we need to handle large amount of information in real life and store it and need to create groups of data which are similar in characteristics, this technique becomes very much useful. For example, data mining needs to handle data with higher dimensions and with a small as well as large number of records. While searching for a topic on the Internet, articles with similar characteristics are grouped together and displayed.

To cluster such types of datasets, a large number of clustering algorithms are available. These algorithms are mainly divided into four categories; namely, hierarchical, partitional, neural network–based and graph theoretic.

The hierarchical clustering algorithm produces a treelike nested sequence of clusters. The whole dataset is represented by the root node of the tree. Each leaf node represents a data object. The extent to which the objects are nearest to each other are represented by the in-between nodes. The distance between each type of object or cluster is measured by the height of the tree.

The partitional clustering algorithm decomposes the dataset into a number of disjointed partitions where each partition represents a cluster. The number of clusters is known in advance in this case. The K-means clustering algorithm is one of the widely used partitional clustering algorithms.

Influenced by the biological neural networks, artificial neural networks are formed. These types of networks are of two types: supervised and unsupervised. In the supervised artificial network, guidelines for the input data vectors are available to produce output data. No guideline or target data are available for unsupervised-type neural networks, which make this field very challenging. In an unsupervised neural network, competitive neural networks are known as winner-takes-all networks [4], and are often used for clustering purposes. The task of these networks is to group the similar patterns based on proximities. These groups function as a single unit and are known as neurons. Input neuron layer and output neuron layer are the two fundamental layers of a basic competitive learning network. The number of output neurons represents the number of classes and generally they are connected to each other. One of the most popular unsupervised neural networks is the self-organizing feature map, popularly called SOFM or SOM.

The SOM was developed by the professor T. Kohonen [5]. SOM generally reduces the number of attributes (dimensions) of a multidimensional dataset and projects the dataset into a lower-dimensional lattice structure. The dimensionality of the SOM neural network is generally one or two [6]. Lattice sizes of different dimensionality are used in SOM. In SOM, it is quite natural to get different output units at different iterations for a particular input pattern. This phenomenon is known as the problem of *stability*. When none of the patterns in the training data changes its types after a finite number of iterations, the system becomes stable. This is known as *plasticity*. As suggested by Jain et al. [7], to overcome the problem of stability, the learning rate is gradually reduced to zero in every iteration [8]. The problem of plasticity is, however, also affected by this effort.

A graph is a good data structure for clustering. For one it can handle various types of distances. Second, a graph has a natural way to represent proximity using edges between nodes. To establish the structural relationship between data objects, the similarity measure does not depend on the closeness between the data objects. Rather, it depends on the vertex closeness through edges between two data objects. Graph algorithms are able to partition a large set of nodes based on attribute similarities. Graph based clustering also has a wide range of applications from the field of data mining [9] to communication networks [10] and has drawn the attention of the researchers as well.

Currently, spectral graph clustering is one of the most promising areas in the field of graph clustering. These methods are extensively used to cluster a very large dataset with more improved performance than the other data clustering methods. Spectral graph theory is mainly related to the eigenvalues and eigenvectors of the adjacency matrix and Laplacian matrix of a graph. The spectral properties of these two matrices are used for graph partitioning. The Laplacian spectrum of a graph gives minimum cut when it is used in graph partitioning.

Another model, which also shows its capability in the field of graph clustering, is the shortest path–based model for constructing a similarity matrix of a graph.

There is a requirement in the different streams of engineering and VLSI design that circuits must be on surface and they should not cross each other. If there are fewer crossings, then the design will be more acceptable. In such cases, planar graphs are very effective. Integration of planar graphs with the shortest path matrix leads towards the goal of successful clustering.

Graph clustering based on planar decomposition has been proposed in this paper. The concept of planar decomposition using Delaunay triangulation has evolved from computational geometry. According to Berg et al. [11], the nearness of the points is preserved by the Delaunay triangulation of the original graph. Due to this property, it is proposed to create the all-pair-shortest-path matrix from the adjacency matrix which is again generated from the Delaunay graph and used as a similarity matrix in graph clustering.

Though there are plenty of existing clustering algorithms, there always remains high demand for innovative computing tools. The search for new ones always goes on. As a result, soft computing techniques are combined with conventional computing techniques. In conventional clustering techniques, one object may belong to only one cluster. On the other hand, in soft computing, the same objects can belong to more than one cluster. Soft computing techniques are mainly based on fuzzy logic, neural networks and genetic algorithms and these techniques produce solutions that can handle uncertainty [12]. Soft computing techniques are generally faster than hard computing techniques though they often end up with a suboptimal solution. Instead, they are practical in many cases where there are trade-offs between time taken to compute and accuracy. For example, neural network–based SOM algorithm is a soft computing technique and the similarity measure of two nodes through the shortest path distance is a hard computing technique. Hard and soft computing techniques are fused with one another with an expectation to get solutions that are faster more accurate, and more robust.

In the present paper, a new data clustering algorithm is proposed which addresses all the problems which have been discussed so far. That is, after obtaining the planar decomposition of the graph, the shortest path matrix is formed before partitioning of the graph. Among the several methods available for partitioning a graph using spectral properties of the Laplacian matrix, in this paper, the method adopted by Shi-Malik [13] has been used to partition the graph, where they have used the eigenvectors corresponding to the first k smallest eigenvalues of the Laplacian matrix. The present paper proposes to use the Laplacian matrix of a correlation-based similarity matrix and to use the SOM algorithm to cluster the Laplacian matrix; i.e., for similarity measure, hardness is applied and to find the clusters, softness is used. By using of SOM as a method of spectral graph clustering, the topological structure of the graph is also preserved. The restriction of SOM is that SOM can be used with vector data only [14]. This restriction has been overcome by the fusion of SOM with graph clustering methods, which is an interesting research area. Integration of SOM with the graph clustering method has a very high demand as it is becoming an increasing challenge when the graphs become larger.

To measure the accuracy level of the proposed method, the results obtained by the proposed method are compared with that of different existing clustering algorithms on various datasets using clustering validity indices.

The rest of the chapter is organized as follows. Section 13.2 presents a literature review. Section 13.3 describes the SOM model and spectral clustering in detail. Section 13.4 discusses some standard clustering algorithms which are used for comparison. Section 13.5 is about the proposed methodology. Section 13.6 analyzes the proposed methodology. Section 13.7 is about the experimental setup. Result and analysis are given in Section 13.8 and the conclusion appears in Section 13.9. At the end, appendices are present. Appendix 13.A is about the basic concepts required to understand the paper. Appendix 13.B contains a list of figures required to explain the basic concepts and to understand the proposed model. Appendix 13.C contains a list of tables required to analyze and compare the results. A list of references comes thereafter.

13.2 Literature Review

An introductory concept of cluster analysis has been given in the book by Xu and Wunsch [15]. The authors have elaborated on some of the most important clustering algorithms. Everitt et al. [16] is also a general reference on clustering. The authors have described the neural networks for clustering along with other related issues. An overview of different clustering methods can also be obtained in the work of Madhulatha [17]. According to Madhulatha, the merging or splitting in a hierarchical method is an irreversible process, which is the main drawback of the hierarchical method. At times, this has an advantage also. If we can accept this drawback the computation is cheaper.

Among the different types of clustering algorithms as mentioned in the previous section, two existing works of hierarchical clustering are given by Yuan et al. [18] and Franti et al. [19]. Yuan et al. [18] have developed an agglomerative MS clustering method, Agglo-MS, which is almost free of parameter tuning and improves the algorithmic speed. A fast agglomerative clustering method has been proposed by Franti et al., which reduces the number of calculations by using an approximate nearest neighbor graph. The quality of results by using these methods is nearer to the full search.

Two existing partitional clustering methods have been developed by Rokach and Maimon [20] and Somasundaram and Rani [21]. In the partitioning method proposed by Rokach and Maimon, partitioning starts from an initial partition and moves data objects from one cluster to another. The cluster size is predetermined in this case and the points are iteratively moved between the k clusters. Somasundaram and Rani have suggested the usage of the eigenvalues in spectral decomposition, which improves the performance of K-means clustering. According to their findings, faster convergence and reduced bit rate than standard K-means are possible through use of the proposed method.

The general reference on neural network–based clustering method SOM is given in Kohonen [5]. A number of research works have also been done on SOM. Vesanto and Alhoniemi [22] have considered different approaches to clustering of the SOM. They have carried out a research on the use of hierarchical agglomerative clustering and partitional clustering using K-means. According to them, after having the prototypes by applying the SOM-based clustering algorithm, the prototypes are further clustered using the hierarchical or K-means algorithm. The authors have concluded that better performance can be achieved compared with direct clustering of the data using the hierarchical or K-means algorithm, which will save lots of computation time in turn. Gunter and Bunke [14] have presented a version of SOM which works in the graph domain. Elghazel and Benabdeslem have also discussed about the different aspects of SOM, like partitioning-, hierarchical-, and graph coloring–based methods [23]. According to them, graph-based SOM methods have the advantage over the other two standard methods because they combine the effect of using dissimilarities and neighborhood relations produced by the map. SOM-based graph clustering can be used to cluster social networks or for document clustering. Ghaemmaghami and Sarhadi [24] have proposed a new method to cluster social networks by using topological structure of SOM. Hagenbuchner et al. [25] have used the capability of SOM to encode structural information for clustering of XML documents.

The proposed method has been developed from the graph-partitioning point of view. Researchers are showing their skills also in this field. Zhou et al. [26] have developed a new graph partitioning method for partitioning a large graph based on attribute similarities. Many graph partitioning methods which are surveyed by Schaeffer [27].

The shortest path distance between two nodes of a graph is considered a similarity measure. Although a number of shortest path algorithms already exists, search for new one never ends. Some of the research works regarding this issue are given in Nawaz et al. [28] and in Zhu and Huang [29]. Nawaz et al. have proposed a new shortcut construction procedure to compute the shortest-path which is extensible, powerful, and whose performance is also cost-effective. Two new shortest-path algorithms are presented by Zhu and Huang. Those are for single-origin and multiple-origins shortest path problems, which are distributed and parallel and often used by many large data networks.

Prior to measuring the shortest path between nodes, planar decomposition of the graph is produced. Details about planar decomposition using Delaunay triangulation is given in Berg et al. [11]. Deng et al. [30] have proposed clustering method using Delaunay triangulation. Busch et al. [31] have presented a new algorithm to construct planar graphs which is nearly optimal and based on the notion of searching for suitable shortest paths in the graph and clustering their proximity. Chen et al. [32] have proposed a method that uses the spectra of Delaunay triangulation.

As proposed in the present paper, the Laplacian spectrum of the graph is used to partition the graph. According to Mohar [33], the Laplacian spectrum of a graph is more important than the adjacency spectrum of the graph and it can be used in different fields of science. General references on the Laplacian matrix are given in Schaeffer [27] and Luxburg [34]. Spectral graph theory can be used to solve large numbers of clustering problems including graphmatching [35] to image segmentation [36,37]. A spectral method which is useful for partitioning a graph into non-overlapping subgraphs and which can be used for graph matching and partitioning has been discussed by Qiu and Hancock [35]. Gou et al. [36] have proposed a new parallel spectral clustering approach that solves the problems associated with the existing spectral clustering methods, which are that they have high computational complexity and use large memory. Gou et al. have also used a distributed parallel computing model which speeds up the computation. Another method which is related to these problems with quantum immune optimization is proposed by Gou et al. [37]. Huang et al. [38] have proposed a spectral clustering method which increases the area of spectral clustering to a range with multiple affinities among the available ones and this method allows the formation of similarity- or distance-based measures for clustering more acceptable. Huang et al. [39] have also proposed another spectral clustering method which uses affinity matrix for extending the spectral clustering algorithm with multiple affinities among the available ones. Dhanjal et al. [40] have derived a new spectral clustering method which can be used for partitioning a graph in more complex situations and this method can be used for exploration of communication networks. Li et al. [41] have developed a scalable constrained spectral clustering method which overcomes the problem associated with existing constrained spectral clustering methods; i.e., this method is able to properly handle medium and large datasets. Jia et al. [42] have summarized the latest research progress of spectral clustering and its main issues. A survey of different spectral clustering methods can be found in Nascimento and Carvalho, 2011 [43].

13.3 SOM and Spectral Clustering

In this section, two models which are proposed to integrate with each other, SOM and spectral graph clustering, are explained in detail.

13.3.1 SOM model

The SOM model is used for clustering data without any knowledge of the class memberships of the input data. In this model, the number of classes of any dataset is predefined, which may not be known at the beginning. The essential characteristics of the input data are identified by the use of this model.

13.3.1.1 SOM architecture

SOM uses two layers of nodes. One is the input layer and the other is the output layer. Each input node represents a vector whose length is same as that of the number of attributes in the corresponding input nodes. All output nodes represents the number of classes or clusters to which the input vectors are classified. For output layer, it uses lattice sizes of different dimensions; generally one or two. The shape of the lattice may be rectangular or hexagonal. All the output nodes are assigned with the corresponding weight vectors. The dimensions of all weight vectors are equal to the number of input nodes. All output nodes are fully connected to all input nodes. Sometimes the output "nodes" are connected to each other but no weights are assigned to these connections; this is used in the algorithm only for updating weights. To refer to a node in the output layer and to calculate the in-between distances of the nodes, each node in the output layer is associated with unique (x, y) co-ordinate. All input nodes map to the nodes of the output layer by maintaining relative distance between the two layers of nodes, i.e., topological order of the nodes are preserved. Nodes that are close to each other in the input layer are mapped to the nearby nodes in the output layer. One- and two-dimensional rectangular arrangements of SOM model are shown in Figure 13.B.1a and b of Appendix 13.B respectively.

To illustrate the architecture with an example, let there be m number of input vectors in an input space and output space is a vector of length n; i.e., number of clusters is n. The input vectors are distinct. Furthermore, it is also assumed that the length of each input vector is d. Vector components may be real numbers. The following is an example of set of input vectors:

$$X : \begin{cases} (x_{11}, x_{12}, \cdots, x_{1d}) \\ (x_{21}, x_{22}, \cdots, x_{2d}) \\ \vdots \\ (x_{m1}, x_{m2}, \cdots, x_{md}) \end{cases}$$

The set of output nodes Y is $(y_1, y_2, \cdots, y_n)$.

Each of $y_1, y_2, \cdots, y_n$ has a corresponding weight vector of dimension d. Here, n may be greater than or less than or equal to m. Each of the m input vectors is assigned to only one of the n clusters. If $d = 5$, $m = 10$ and $n = 3$, then all the 10 input vectors are of dimension five and all the three clusters have a weight vector of dimension five. Ten input vectors are classified among the three clusters.

Two-dimensional representation of lattice may not produce the desired number of clusters. It may be greater than the actual requirements. So, we need to merge the output nodes to meet the actual requirements. Some other algorithms are needed to merge the output nodes. If we need three clusters and the lattice size is 3×3, then the number of output nodes will be nine, which is much greater than the actual requirements. We need to merge these nine nodes to produce three clusters.

In the proposed model, it is assumed that the output layer is a rectangular lattice and while mapping the input data vector into lower dimensional neural networks, the SOM uses a two dimensional lattice of nodes. There exists no connection between the nodes in the output layer. Furthermore, partitional K-means clustering is used to merge the SOM outputs.

13.3.1.2 Working principle of SOM model

The SOM model used in the proposed method is basically a two-pass model.

The first pass of the model starts with *normalization* of the data vectors. The details of the *normalization* process are given in Appendix 13.A. Next, the SOM algorithm is applied to this transformed data. The SOM algorithm is a two-phase algorithm. The first phase of the algorithm trains the SOM network and the second phase is the testing phase. The training employs competitive learning. As a result, different parts of the network respond similarly to certain input vectors and the network starts to learn. After completion of learning, the input vectors are tested by using the same network. The working algorithm of a standard SOM is given below.

13.3.1.2.1 SOM Algorithm

The convergence of the SOM algorithm depends on different parameters like the learning rate and a neighborhood of the winning node. At first, initial weights are assigned randomly and the learning rate is assigned an initial value near to one. Initial neighborhood size is assigned to the lattice radius. After setting the initial weights, learning rate (α), and the initial neighborhood size, the transformed input vectors to be clustered are presented to the network. A large number of transformed input vectors, which are close enough, are presented to the network. Generally, the samples are presented several times to the network.

When a sample input vector is fed to the network, its Euclidean distance to all weight vectors are computed by using Equation 13.1.

$$Diatance\ from\ input = \sqrt{\sum_{j=0}^{d-1} (X_i - W_j)^2} \tag{13.1}$$

where X_i is i-th input vector, W_j is j-th weight and d is the dimension of weight vectors.

The weight vector with minimum Euclidean distance from i-th input vector is considered as most similar to i-th input. This weight vector is known as the best matching unit (BMU) and the weight of this BMU is updated. Let us consider an example to understand the process of BMU computation. Suppose an input vector is $(1, 2, 1)$ and for a 2×2 rectangular lattice, the initial weight vectors for the four output nodes, are $(2.1, 3.4, 1.1)$, $(3.2, 1.5, 1.7)$, $(1.2, 2.4, 2.1)$ and $(2.2, 1.3, 3.1)$.

The square of Euclidean distance of the input vector from weight vector1 $= (1-2.1)^2 + (2-3.4)^2 + (1-1.1)^2 = 3.18$

The square of Euclidean distance of the input vector from weight vector2 $= (1-3.2)^2 + (2-1.5)^2 + (1-1.7)^2 = 5.58$

The square of Euclidean distance of the input vector from weight vector3 $= (1-1.2)^2 + (2-2.4)^2 + (1-2.1)^2 = 1.41$

The square of Euclidean distance of the input vector from weight vector4 $= (1-2.2)^2 + (2-1.3)^2 + (1-3.1)^2 = 6.34$

As the minimum distance is corresponding to the third weight vector, the node corresponding to this weight vector is the BMU.

The weight of the BMU is updated by using Equation 13.2.

$$W_{new} = W_{old} + \alpha[X_i - W_{old}] \tag{13.2}$$

Next, the weights of the nodes, which are in the same neighborhood of the BMU, are considered and their weights are adjusted towards the i-th input vector by using Equation 13.3.

$$W_{new} = W_{old} + \alpha\beta[X_i - W_{old}] \tag{13.3}$$

where β is spatial distance of the neighboring node from BMU.

When all the input vectors are presented to the network, one epoch completes. Several epochs of training may be performed by updating the learning rate and neighborhood radius of BMU. The amounts of learning rate and neighborhood radius are reduced after each epoch. Learning rate and neighborhood radius are updated by using Equations 13.4 and 13.5 respectively.

$$\alpha_{new} = \alpha_{old} \times 0.5 \tag{13.4}$$

$$\delta(t) = \delta_0 \times e^{(-t/\lambda)} \tag{13.5}$$

where t is the current pass, λ is the maximum number of passes and δ_0 is the radius of the lattice.

This ends the training phase. Next, the testing phase starts. In this phase, sample input vectors are tested with the training weights and the same steps are repeated.

13.3.1.2.2 Steps of the SOM algorithm

Algorithm 13.1

---

1. Initialize weight vectors $W_1, W_2, \cdots, W_n$ randomly.
2. Set Learning rate α.
3. Set topological neighborhood to lattice radius.
4. Normalize the input vectors by using Equation 13.A.1 of Appendix 13.A.
5. Choose next transformed input vector X_i and do the following.
6. Calculate the distance of X_i from all nodes of the lattice by using Equation 13.1.
7. Find the Best Matching Unit (BMU)
8. Update BMU by using Equation 13.2.
9. Consider all nodes inside the radius of the BMU and update their weights by using Equation 13.3.
10. Repeat Steps (5) to (9) for m input vectors.
11. Update learning rate α by using Equation 13.4.
12. Update the radius of neighborhood of the BMU by using Equation 13.5.
13. Stop if the stopping condition is achieved. Else repeat steps (5) to (12).

The second pass of the model involves merging of partitions. At the end of testing phase, the SOM prototypes are further merged to get the final clusters. For this purpose, the partitional K-means clustering has been opted which will be discussed in Section 13.4.

13.3.2 Spectral clustering

Spectral clustering is basically the problem of partitioning a dataset represented by graph data structure based on the properties derived from the data. It is the study of the properties of a graph via eigenvalues and eigenvectors of graph matrices. The fundamental concepts of graph theory, such as what a graph is and what its different types are, have been discussed in Appendix 13.A. Before moving into the in-depth discussions of spectral clustering, let us discuss what graph partitioning is.

In the problem of graph partitioning, the entire dataset is expressed as a graph $G = (V, E)$, where V is the set of vertices and E is the set of edges. Then the vertex set V is divided into a number of smaller components k, with some specific properties. Graph clustering algorithms are mostly based on the graph partitioning problems. The partition in which the number of edges connecting two separate components is small, i.e., a minimum cut can be obtained, is considered as a good partition. There are various methods like ratio cut and normalized cut available to find the minimum cut which are NP-hard problems. In the proposed method, a shortest path based planar graph cut method is used. For this, planar decomposition of the original graph representing the input vectors is made. This is also known as Delaunay triangulation. For planar graph, Delaunay triangulation, and shortest path see Appendix 13.A. The spectral properties of the graph matrices are used to solve the graph cut problems. Actually, through the use of spectral clustering methods these NP-hard graphs cut problems can be solved more easily. Spectral clustering represents data in lower-dimensional space which can be clustered easily. In spectral clustering, a similarity graph is used. A similarity graph is a weighted graph in which each data object is expressed as a node of the graph and the connection between two data objects is made if the similarity between the corresponding data objects is positive or exceeds a certain amount of threshold value. The weight of any edge is assigned with the amount of similarity between two data objects. As distance decreases, the similarity increases. Now, if this graph is partitioned, the weight of an edge connecting two data objects, each one from two different groups has very low weight. When two data objects are chosen from the same group, the edge between them has high weight. Figure 13.B.3b shows a similarity graph.

From the similarity graph, other graph matrices like the adjacency matrix (see Appendix 13.A), the weighted adjacency matrix, the degree diagonal matrix and the Laplacian matrix are formed and used in spectral clustering. Each of these matrices is defined below.

Weighted adjacency matrix/Similarity matrix: In weighted adjacency matrix, the ij-th entry W_{ij} is set to the weight assigned to the corresponding edge if there is an edge between the i-th and the j-th node; otherwise, it is set to zero. Again, if the graph is undirected, then $W_{ij} = W_{ji}$ and if the graph has no self-loops, then $W_{ii} = 0$. When the assigned weights are the amount of similarity of the associated nodes, the matrix is known as similarity matrix. It is a symmetric matrix. For example, the similarity matrix of the graph given in Figure 13.B.3b is as follows.

$$W = \begin{bmatrix} 0 & 0.9 & 0.9 & 0 & 0 & 0 & 0 & 0 & 0 \\ 0.9 & 0 & 0.7 & 0.8 & 0 & 0 & 0 & 0 & 0 \\ 0.9 & 0.7 & 0 & 0.8 & 0.3 & 0 & 0 & 0 & 0 \\ 0 & 0.8 & 0.8 & 0 & 0 & 0.1 & 0 & 0 & 0 \\ 0 & 0 & 0.3 & 0 & 0 & 0.7 & 0.9 & 0.6 & 0 \\ 0 & 0 & 0 & 0.1 & 0.7 & 0 & 0.8 & 0 & 0.7 \\ 0 & 0 & 0 & 0 & 0.9 & 0.8 & 0 & 0.9 & 0.6 \\ 0 & 0 & 0 & 0 & 0.6 & 0 & 0.9 & 0 & 0.8 \\ 0 & 0 & 0 & 0 & 0 & 0.7 & 0.6 & 0.8 & 0 \end{bmatrix}$$

Degree diagonal matrix: The degree diagonal matrix D is a diagonal matrix whose diagonal entries are the degrees of the corresponding nodes, i.e., the entry D_{ii} is the degree of the i-th node. For example, the degree diagonal matrix of the similarity graph shown in Figure 13.B.3b is as follows.

$$W = \begin{bmatrix} 1.8 & 0 & 0 & 0 & 0 & 0 & 0 & 0 & 0 \\ 0 & 2.4 & 0 & 0 & 0 & 0 & 0 & 0 & 0 \\ 0 & 0 & 2.7 & 0 & 0 & 0 & 0 & 0 & 0 \\ 0 & 0 & 0 & 1.7 & 0 & 0 & 0 & 0 & 0 \\ 0 & 0 & 0 & 0 & 2.5 & 0 & 0 & 0 & 0 \\ 0 & 0 & 0 & 0 & 0 & 2.3 & 0 & 0 & 0 \\ 0 & 0 & 0 & 0 & 0 & 0 & 3.2 & 0 & 0 \\ 0 & 0 & 0 & 0 & 0 & 0 & 0 & 2.3 & 0 \\ 0 & 0 & 0 & 0 & 0 & 0 & 0 & 0 & 2.1 \end{bmatrix}$$

Laplacian Matrix: The Laplacian matrix of a graph, also known as Unnormalized Laplacian Matrix, is a matrix whose $(i, j) - th$ entry L_{ij} is assigned to d_{ii}, when i and j have equal value, otherwise it is assigned to $-W_{ij}$, where d_{ii} represents the degree of the i-th node or it can be defined by using Equation 13.6.

$$L = D - W \tag{13.6}$$

where D and W represent the degree diagonal matrix and the weight matrix respectively. As an illustration, the Laplacian matrix of the similarity graph shown in Figure 13.B.3b is as follows.

$$W = \begin{bmatrix} 1.8 & -0.9 & -0.9 & 0 & 0 & 0 & 0 & 0 & 0 \\ -0.9 & 2.4 & -0.7 & -0.8 & 0 & 0 & 0 & 0 & 0 \\ -0.9 & -0.7 & 2.7 & -0.8 & -0.3 & 0 & 0 & 0 & 0 \\ 0 & -0.8 & -0.8 & 1.7 & 0 & -0.1 & 0 & 0 & 0 \\ 0 & 0 & -0.3 & 0 & 2.5 & -0.7 & -0.9 & -0.6 & 0 \\ 0 & 0 & 0 & -0.1 & -0.7 & 2.3 & -0.8 & 0 & -0.7 \\ 0 & 0 & 0 & 0 & -0.9 & -0.8 & 3.2 & -0.9 & -0.6 \\ 0 & 0 & 0 & 0 & -0.6 & 0 & -0.9 & 2.3 & -0.8 \\ 0 & 0 & 0 & 0 & 0 & -0.7 & -0.6 & -0.8 & 2.1 \end{bmatrix}$$

There exist two other variations of the Laplacian matrix which are also used in the field of clustering. These are known as normalized symmetric Laplacian matrix and random walk Laplacian matrix. The second one is related to the random walk property of the nodes of the graph. These two types of Laplacian matrices are defined by using Equations 13.7 and 13.8 respectively.

$$L_{sym} = I - D^{-1/2} W D^{-1/2} \tag{13.7}$$

$$L_{rw} = I - D^{-1} W \tag{13.8}$$

where D and W are defined as above.

These three types of Laplacian matrices are positive semidefinite matrices and L_{sym} is symmetric. The row sum and column sum of a Lalacian matrix are always zero.

The idea behind the spectral geometry leads towards the deep understanding of the study of graph eigenvalues. By the term "spectra of Laplacian matrix", we mean the eigenvalues and their corresponding eigenvectors of the Laplacian matrix. There are m eigenvalues of an $m \times m$ Laplacian matrix and m eigenvectors corresponding to these eigenvalues. If the assigned weights are all non-negative, then the Laplacian matrix has all real eigenvalues. The eigenvalues are all positive and 0 is the smallest eigenvalue and the corresponding eigenvector is $(1, 1, 1, \cdots, 1)^T$. These m eigenvalues are not all distinct.

Among the m eigenvalues different eigenvalues hold some specific spectral properties and the corresponding eigenvectors can be used to cluster the graph. The multiplicity of these eigenvalues corresponds to the number of connected components of the graph. For example, the smallest non-null eigenvectors of the unnormalized Laplacian matrix returns the ratio cut and the smallest non-null eigenvector of the random walk-based Laplacian returns the normalized cut. The second smallest eigenvalue is known as Fiedler value. The eigenvector corresponding to the Fiedler value is known as the Fiedler vector. The multiplicity of this vector is always one. The Fiedler vector is used for spectral bi-partitioning. The first non-zero eigenvalue is known as eigengap. Sometimes the eigenvector corresponding to this eigenvalue is also used for partitioning. According to Shi-Malik [13], the eigenvectors corresponding to the first k smallest eigenvalues of the Laplacian matrix are used for clustering in the proposed model.

13.4 Some Standard Clustering Algorithms

This section discusses about some standard clustering algorithms like K-means, (PAM), fuzzy K-means with polynomial fuzzifier (FKMPF), and hierarchical. These algorithms are used to compare the performance of the proposed model.

The K-Means clustering algorithm is used in the proposed model to compare the results obtained by applying this algorithm and the proposed one. This algorithm is also used to merge the SOM prototypes to produce the desired number of clusters. In this method, the number of clusters K is predetermined. K number of cluster centers are chosen randomly. For merging SOM prototypes, all these prototypes are supplied as input vectors. The algorithm proceeds by iteratively minimizing the error function given in Equation 13.9 and updating the cluster centers accordingly. The error function is defined by Equation 13.9.

$$E = \sum_{i=1}^{C} \sum_{j=1}^{N} \|X_j - K_i\|^2 \tag{13.9}$$

where X_j is the j-th input vector, K is the number of clusters, and K_i is the center of i-th cluster.

Until the error value E becomes less than the desired criterion, i-th input vector X_i is assigned to its nearest cluster by calculating the Euclidean distance between the input vector and cluster centers by using Equation 13.1. In the same way, the other input vectors are also assigned to their nearest clusters. Then the cluster centers are updated by calculating the mean of the clusters. Again, the error function E is calculated for updated centers by using Equation 13.9. The whole process is repeated until the partitioning is unchanged or the algorithm has converged.

Although, we have used the K-means clustering algorithm, this algorithm suffers from several disadvantages. One of the disadvantages of K-means clustering is that the output is very much influenced by the choice of initial cluster centers. It is also not suitable for finding the clusters which are of different sizes. This algorithm can only be applied for the defined mean value. It is also sensitive to noise and outliers which may influence the mean value.

Another partitional algorithm which is related to the K-means algorithm and used for comparison is the K-medoids or PAM algorithm. In this algorithm, the number of clusters is known a priori, but instead of cluster means individual data points are chosen as cluster centers. This algorithm is more robust than the K-means algorithm but is not well-suited for large datasets. This algorithm is also sensitive to noise and outliers.

The fuzzy K-means FKMPF algorithm is also a partitional algorithm which performs the fuzzy K-means algorithm with polynomial fuzzifier function that produces the soft clusters. Unlike hard clusters, in a soft cluster, any point can belong to more than one cluster with fuzzy membership degrees. The disadvantages associated with this algorithm are that this algorithm is unable to handle noise and outliers and cannot handle high-dimensional datasets and large numbers of prototypes.

An hierarchical clustering algorithm groups the data in one of two ways: either by top-down approach or by bottom-up approach. At first, all data objects are considered as a single cluster (root) and next, the root is divided into a number of small clusters. This method is known as the divisive clustering algorithm. In the second case, grouping of objects is done just in the reverse order of the first case and this technique is known as the agglomerative clustering algorithm.

Single linkage, complete linkage, and average linkage are three types of linkage methods available in the hierarchical clustering algorithm. Single linkage and complete linkage measure the distance between two clusters based on the smallest or largest distance between two data points each from a different cluster respectively. All distance values between the pair of data points from different clusters are averaged in average linkage method.

Hierarchical clustering algorithm also has sensitivity to noise and outliers. This algorithm faces difficulty in handling different-sized clusters and of irregular shapes.

To overcome the problems associated with K-means, PAM, FKMPF, and hierarchical clustering algorithms, SOM has been proposed to cluster the Laplacian eigenvectors.

13.5 Proposed Methodology

In this section, the motivation behind the proposed model and the proposed model itself are discussed in detail.

13.5.1 Motivation

Motivation behind the proposed model comes from the random-walk nature of the graph. A random walk on a graph is a stochastic process which randomly jumps from one subgraph to the other [34]. Its time to stay within a dense subgraph lasts for long and seldom jumps from one subgraph to the other.

It is also observed that there exists a strong correlation between nodes of a graph when a shortest path is drawn from two nodes of a weighted undirected graph, which are closed enough. That means, if the shortest path from an arbitrary node to a first node is positive, then the shortest path from a second node to the arbitrary node will also be the same and those nodes will form a cluster. And if the two nodes are far away, then the shortest path between those two nodes will have negative correlation and they will belong to a different cluster. This observation has also motivated the existence of the proposed model.

13.5.2 Proposed model

A novel spectral clustering–based model is proposed in this paper, in which spectral properties of Delaunay triangulation are used as similarity measure.

As the random-walk nature of graph is the main motivation, it is applied to compute the Laplacian matrix of the Delaunay graph. The idea of random walk–based clustering, proposed in the present paper, comes from the random-walk nature of a graph. The transition matrix T of the random walk is defined by using Equation 13.10:

$$T = D^{-1}W \tag{13.10}$$

As the nearness of the points remains unaltered in a planar graph, such types of graphs are chosen for the proposed model.

A new shortest path–based measurement method has been proposed which comes from the fact that if we have to measure the similarity between two nodes of a weighted undirected graph, then we need to calculate the all-pair-shortest-path matrix of the graph based on by which method the shortest path distance from those two nodes to all other nodes will be measured. If the shortest paths from those two nodes to all other nodes will be almost same, then they will be nearer; otherwise they are at a great distance.

To construct the model, a planar decomposition, i.e., the Delaunay triangulation of the original graph, is obtained followed by the creation of adjacency matrix. The all-pair-shortest-path matrix is formed from this adjacency matrix. Correlations between the nodes are measured thereafter to assign weight to the edges and it is assumed that no loops exist in the graph, i.e., the weights for these entries in the correlation matrix are assigned to zero. A threshold value is set so that if the correlation falls below this threshold value, the correlation will also be set to zero. In this way, a correlation-based similarity matrix is formed. Next, the random walk–based Laplacian matrix of the graph is obtained from this matrix by using Equation 13.8. The eignvectors corresponding to the first K smallest eigenvalues of the Laplacian matrix are clustered thereafter by using SOM clustering algorithm. The number K is the number of clusters and is assumed to be known in advance. As in the present work the structural property of the clusters is not considered, the knowledge of K is mandatory. The algorithm of the proposed model is stated below.

Algorithm 13.2

[The algorithm assumes that the number of clusters K is known in advance.]

input: Data Vector Set D, cut-off threshold value for selecting correlation of data vectors and Number of Clusters K.

output: K number of clusters.

1. Compute the planar decomposition PD of the graph G.
2. Create the adjacency matrix AM from this planar decomposition.
3. Compute the all-pair-shortest-path matrix $APSP$ from the adjacency matrix AM and hence form the correlation matrix CM from the matrix $APSP$. The correlation, less than the threshold value, and the correlation of a data vector to itself are set to zero.

4. Compute the Laplacian matrix LM from this correlation matrix CM by using Equation 13.8.
5. Calculate the eigenvalues and eigenvectors of the Laplacian matrix LM.
6. Choose the eigenvectors corresponding to the first K smallest eigenvalues of the Laplacian matrix LM and apply SOM algorithm to these eigenvectors to produce K number of clusters.

13.6 Analysis of the Proposed Methodology

This section analyzes the proposed methodology. It is proposed to convert the original graph to its planar representation. This is done by finding the Delaunay triangulation of the graph. In the Delaunay graph, the nearness of the points also remains as it is in the original graph; i.e., if two points A and B are in proximity in the original graph, they will also remain in proximity to each other in the Delaunay graph.

This property is also maintained while computing the all-pair-shortest-path from the adjacency matrix of the Delaunay graph. If two points P and Q are close to each other and a third point R is far away from the point P, then the point R is also far away from the point Q. That is to say, two points which are closed in the Delaunay graph will possess high correlation when the shortest path from these two points are drawn. If the points are far away, then the correlation between the points will be low.

The time complexity associated with the divide-and-conquer algorithm to find Delaunay triangulation is $O(m \log m)$, m being the number of vertices of the graph, i.e., the number of input vectors. On the other hand, Floyd–Warshall's algorithm to compute all-pair-shortest-path has a worst case time complexity of $O(m^3)$, m being the number of vertices. Although an additional complexity of $O(m \log m)$ will be added to the complexity of $O(m^3)$ for Delaunay triangulation prior to computing all-pair-shortest-path matrix, it is negligible when m is very large. So, it is advantageous to create an all-pair-shortest-path matrix from the adjacency matrix of the Delaunay graph.

By the use of random walk–based Laplacian matrix, the random walk property of the graph, as explained in the last section, can also be obtained in the proposed model. The spectral properties of the Laplacian matrix are used to cluster the Delaunay as well as the original graph. By considering the eigenvectors corresponding to the first k smallest eigenvalues of the Laplacian matrix, the dimensionality of the matrix, which is to be clustered, is also reduced; i.e., the problem of clustering an $m \times m$ Laplacian matrix reduces to the problem of clustering an $m \times k$ matrix, where k is the desired number of clusters.

Although any standard clustering algorithm can be applied to cluster the Laplacian eigenvectors into a desired number of clusters, it is proposed to use SOM for this purpose. The orders of the input vectors are also maintained by the use of SOM, as this algorithm is a topology-preserving algorithm. At a time, SOM can handle large datasets and help to reduce the dimensionality of the participating matrix. The disadvantages associated with other standard algorithms can also be avoided by the use of SOM.

The computational complexity of the SOM algorithm is only linear in the number of samples [44]. The sample size is again proportional to the number of prototypes. The complexity increases as the number of prototypes increases in the SOM algorithm. On the other hand, the computational complexity of the K-means algorithm is $O(mni)$ and that of PAM is $O(n(m-n)^2 i)$, m, n, i being the sample size, number of clusters and number of iterations respectively. When m is very large, i.e., $n << m$ and $i << m$, the complexity of PAM is almost $O(m^2 n)$. The computational complexity of the hierarchical clustering algorithm is also at least $O(m^2)$. So, the choice of SOM algorithm for clustering Laplacian eigenvectors is effective from the computational end also.

13.7 Experimental Setup

This section describes the different datasets which are considered for experimental purposes and also the different clustering validity indices which are considered for measuring the similarity between two clusters.

Seven standard datasets have been chosen for this purpose. These are selected for their inherent features; i.e., the datasets differ in shapes and sizes. The datasets are *iris* and *seeds* datasets of UCI Machine Learning Repository [R. A. Fisher, 1936] [45] and *flame*, *Jain*, *spiral*, *compound* and *aggregation* datasets of Speech and Image Processing Unit of the School of Computing [46]. The proposed algorithm is applied on these datasets. The *flame*, *Jain*, *spiral*, *compound* and *aggregation* datasets are proposed by Fu and Medico, 2007 [47], Jain and Law, 2005 [7], Chang and Yeung, 2008 [48], Zahn, C. T., 1971 [49] and Gionis, et al., 2007 [50] respectively. The properties which are considered for selecting these datasets for experimental purposes are given below.

The *iris*, *seeds*, and *spiral* datasets consist of three classes but one of the classes of the *iris* dataset is linearly separable from the other two, whereas none of the classes of the *seeds* dataset are linearly separable, which is clear from Figure 13.B.5a and b of Appendix 13.B. The shape of the *spiral* dataset is more complex than the other two datasets and holds a combination of sparse and dense data points, again, which is clear from Figure 13.B.6a of Appendix 13.B.

Both the *flame* and *Jain* datasets consist of two classes with different features. The classes of the *flame* dataset are touching whereas the *Jain* dataset is touching but one class consists of dense data and the other of sparse data. The distributions of these datasets are given in Figures 13.B.6b and 13.B.7a of Appendix 13.B respectively.

The classes of *compound* and *aggregation* datasets are greater in number than the other five datasets that have been discussed earlier. The *compound* dataset is again consisting of dense and sparse data points and the shape is more complex than the *aggregation* dataset which can be seen in Figures 13.B.7b and 13.B.8a of Appendix 13.B respectively. The size of the *aggregation* dataset is also larger than the other datasets.

It is clear from Figures 13.B.5a through 13.B.8a that all the datasets have overlapping classes, contain dense as well as sparse data, and are of complex shapes. Because of these distinct properties, all of these datasets are considered as test data.

The similarity between two clusters is measured by using clustering validity indices. Such a measure is used to compare the different data clustering algorithms on a dataset.

Among the available clustering validity indices, external and internal clustering indices are used for comparison. External indices are mainly used to measure the similarity of the computed clustering structure ($C1$) with an existing clustering structure ($C2$). Only distribution of points in the different clusters is considered in these comparison indices. These indices do not measure the quality of the distribution. These indices lie between zero and one; one means the two partitions are perfectly similar. There are other indices for measuring different attributes of clusters but the present paper emphasizes the external indices as the distribution of the sample datasets are known and are used for comparison.

To measure the quality of the distribution of data points, internal indices are used. Saha and Bandyopadhyay [51] and Roy and Mandal [52] have measured the performance of some of these indices.

Some of the external clustering indices which are used to test datasets for comparison are Jaccard, Rand, Recall, and Sokal–Sneath.

The Jaccard index is defined using Equation 13.11.

$$I = \frac{a}{(a+b+c)} \tag{13.11}$$

The Rand index is defined by using Equation 13.12.

$$I = \frac{a+d}{(a+b+c+d)} \tag{13.12}$$

The Recall co-efficient is defined by using Equation 13.13.

$$I = \frac{a}{(a+b)} \tag{13.13}$$

The two versions of Sokal–Sneath index are defined by using Equations 13.14 and 13.15.

$$I = \frac{a}{(a+b)} \tag{13.14}$$

$$I = \frac{a}{(a+b)} \tag{13.15}$$

where a is the number of pairs of samples which are assigned to the same cluster in $C1$ as well as in $C2$, b is the number of pairs of samples which are in the same cluster in $C1$ only, c is the number of pairs of samples which are in the same cluster in $C2$ only, and d is the number of pairs of samples that are assigned to different clusters in $C1$ and $C2$. The quantities a and d are known as agreements, and b and c as disagreements.

Among the available internal clustering indices, the Ray–Turi index is used to measure the quality of distribution. If the index value is minimum, in this case, then the result is more acceptable.

13.8 Results and Analysis

This section analyzes the results obtained by applying the proposed algorithm on test datasets as mentioned in the previous section. Descriptions of each dataset are given below.

A total of 150 instances of three types of flowers, namely, *Iris setosa*, *Iris versicolor*, and *Iris virginica*, are in the *iris* dataset. All of these three types of flowers have 50 instances in each. Among these three types of flowers, only one type of flower is linearly separable from the other two. Each instance has four attributes.

The *seeds* dataset consists of a group of kernels which belong to a total of 210 instances of three different varieties of wheat, Kama, Rosa, and Canadian, with 70 elements in each. Each instance has seven attributes.

A total of 312 instances are there in the *spiral* dataset with three classes comprised of two attributes.

The *flame* dataset comprises a total of 240 instances and consists of two classes of different sizes. Each instance consists of two attributes.

The *Jain* dataset contains a total of 373 instances with two classes of different sizes. The number of attributes in each instance is two.

The *compound* dataset has 399 instances of six classes with two attributes for each instance.

The *aggregation* dataset is consisting of 788 instances of seven classes. Again, in this dataset each instance has two attributes.

No dataset has missing values. The number of classes of each dataset is clear from Figures 13.B.5a through 13.B.8a.

The planar decompositions of the *iris*, *spiral*, *flame*, *Jain*, *compound*, and *aggregation* datasets are shown in Figure 13.B.8b, 13.B.9a and b, 13.B.10a and b, 13.B.11a of Appendix 13.B respectively. The dotted lines in the figures show the good cuts which are possible in each of the decomposition.

By observing Figures 13.B.5a through 13.B.8a and Figures 13.B.8b through 13.B.11a, it is clear that the proximities of the data points are maintained in their planar decompositions also, for all the test datasets.

The clustering results obtained by applying the proposed algorithm is compared with the results obtained by applying the K-means, PAM, FKMPF, single-link hierarchical, complete-link hierarchical, average-link hierarchical, and SOM algorithms on the test datasets using external and internal clustering indices.

The graphical view of the results obtained by clustering the seven test data *iris*, *seeds*, *spiral*, *flame*, *Jain*, *compound*, and *aggregation* by the proposed algorithm using Recall and second version of Sokal–Sneath are given in Figures 13.B.11b, 13.B.12a and b, 13.B.13a and b, 13.B.14a and b of Appendix 13.B respectively. A comparative study of the above-mentioned seven clustering methods including the proposed one, when they are applied on seven test data respectively, using Jaccard and Rand indices, are given in Tables 13.C.1 through 13.C.7 of Appendix 13.C. The result after comparison using internal clustering index *Ray–Turi* can be found in Figure 13.B.15 of Appendix 13.B.

By inspecting the results given in Tables 13.C.1 through 13.C.7, it is clear that our proposed algorithm can cluster the above-mentioned data almost with the same accuracy as that of other clustering algorithms. From Table 13.C.4, it is seen that the proposed method is powerful enough to cluster the *flame* data with far greater accuracy than the other methods. Although the proposed method is unable to cluster the *aggregation* data with more accuracy than the other methods, which can be seen from Table 13.C.6, comparatively this inability is negligible.

By examining the results represented by Figures 13.B.11b through 13.B.14b, it is clear that the proposed method is able to cluster the test data with nearly perfect accuracy, which is more or less the same as that of other clustering algorithms. Although in Figure 13.B.12b, the single-link hierarchical clustering method is showing its capability to cluster the *spiral* data with 100% accuracy, the accuracy level of the clustering by the proposed method is almost the same as that of the remaining clustering algorithms, as mentioned earlier.

So, by considering all the comparison results given in the figures and tables, it is shown that the accuracy level of the proposed method is sometimes better and sometimes slightly less than that of some of the data.

Figure 13.B.15 shows the comparative study of measuring the quality of clustering methods by using the Ray–Turi index and it can be seen that the quality of proposed algorithm is the same as that of almost all test data.

13.9 Conclusion

This paper proposes a new data clustering method by using all-pair-shortest-path distances of graph and its Laplacian spectra. As the new algorithm is capable of clustering the different datasets with almost the same accuracy as the other existing clustering algorithms, and gives more or less the same output when the proposed algorithm is used, we can conclude that this new algorithm is able to cluster large, non-categorical datasets successfully. As a result, it overcomes the disadvantages of the K-means, PAM, FKMPF, and hierarchical clustering algorithms. There is also a scope for further improvement of the proposed algorithm. Instead of using the adjacency matrix, other graph matrices can be used; instead of using Laplacian spectra, the generalized inverse of Laplacian matrix, and the commute distance of the graph can also be used for clustering purpose.

Acknowledgment

The authors are thankful to the Department of Computer Science, University of Burdwan; the Department of Computer Science and Engineering of Kalyani University; and DST PURSE of the Govt. of India at Kalyani University for their help and support, and also to UCI Machine Learning Center and to the Speech and Image Processing Unit of School of Computing for their online standard dataset.

Appendix 13.A

Essential Mathematics

Normalization: Scaling is a problem due to which there exists a mismatch in the range for different attribute parameters in a dataset. Due to this, one attribute parameter may dominate over the other. To avoid the problem of scaling, a technique, known as normalization, is adopted. In this technique, the data vectors are normalized to [0, 1] by using Equation 13.A.1.

$$X_i = \frac{X_i^j - min\{X^j\}}{max\{X^j\} - min\{X^j\}}, \forall X_i \in S \tag{13.A.1}$$

where X_i is the i-th data vector, X_i^j is the j-th parameter of the i-th data vector X_i and X_j is the j-th parameter values for all of X_i.

Graph: A graph G is an ordered pair $G = (V, E)$ where V is the set of vertices or nodes and E is the multiset of edges. The data vectors are represented by nodes of a graph and the connection between two nodes are established by a number of edges.

Subgraph: A subgraph S of a graph G is a graph, each of whose vertices belongs to the set V and each edge belongs to the set E.

Different types of Graphs: Sometimes, the edges between two nodes of a graph may be directional. If it is directional, then the graph is directed; otherwise it is undirected. It may be possible to have multiple edges between two nodes of a graph. When a node points to itself, it is known as loop. Figure 13.B.2a of Appendix 13.B shows the picture of a directed graph with self-loop and multiple edges.

Sometimes, a number of weights are assigned to each edge of the graph and this type of graph is known as a weighted graph. If no weights are assigned, then the graph is unweighted. The graph in Figure 13.B.2a of Appendix 13.B is an unweighted graph.

A simple graph is an undirected graph, in which only a single edge between two nodes exists and no self-loops exist. The picture of a simple weighted graph is shown in Figure 13.B.2b of Appendix 13.B.

A planar graph is a graph in which no two edges cross each other. Figure 13.B.3a of Appendix 13.B is the planar representation of the simple graph shown in Figure 13.B.2b.

Adjacency: Two vertices u and v of the graph G are said to be adjacent if there is an edge uv connecting the two vertices exists. The vertices u and v are said to be incident with the edge uv.

Adjacency Matrix: The adjacency matrix A of a graph G with m vertices is a $m \times m$ matrix in which the ij-th entry A_{ij} is set to one if there is an edge between the i- and the j-th node; otherwise, it is set to zero. If the graph is undirected, then $A_{ij} = A_{ji}$. If the graph has no self-loops, then $A_{ii} = 0$. The adjacency matrix is a symmetric matrix. For example, the adjacency matrix of the similarity graph given in Figure 13.B.3b of Appendix 13.B is as follows.

$$A = \begin{bmatrix} 0 & 1 & 1 & 0 & 0 & 0 & 0 & 0 & 0 \\ 1 & 0 & 1 & 1 & 0 & 0 & 0 & 0 & 0 \\ 1 & 1 & 0 & 1 & 1 & 0 & 0 & 0 & 0 \\ 0 & 1 & 1 & 0 & 0 & 1 & 0 & 0 & 0 \\ 0 & 0 & 1 & 0 & 0 & 1 & 1 & 1 & 0 \\ 0 & 0 & 0 & 1 & 1 & 0 & 1 & 0 & 1 \\ 0 & 0 & 0 & 0 & 1 & 1 & 0 & 1 & 1 \\ 0 & 0 & 0 & 0 & 1 & 0 & 1 & 0 & 1 \\ 0 & 0 & 0 & 0 & 0 & 1 & 1 & 1 & 0 \end{bmatrix}$$

Degree: The degree of a vertex u of the graph G is determined by the number of vertices incident with u. It is the row sum or column sum of the adjacency matrix.

Walk: Walk is to travel through vertex to vertex along edges. In a walk, any vertex and any edge may repeat any number of times. The starting and ending may be same or different in a walk.

Path: A path is a simple walk where none of the vertex is repeated except the first one.

Graph Cut: A cut partitions the set of vertices of a graph into two separate subsets. The set of edges that have one endpoint in each subset of the partitions is considered to be the cut set. A cut set $S = (A, B)$ is a partition which divides the vertex set V of a graph $G = (V, E)$ into two groups. Any edge of the cut set S belongs to the edge set E, and one end point of the edge belongs to the set A and the other end point belongs to the set B. An $a - b$ cut of the graph G, where $a \in A$ and $b \in B$, is a set of edges which when removed from G there remain no edges connecting the two groups. The cost of a cut is the sum of edge weights in the cut set. A cut is considered to be minimum cut when the cost of the cut is smallest among all cuts. If the cost is not smaller than any other cut, then it is considered as maximum cut. In Figure 13.B.3b, the dotted line shows a minimum cut.

Delaunay triangulation: In graph theory, maximal planar graphs are known as triangulations. A maximal planar subdivision can be defined as a subdivision such that no edge connecting two nodes can be added to this subdivision without destroying its planarity [11].

If $P := p_1, p_2, \cdots, p_m$ be a set of m distinct points in the plane, then the Voronoi diagram of P can be defined as the subdivision of the plane into m cells, such that a point q lies in the cell corresponding to a point p_i if and only if,

$dist(q, p_i) < dist(q, p_j), \forall p_j \in P \;\&\; j \neq i$

A Delaunay triangulation of a set of points P in a plane can be defined as a triangulation so that none of the point in P is inside the circle that circumscribes any triangle in that triangulation. The necessary condition for a triangle to be a Delaunay triangle is that the inside of the circumcircle of the triangle should be empty.

No Delaunay triangulation exists for the collinear points, as the necessary condition for triangulation, violets in this case. It is not possible to have a unique Delaunay triangulation for more than three points lying on the circumference of the same circle. Two or more triangles satisfy the condition for Delaunay triangulation in this case.

In general, Voronoi diagram is the dual graph of Delaunay triangulation. Figure 13.B.4a of Appendix 13.B is the Delaunay triangulation of a set of points. All the circumcircles and their centers are shown in Figure 13.B.4b of Appendix 13.B shows Voronoi diagram corresponding to the Delaunay triangulation shown in Figure 13.B.4a by connecting centers of the circumcircles shown in this figure.

Although, the minimum angle is maximized by Delaunay triangulation in the plane but the reverse is not true; i.e., it is not necessary that the maximum angle is minimized by the Delaunay triangulation. The length of the edges also does not necessarily minimized by the Delaunay triangulation. The point which is nearest to any of the given input points lies on an edge of the Delaunay triangulation.

One of the popular algorithms that compute the Delaunay triangulation is the divide-and-conquer algorithm. The worse case time complexity of this algorithm is $O(m \log m)$, m being the number of input vectors.

Shortest Path: The shortest path distance is the distance between two nodes in a graph so that the sum of the weights of its component edges is minimized. The shortest path distance can be defined for both of undirected and directed graphs.

There are three different types of problems related to shortest paths. These are single source, single destination, and all-pair shortest path problems. Single source shortest path finds the shortest paths from a source node u to all other nodes in the graph. Single-destination shortest path problem finds shortest paths from all nodes in a graph to a single destination node. The all-pair-shortest-path problem finds the shortest path between each and every pair of nodes of a graph. The graph may contain negative edges but no negative cycles. The all-pair-shortest-path problem can be represented as follows:

$$W_{ij} = \left\{ \begin{array}{lll} 0 & ; & \text{if } i = j \\ \infty & ; & \text{if } \exists \text{ no edge between } i \text{ and } j \\ \delta_{i,j} & ; & \text{Otherwise} \end{array} \right\}$$

where W represents the weight matrix.

Numbers of algorithms are available to compute the all-pair-shortest-path of a graph like Floyd–Warshall algorithm, Dijkstra's algorithm, Bellman–Ford's algorithm, etc. Among these algorithms, the Floyd–Warshall algorithm has a time complexity of $O(m^3)$, m being the number of input vectors.

Eigenvalue and Eigenvector: An eigenvector of a linear transformation can be defined as an inclination that remains unchanged under the transformation. A scalar quantity λ is said to be an eigenvalue of the $m \times m$ matrix A if there exists a nontrivial solution x of $Ax = \lambda x$ where x is known as the eigenvector corresponding to the eigenvalue λ. There are m eigenvalues of an $m \times m$ matrix and m eigenvectors corresponding to these eigenvalues.

Appendix 13.B

Figures

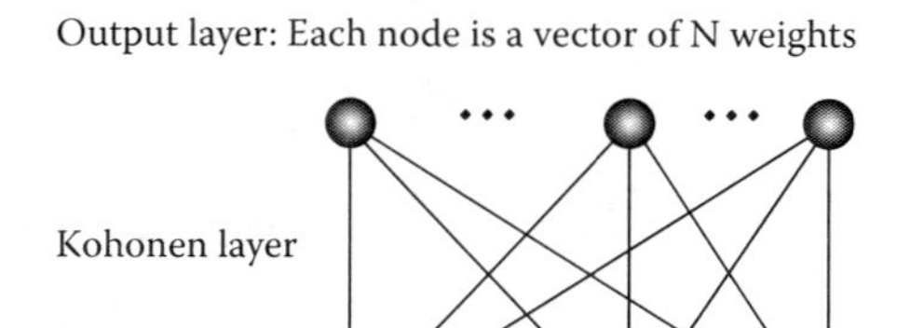

FIGURE 13.B.1: (a)1-dimensional SOM model and (b) 2-dimensional SOM model.

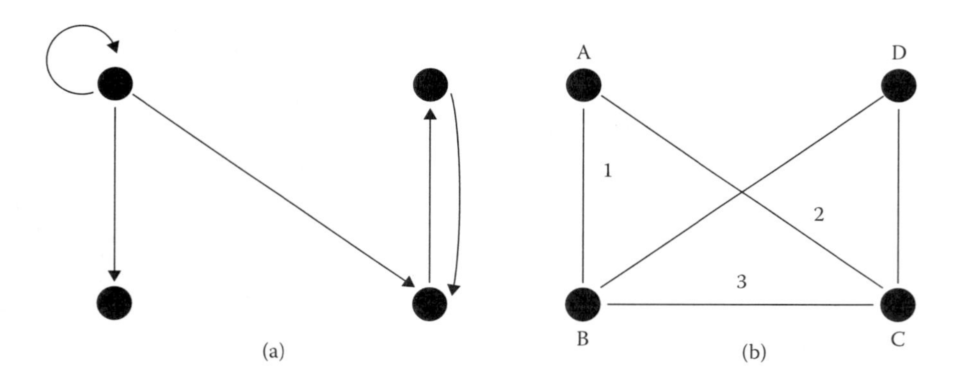

FIGURE 13.B.2: (a) Directed graph and (b) simple graph.

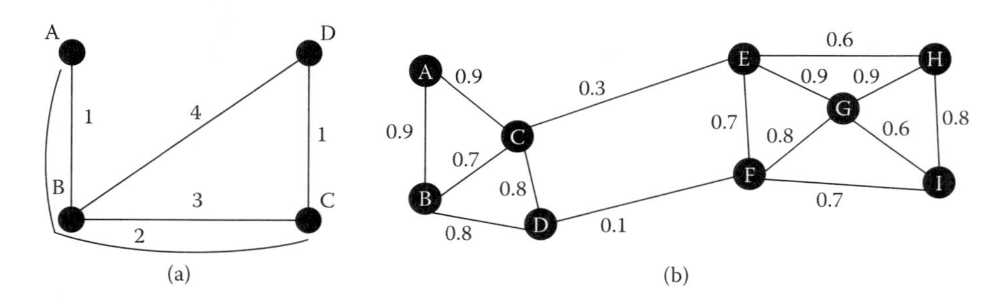

FIGURE 13.B.3: (a) Planar graph and (b) similarity graph.

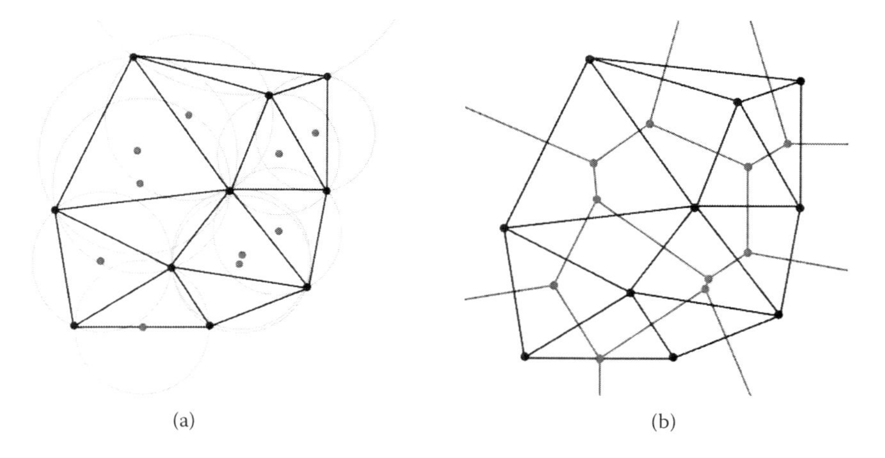

FIGURE 13.B.4: (a) Delaunay triangulation and (b) Delaunay–Voronoi diagram.

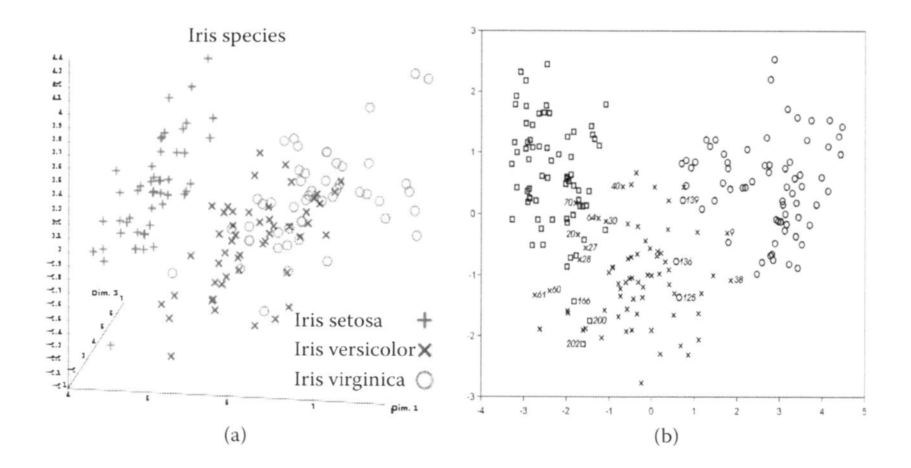

FIGURE 13.B.5: (a) The iris dataset and (b) the seeds dataset.

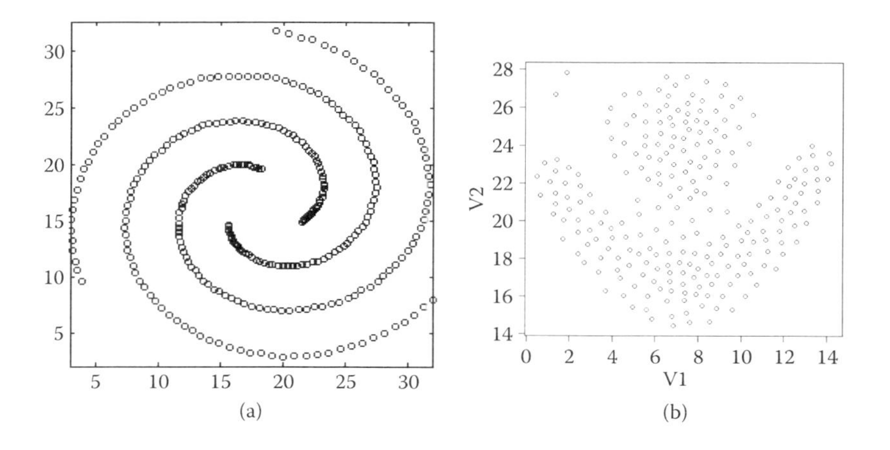

FIGURE 13.B.6: (a) The spiral dataset and (b) the flame dataset.

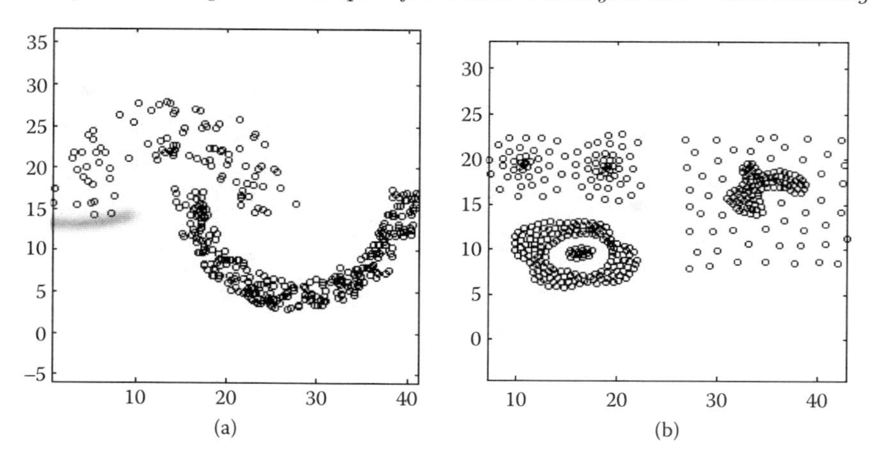

FIGURE 13.B.7: (a) The Jain dataset and (b) the compound dataset. (From Speech and image Processing Unit of the School of Computing, cs.joensuu.fi/sipu/datasets/, accessed on March 20, 2015.)

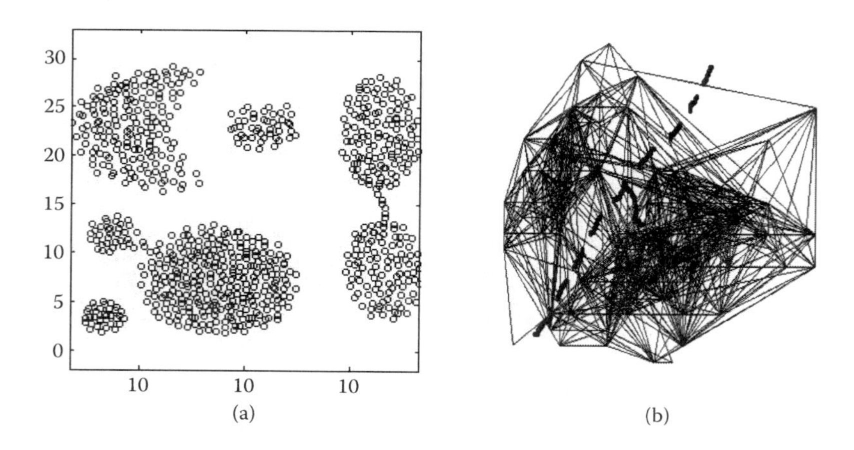

FIGURE 13.B.8: (a) The aggregation dataset (b) Delaunay triangulation of iris dataset; the Min-Cut is shown in dotted line. (From Speech and Image Processing Unit of the School of Computing, cs.joensuu.fi/sipu/datasets/, accessed on March 20, 2015.)

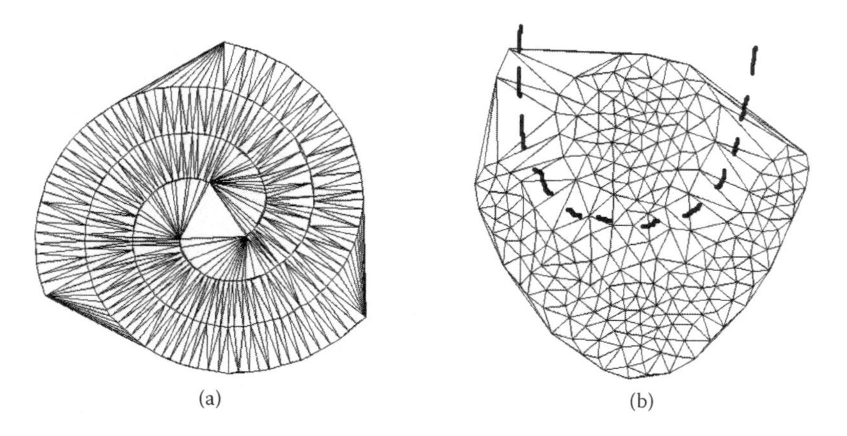

FIGURE 13.B.9: (a) Delaunay triangulation of spiral dataset. (b) Delaunay Triangulation of flame Dataset; the Min-Cut is shown in dotted line.

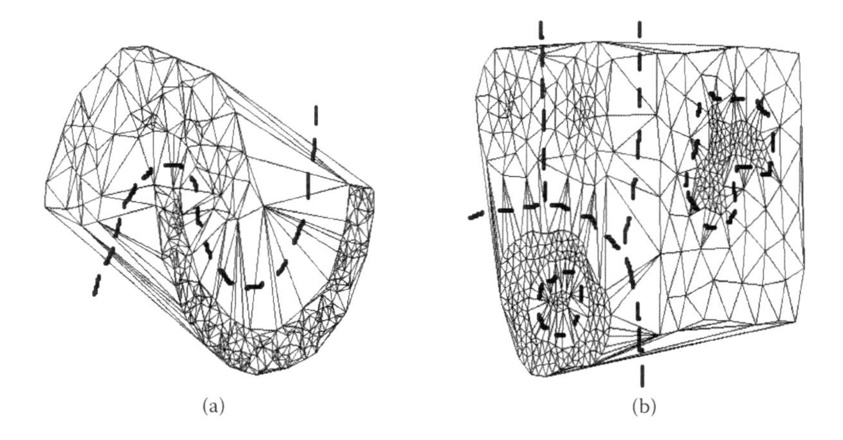

FIGURE 13.B.10: (a) Delaunay triangulation of Jain dataset; the Min-Cut is shown in dotted line. (b) Delaunay triangulation of compound dataset; the Min-Cuts are shown in dotted line.

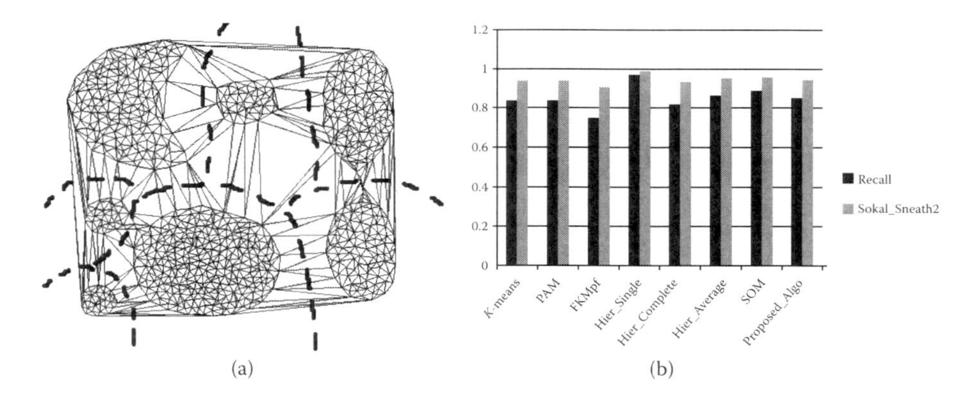

FIGURE 13.B.11: (a) Delaunay triangulation of aggregation dataset; the Min-Cuts are shown in dotted line. (b) Graphical view of clustering results for the iris dataset using Recall and Sokal–Sneath indices.

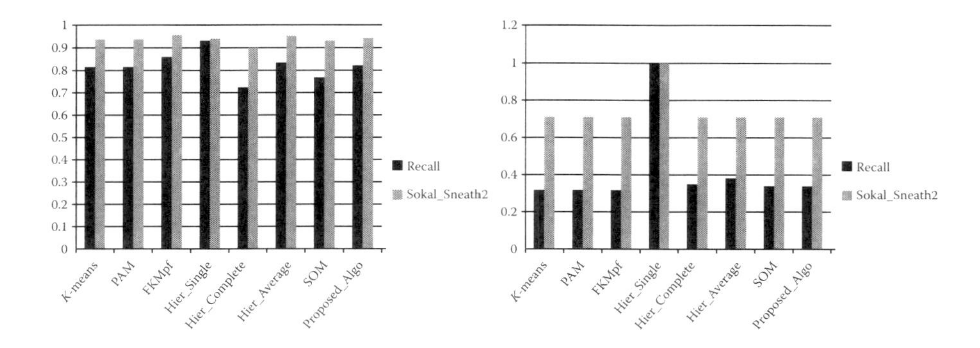

FIGURE 13.B.12: (a) Graphical view of clustering results for the seeds dataset using Recall and Sokal–Sneath indices. (b) Graphical view of clustering results for the spiral dataset using Recall and Sokal–Sneath indices.

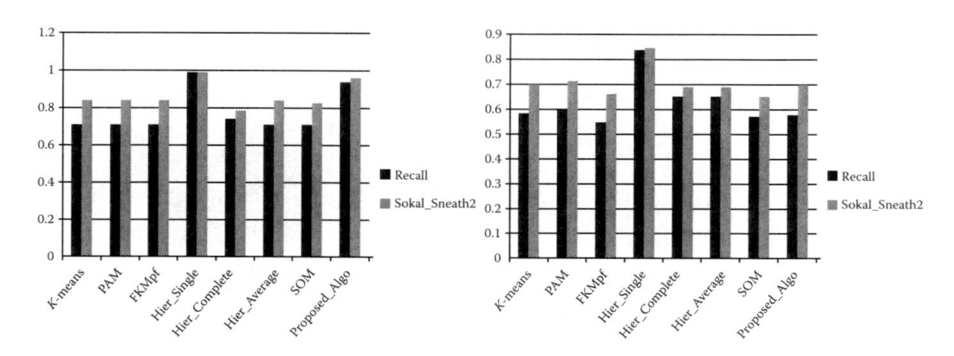

FIGURE 13.B.13: (a) Graphical view of clustering results for the flame dataset using Recall and Sokal–Sneath indices. (b) Graphical view of clustering results for the jain dataset using Recall and Sokal–Sneath indices.

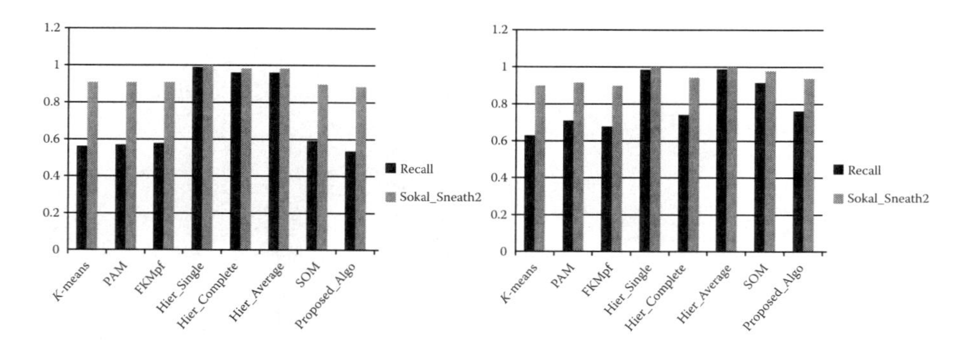

FIGURE 13.B.14: (a) Graphical view of clustering results for the compound dataset using Recall and Sokal–Sneath indices. (b) Graphical view of clustering results for the aggregation dataset using Recall and Sokal–Sneath indices.

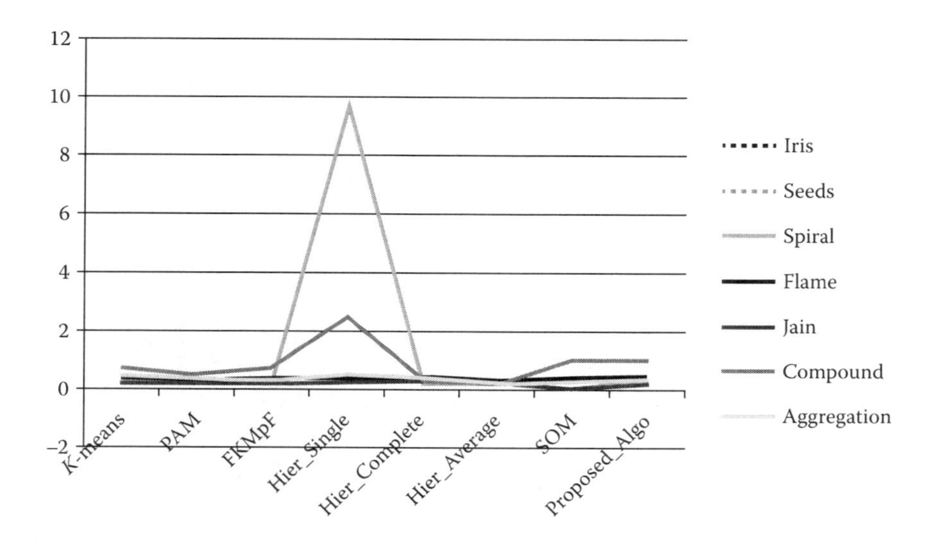

FIGURE 13.B.15: Graphical view of clustering results using Ray–Turi index.

Appendix 13.C

Tables

TABLE 13.C.1: Comparative study of clustering results of iris

External Index	K-Means	PAM	FKMPF	Hierarchical (Sngl-Link)	Hierarchical Compl-Link	Hierarchical Avg-Link	SOM	Proposed Model
Jaccard	0.6958588	0.6958588	0.5938922	0.5891358	0.622282	0.7248	0.8000976	0.7100295
Rand	0.8797315	0.8797315	0.8322148	0.7766443	0.8367785	0.8922595	0.9267114	0.885906

Source: R. A. Fisher. UCI machine learning repository, [online] http://archive.ics.uci.edu/ml, 1936. *Note*: Data by using Jaccard and Rand indices.

TABLE 13.C.2: Comparative study of clustering results of seeds

External Index	K-Means	PAM	FKMPF	Hierarchical (Sngl-Link)	Hierarchical Compl-Link	Hierarchical Avg-Link	SOM	Proposed Model
Jaccard	0.6760758	0.6760758	0.7483116	0.3227432	0.5389135	0.7081771	0.6316851	0.6815294
Rand	0.8713602	0.8713602	0.9048986	0.3569378	0.7961723	0.886489	0.8503076	0.8743677

Source: R. A. Fisher. UCI machine learning repository, [online] http://archive.ics.uci.edu/ml, 1936. *Note*: Data by using Jaccard and Rand indices.

TABLE 13.C.3: Comparative study of clustering results of spiral

External Index	K-Means	PAM	FKMPF	Hierarchical (Sngl-Link)	Hierarchical Compl-Link	Hierarchical Avg-Link	SOM	Proposed Model
Jaccard	0.1957467	0.1961039	0.1961134	1	0.2036921	0.216655	0.1958195	0.1970835
Rand	0.5541265	0.5542914	0.5540646	1	0.553673	0.5366065	0.5543326	0.5551158

Source: H. Chang and D. Y. Yeung, *Pattern Recognition*, 41, 191–203, 2008. *Note*: Data by using Jaccard and Rand indices.

TABLE 13.C.4: Comparative study of clustering results of flame

External Index	K-Means	PAM	FKMPF	Hierarchical (Sngl-Link)	Hierarchical Compl-Link	Hierarchical Avg-Link	SOM	Proposed Model
Jaccard	0.6104777	0.60322	0.6103508	0.5356665	0.4425022	0.5755743	0.5485405	0.8740916
Rand	0.7498257	0.7439331	0.7498257	0.5406206	0.4984658	0.72106	0.6888424	0.9275104

Source: L. Fu and E. Medico, *BMC Bioinformatics*, 8, 3, 2007. *Note*: Data by using Jaccard and Rand indices.

TABLE 13.C.5: Comparative study of clustering results of jain

External Index	*K*-Means	PAM	FKMPF	Hierarchical (Sngl-Link)	Hierarchical Compl-Link	Hierarchical Avg-Link	SOM	Proposed Model
Jaccard	0.4780714	0.503857	0.4240342	0.5400562	0.4406952	0.4406952	0.4149817	0.4690857
Rand	0.6065035	0.6328807	0.5383839	0.5567759	0.5009081	0.5009081	0.5041367	0.5967598

Source: A. Jain and M. Law, *Lecture Notes in Computer Science*, 3776, 1–10, 2005. *Note*: Data by using Jaccard and Rand indices.

TABLE 13.C.6: Comparative study of results of compound

External Index	*K*-Means	PAM	FKMPF	Hierarchical (Sngl-Link)	Hierarchical Compl-Link	Hierarchical Avg-Link	SOM	Proposed Model
Jaccard	0.447631	0.4525963	0.4990489	0.6911561	0.7386037	0.7495113	0.4524205	0.398263
Rand	0.827773	0.8292591	0.8540698	0.890354	0.9165879	0.9209204	0.821501	0.7940706

Source: C. T. Zahn, *IEEE Transactions on Computers*, C-20, 68–86, 1971. *Note*: Data by using Jaccard and Rand indices.

TABLE 13.C.7: Comparative study of clustering results of aggregation

External Index	*K*-Means	PAM	FKMPF	Hierarchical (Sngl-Link)	Hierarchical Compl-Link	Hierarchical Avg-Link	SOM	Proposed Model
Jaccard	0.5741431	0.6853195	0.6338657	0.7430394	0.6907042	0.9898142	0.9191762	0.5939304
Rand	0.8975838	0.9300692	0.9154277	0.92568	0.9297048	0.9977844	0.9823077	0.8963487

Source: A. Gionis, H. Mannila, and P. Tsaparas, *ACM Transactions on Knowledge Discovery from Data (TKDD)*, 1, 1–30, 2007. *Note*: Data by using Jaccard and Rand indices.

References

1. R. O. Duda, P. E. Hart, and D. G. Stork. *Pattern Classification.* New York, NY: John Wiley & Sons Inc., 2nd edition, 2001.

2. S Wang, M. Zhou, and G. Geng. Application of fuzzy cluster analysis for medical image data mining. In *IEEE International Conference on Mechatronics and Automation, 2005,* Niagara Falls, Canada, vol 2, pp. 631–636, July 2005.

3. V. K. Dehariya, S. K. Shrivastava, and R. C. Jain. Clustering of image data set using K-means and fuzzy K-means algorithms. In *2010 International Conference on Computational Intelligence and Communication Networks (CICN),* Bhopal, India pp. 386–391, November 2010.

4. A. K. Jain, J. Mao, and K. M. Mohiuddin. Artificial neural networks: A tutorial. *Computer*, 29(3):31–44, March 1996.

5. T. Kohonen. *Self-Organizing Maps.* New York, NY: Springer-Verlag, 3rd edition, 2001.

6. K. A. Han, J. C. Lee, and C. J. Hwang. Image clustering using self-organizing feature map with refinement. In *Proceedings of IEEE International Conference on Neural Networks, 1995,* Perth, Australia vol. 1, pp. 465–469, November 1995.

7. A. Jain and M. Law. Data clustering: A user's dilemma. *Lecture Notes in Computer Science*, 3776:1–10, 2005.

8. A. K. Jain, M. N. Murty, and P. J. Flynn. Data clustering: A review. *ACM Computing Surveys*, 31(3):264–323, September 1999.

9. K. Kameshwaran and K. Malarvizhi. Survey on clustering techniques in data mining. *International Journal of Computer Science and Information Technologies (IJCSIT)*, 5(2):2272–2276, 2014.

10. S. Deswal and A. Singhrova. Application of graph theory in communication networks. *International Journal of Application or Innovation in Engineering & Management (IJAIEM)*, 1(2):66–70, October 2012.

11. M. D. Berg, O. Cheong, M. V. Kreveld, and M. Overmars. *Computational Geometry.* Berlin, Heidelberg: Springer-Verlag, 3rd edition, 2008.

12. K. Vijaya Sri and K. Usha Rani. Ingenious tools of soft computing. *International Journal of Engineering Sciences Research (IJESR)*, 4(1):1305–1312, October 2013.

13. J. Shi and J. Malik. Normalized cuts and image segmentation. *IEEE Transactions on Pattern Analysis and Machine Intelligence*, 22(8):888–905, August 2000.

14. S. Gunter and H. Bunke. Self-organizing map for clustering in the graph domain. *Pattern Recognition Letters*, 23(4):405–417, February 2002.

15. R. Xu and D. Wunsch. *Clustering.* Piscataway, NJ: Wiley-IEEE Press, October 2008.

16. B. Everitt, S. Landau, M. Leese, and D. Stahl. *Cluster Analysis.* West Sussex, UK: Wiley-IEEE Press, 5th edition, February 2011.

17. T. Soni Madhulatha. An overview on clustering methods. *IOSR Journal of Engineering*, 2(4):719–725, April 2013.

18. X. T. Yuan, B. G. Hu, and R. He. Agglomerative mean-shift clustering. *IEEE Transactions on Knowledge and Data Engineering*, 24(2):209–219, February 2012.

19. P. Franti, O. Virmajoki, and V. Hautamaki. Fast agglomerative clustering using a K-nearest neighbor graph. *IEEE Transactions on Pattern Analysis and Machine Intelligence*, 28(11):1875–1881, November 2006.

20. L. Rokach and O. Maimon (eds.). *Data Mining and Knowledge Discovery Handbook*. New York, NY: Springer, 2nd edition, 2010.

21. K. Somasundaram and M. Mary Shanthi Rani. Eigen value based K-means clustering for image compression. *International Journal of Applied Information Systems (IJAIS)*, 3(7):21–24, August 2012.

22. J. Vesanto and E. Alhoniemi. Clustering of the self-organizing map. *IEEE Transactions on Neural Networks*, 11(3):586–600, May 2000.

23. H. Elghazel and K. Benabdeslem. Different aspects of clustering the self-organizing maps. *Neural Process Letters*, 39(1):97–114, February 2014.

24. F. Ghaemmaghami and R. M. Sarhadi. Somsn: An effective self organizing map for clustering of social networks. *International Journal of Computer Applications*, 84(5):975–987, December 2013.

25. M. Hagenbuchner, A. Sperduti, A. C. Tsoi, F. Trentinil, F. Scarselli, and M. Gori. Clustering XML documents using self-organizing maps for structures. *Lecture Notes in Computer Science*, 3799(5):481–496, 2006.

26. Y. Zhou, H. Cheng, and J. X. Yu. Graph clustering based on structural/attribute similarities. *VLDB*, 2(1):718–729, August 2009.

27. S. E. Schaeffer. Graph clustering. *Computer Science Review*, 1(1):27–64, August 2007.

28. W. Nawaz, K.-U. Khan, and Y.-K. Lee. Spore: Shortest path overlapped regions and confined traversals towards graph clustering. *Applied Intelligence*, 43(1):208–232, July 2015.

29. S. Zhu and M. Garng Huang. A new parallel and distributed shortest path algorithm for hierarchically clustered data networks. *IEEE Transactions on Parallel and Distributed Systems*, 9(9):841–855, September 1998.

30. M. Deng, Q. Liu, T. Cheng, and Y. Shi. An adaptive spatial clustering algorithm based on Delaunay triangulation. *Computers, Environment and Urban Systems*, 35(4):320–332, July 2011.

31. C. Busch, R. LaFortune, and S. Tirthapura. Sparse covers for planar graphs and graphs that exclude a fixed minor. *Algorithmica*, 69:658–684, 2014.

32. R. Chen, Y. Xu, C. Gotsman, and L. Liu. A spectral characterization of the Delaunay triangulation. *Algorithmica*, 27(4):295–300, May 2010.

33. B. Mohar. The Laplacian spectrum of graphs. In Y. Alavi, G. Chartrand, O. R. Oellermann, A. J. Schwenk, (Eds.), *Graph Theory, Combinatorics, and Applications*, vol. 2, New York, NY: Wiley, New York, pp. 871–898, 1991.

34. U. von Luxburg. A tutorial on spectral clustering. *Statistics and Computing*, 17(4):1–32, 2007.

35. H. Qiu and E. R. Hancock. Graph matching and clustering using spectral partitions. *Pattern Recognition*, 39(1):22–34, January 2006.

36. S. Gou, X. Zhuang, H. Zhu, and T. Yu. Parallel sparse spectral clustering for SAR image segmentation. *IEEE Journal of Selected Topics in Applied Earth Observations and Remote Sensing*, 6(4):1949–1963, August 2013.

37. S. P. Gou, X. Zhuang, and L. C. Jiao. Quantum immune fast spectral clustering for SAR image segmentation. *IEEE Geoscience and Remote Sensing Letters*, 9(1):8–12, January 2012.

38. H. C. Huang, Y. Y. Chuang, and C. S. Chen. Affinity aggregation for spectral clustering. In *2012 IEEE Conference on Computer Vision and Pattern Recognition (CVPR)*, Providence, RI, pp. 773–780, June 2012.

39. H. C. Huang, Y. Y. Chuang, and C. S. Chen. Multi-affinity spectral clustering. In *2012 IEEE International Conference on Acoustics, Speech and Signal Processing (ICASSP)*, Kyoto, Japan, pp. 2089–2092, March 2012.

40. C. Dhanjal, R. Gaudel, and S. Clemenon. Efficient eigen-updating for spectral graph clustering. *Neurocomputing*, 131:440–452, 2014.

41. J. Li, Y. Xia, Z. Shan, and Y. Liu. Scalable constrained spectral clustering. *IEEE Transactions on Knowledge and Data Engineering*, 27(2):589–593, February 2015.

42. H. Jia, S. Ding, X. Xu, and R. Nie. The latest research progress on spectral clustering. *Neural Computing and Applications*, 24(7–8):1477–1486, June 2014.

43. M. C. V. Nascimento and A. C. P. L. F. de Carvalho. Spectral methods for graph clustering—A survey. *European Journal of Operational Research*, 211(2):221–231, June 2011.

44. A. S. Drigas and J. Vrettaros. Using the self-organizing map (SOM) algorithm, as a prototype E-content retrieval tool. In *Proceedings of the International Conference on Computational Science and its Applications, ICCSA 2008, Part II*, Springer-Verlag, Berlin, Heidelberg, pp. 14–23, 2008.

45. R. A. Fisher. UCI machine learning repository, [online] http://archive.ics.uci.edu/ml, 1936.

46. P. Fränti, J. Saastamoinen, I. Kärkkäinen, T. Kinnunen, V. Hautamäki, and I. Sidoroff. Clustering datasets, speech, and image processing unit, University of Eastern Finland, [Online]. Available at http://cs.joensuu.fi/sipu/datasets/. Accessed on 20 March 20, 2015.

47. L. Fu and E. Medico. A novel fuzzy clustering method for the analysis of DNA microarray data. *BMC Bioinformatics*, 8(1):3, 2007.

48. H. Chang and D. Y. Yeung. Robust path-based spectral clustering. *Pattern Recognition*, 41(1):191–203, 2008.

49. C. T. Zahn. Graph-theoretical methods for detecting and describing gestalt clusters. *IEEE Transactions on Computers*, C-20(1):68–86, January 1971.

50. A. Gionis, H. Mannila, and P. Tsaparas. Clustering aggregation. *ACM Transactions on Knowledge Discovery from Data (TKDD)*, 1(1):1–30, 2007.

51. S. Saha and S. Bandyopadhyay. Performance evaluation of some symmetry-based cluster validity indexes. *IEEE Transactions on Systems, Man, and Cybernetics, Part C (Applications and Reviews)*, 39(4):420–425, July 2009.

52. P. Roy and J. K. Mandal. Performance evaluation of some clustering indices. In L. Jain, H. Behera, J. Manda, D. Mohapatra (eds.), *Proceedings of Computational Intelligence in Data Mining*, Volume 3: *Proceedings of the International Conference on CIDM, 20–21 December 2014, Sambalpur, India,* New Delhi, India: Springer, pp. 509–517, 2015.

Chapter 14

Quantum Algorithm for Sequence Clustering

Arit Kumar Bishwas, Ashish Mani, and Vasile Palade

14.1 Introduction

It is observed in recent years that we are living in a world increasingly driven by data with more information about individuals, corporates, institutions, and governments available than ever before. Data has always played an important role in business research but in the current era the concept of big data is changing the way people use data for conclusions. Whether it is astronomy, genome modeling, or stock marketing, big data is playing a big role in future decision making. Recently, the increasing of importance of clustering with big sequential data sets has been noticed due to its wide application areas like DNA genome modeling and financial data analysis. Handling big data for machine learning activities such as classification and clustering is still very cumbersome and complex. One such problem is the clustering of sequences. Clustering is a difficult problem and this intrinsic difficulty worsens with huge sequential data (which can be termed *big sequences*). The structure of the underlying process is often difficult to deduce. Because of the typically different lengths, sequences need to be dealt with and very large sequences set cardinally. In this chapter we have proposed an interesting approach for clustering sequencing data in a quantum paradigm.

The proposed approach is based on the *hidden Markov model* (HMM) and quantum K-means clustering. In this approach at first, a new representation space of input data objects (sequences) is developed. The vector of each object's similarities with respect to a predeterminated set of other objects describes the respective object. These similarities are determined using the HMM. Then the new representation space of input sequences will be mapped to quantum states with the help of quantum RAM (QRAM). Clustering is then performed using quantum K-means clustering over the new transformed input data set. We have observed around an exponential speed-up gain in computational complexity in our approach, as compared to the classical approaches. Before discussing the new proposed approach, we will briefly walk through some of the important and relevant concepts in the following sections, which will develop a base to understand our proposed approach.

14.2 Clustering

Clustering is the process of categorizing a set of entities in such a way that entities in the same category share more similarities with one an other than with those in other categories and in this sense, the categories are called clusters. This is an unsupervised learning process where the task is to find the pattern with unlabeled data. There are many classical clustering methods that have been proposed but the classical K-means clustering method (used for classical computers) is the most widely used. We will talk about the K-means clustering method in detail in Section 14.3. First, we will discuss some other important types of clustering techniques in brief.

14.2.1 Hierarchical clustering

The basic idea of this type of clustering is to start with a large data set and divide it into some criterion-based partitions. Hierarchical-based clustering is a connectivity-based clustering method in which the more nearby objects in a set of "objects to cluster" are put together to form a group or cluster based on their inter-object distances. The process of clustering can be represented by a *dendrogram*, as at different distances, different clusters will be formed. In this type of clustering process, we don't develop single clusters of data sets but a hierarchy of clusters where each cluster is connected to some other cluster hierarchically so we can say that they merge somewhere in the hierarchy. In a dendrogram, the distance at which the clusters merge is represented by the *y-axis*, and the *x-axis* represents the cluster of objects that don't combine.

14.2.2 Centroid-based clustering

Centroid-based clustering is a method of clustering in which clusters are represented by a central vector that may not necessarily be a member of the data set [1]. The K-means clustering method is a widely known example of this type of clustering, where the number of clusters K is defined before starting the clustering process. And, after the clustering process, it outputs K clusters. In the K-means clustering method, at first we define the number of clusters required. Then we allocate the initial K-means in the given data set. Then, by using any distance-based method, we update the means till the time it converges, and no further change of position of the means is likely.

14.2.3 Density-based clustering

In density-based clustering, the higher-density areas are categorized as clusters, and the objects which are required to separate the clusters are usually considered to be border points and noises [2,3]. Density-based spatial clustering of applications with noise (DBSCAN) is a well-known density-based clustering method. It works by connecting the objects which are within certain a distance threshold value, based on some density criterion. DBSCAN requires a linear number of range queries in the database so the complexity of DBSCAN is low. Apart from some good properties, DBSCAN has some drawbacks too. DBSCAN cannot detect intrinsic cluster structures and also has difficulty in detecting cluster borders, which forces some kind of density drop.

14.2.4 Clustering with big data

In the past few years, big data has started to play a vital role in the industry for large amounts of data processing and analytical purposes in business. Dealing with large volumes of data for business needs is a growing trend for organizations. *Gartner (earlier META Group)* analyst *Doug Laney* discussed and defined data growth challenges and its opportunities with the *3V's* model, *volume*, *velocity*, and *variety*, in a research report in 2001 [4]. Over the years, new challenges and opportunities added some more parameterized characteristics to the 3Vs model.

Volume: This characteristic defines the amount of data generated and stored. The amount of data determines the value of the data to be processed for business forecasting or analytical purposes. How large an amount of data is required to be considered big data is a relative decision of the organization that will work with it.

Velocity: This is in the context of the speed of the data generated and processed for strategic, analytical, etc. business or non-business decisions and demand fulfillment.

Variety: This is in the context of the nature of the data. In today's world, heterogeneous data analysis is very important because it helps in analyzing such a variety of data.

Veracity: It is not only the quantity of data but the quality of data that is also very important for accurate analysis. It helps in maintaining the reputation of high quality of large volumes of data.

Variability: Inconsistent or corrupted data can hamper and damage the purpose of the data for analysis. Analysis with wrong data can lead to disasters.

With classical computers, there are many solutions that have been developed to deal with the big data issue. Among them, Hadoop [5] is one of the most popular.

Clustering with big data is a challenging and complex task. Recently, many techniques and technologies have been developed to handle these kind of machine learning tasks. One of the most common ways is by using the Hadoop framework where big data can be stored in multiple computers in Hadoop clusters by using the map-reduce technique. Then we can use any clustering technique to do clustering on these big data sets (obviously with some changes to make them compatible to work with the map-reduce technique). But these types of techniques suffer in run time due to the classical computing limitations. In the next section we will discuss a quantum clustering technique which works for quantum computers and is exponentially faster than existing classical clustering techniques when dealing with big data.

14.2.5 Recent developments—quantum clustering

In very recent years, researchers have started to work with efficient quantum-clustering algorithms. In [6], the authors demonstrated a quantum K-means clustering algorithm and claimed an exponential speed-up in run time in comparison to the classical K-means clustering algorithm. We have used this quantum K-means clustering algorithm to address the sequence-clustering problem; we will discuss the approach in a later section.

14.3 Classical K-Means Clustering

The K-means clustering algorithm is an unsupervised machine-learning algorithm, which helps in solving clustering problems. It classifies a given data set through a certain number of clusters. In this approach initially k centroids are defined, one for each cluster. Then, the centroids are placed, in a calculating way as much as possible, far away from each other because a different location causes a different result. After that, each point belonging to the data set is associated with the nearest available centroid. The set of data points associated with each mean creates one cluster, hence results the initial k clusters.

We now need to calculate the new means, which are to be the centroids of the observations in the new clusters. And the iterations keep on going until the centroids do not change their positions in the data set. Finally, the algorithm aims at minimizing an objective function, which may be a squared error function.

Let us suppose $\{O_1, O_2, \cdots, O_n\}$ is the set of observations and each observation is a d-dimensional real vector, i.e.,

$$O_n = \{O_{n1}, O_{n2}, \cdots, O_{nd}\}.$$

The aim is to partition n observations into k groups $C = \{C_1, C_2, \cdots, C_k\}$ in such a way that the sum of the distance functions of each point in the group to the k center is at a minimum, i.e.,

$$argmin \sum_{i=1}^{k} \sum_{x \in C_i} \|O - \mu_i\|^2,$$

where μ_i is the mean of the points in group C_i. The algorithm is as follows:

Step 1: Define k, the number of clusters required.

Step 2: Select initial means $\mu_1, \mu_2, \ldots, \mu_k$ in each k clusters $C_1, C_2, \ldots, C_k$.

Step 3: Assign each observation to the cluster, for which the square sun of the distance between observation and the mean of that cluster is least:

$$C_i^t = \{O_j : \|O_j - \mu_i^t\|^2 \leq \|O_j - \mu_l^t\|^2 \ \forall\ l\},$$

where $1 \leq l \leq k$.

Step 4: Calculate the means in the cluster of newly assigned observations to get the centroid of the observations, i.e.,

$$\mu_l^{t+1} = |C_i^t|^{-1} \sum_{O_l \in C_i^t} O_l.$$

Step 5: Keep on iterating step 3 and step 4 until the algorithm reaches convergence, which means that means no new mean is possible to find any more, as the further assignments no longer changes the means.

14.4 Introduction to Quantum Computing

In this section we will discuss some very important basic concepts of quantum computing. Proper grasping of these concepts will help us to understand the proposed quantum algorithm for sequence clustering.

14.4.1 Quantum bits and quantum gates

A *qubit* [7] is a quantum bit in the quantum computing paradigm. In a similar way as, the binary digit or bit is defined in classical computing. A qubit is regarded as the basic unit of information in the quantum computer.

For manipulating any information in a qubit we use quantum gates [7]. A quantum gate works in a similar way as a classical logic gate. Classical logic gates take bits as input then evaluate and process the input to produce new output bits. In the same way, quantum gates take input as qubits, but the qubit can exist in a state of superposition. It is interesting to note that quantum gates are reversible in nature, unlike many classical logic gates. This means that the quantum gate's outputs can be transformed back into the original input, which is actually essential in order to preserve the quantum state. And, the number of outputs must be the same as the number of inputs in order for the gates to be reversible.

14.4.2 Entanglement

This is a state of a composite quantum system that embroils unusually strong correlations between parts of the system [8]. It describes the way that particles of physical parameters like energy/matter can be correlated to predictably interact with one another, regardless of how far they are separated.

The qubits that have interacted with each other retain a type of connection and can be entangled with each other in pairs in the process known as correlation. For example, knowing the spin state of one of the entangled particles (the direction of the spin may be up or down) allows us to know that the direction of the spin of its other entangled particle is in the reverse direction. Even more amazing is the knowledge that the measured particle has no single spin direction before being measured, but is simultaneously in both a spin-up and spin-down state due to the phenomenon of superposition. This gives a way to define quantum channels through which the quantum information can be traveled.

Quantum entanglement permits qubits that are separated by inconceivable distances to interact with one another immediately, and communication that is not limited to the speed of light. No matter how far the distance between the linked particles, as long as they are isolated, they will stay entangled.

We are considering two non-interacting systems A and B, with respect to finite dimensional Hilbert spaces H_A and H_B. The Hilbert space of the composite system is the tensor product,

$$H_A \otimes H_B.$$

So, if the first system is in state $|\psi_A\rangle$ and the second in the state $|\varphi_B\rangle$, the state of the composite system will be in

$$|\psi_A\rangle \otimes |\varphi_B\rangle.$$

The above-mentioned composite system is called separable states, or (in the simplest case) product states.

All states are not, in general, the separable states (and thus product states). By fixing a basis $|i_A\rangle$ for H_A and a basis $|j_B\rangle$ for H_B, the most general state is of the form

$$|\psi_{AB}\rangle = \sum_{i,j} c_{ij}|i_A\rangle \otimes |j_B\rangle.$$

The above state is separable if

$$c_{ij} = c_i^A c_j^B,$$

which yields

$$|\psi_A\rangle = \sum_i c_i^A|i_A\rangle \text{ and } |\varphi_B\rangle = \sum_j c_j^B|j_B\rangle$$

If a state is inseparable, it is called an entangled state and in that case,

$$c_{ij} \neq c_i^A c_j^B.$$

14.4.3 Superposition

Quantum superposition [9] is the phenomenon in the quantum space system in which quantum particles appear to exist in all states simultaneously. It is the fundamental principle of quantum mechanics. As per the principle, a particle (a subatomic particle like an electron) exists partly in all its particular, theoretically possible states simultaneously; but, when observed, it gives an outcome corresponding to only one of the possible configurations.

Thus, at any instance a qubit can be in a superposition of both a $|0\rangle$ and a $|1\rangle$ simultaneously, i.e.,

$$|\psi\rangle = a|0\rangle + b|1\rangle,$$

where a and b are complex numbers and have the property

$$|a|^2 + |b|^2 = 1.$$

Here, qubit is properly normalized, and the coefficients a and b are known as the amplitudes of the $|0\rangle$ and $|1\rangle$ component respectively.

14.4.4 Hamiltonian

Hamiltonian is a function used to express the energy of a system in terms of its momentum and positional coordinates [10]. This is the sum of its kinetic and potential energies in general cases. In Hamiltonian equations, the usual equations used in mechanics are replaced by equations expressed in terms of momentum.

In quantum mechanics, operators describe observables, and these operators can be represented as *Hermitian* matrices. The permissible values for an observable are the *eigenvalues* of its associated Hermitian matrix. The Hamiltonian, in general represented as H, is the observable corresponding to the total energy of the system. When one measures the total energy of the system, the possible obtained values are its eigenvalues.

14.4.5 Quantum algorithm

A quantum algorithm runs on a realistic model of quantum computation; the quantum circuit models of computation are the most commonly used models, as is the recently explored adiabatic quantum computation model. The quantum algorithm is a step-by-step procedure where each step is performed on a quantum computer. The term quantum algorithm is usually associated with those algorithms which use some essential features of quantum computation like quantum superposition or quantum entanglement. Although it is to note that all classical algorithms can also be executed on a quantum computer. Some of the famous quantum algorithms are Shor's algorithm [11] for factorization, and Grover's algorithm [12] used for searching an unstructured database. Shor's algorithm is exponentially faster than the best known classical algorithm for factorization. Grover's algorithm runs with quadratic speed improvement compared to the best possible counter classical algorithm for the similar task.

14.4.6 Adiabatic quantum computation

Adiabatic quantum computation is a method used to perform quantum computation, which is based on the adiabatic process [13,14]. The adiabatic quantum computer based on adiabatic quantum computation is considered a polynomial equivalent to the quantum computer based on a circuit model. In the adiabatic process, the system adapts its configuration during the process with gradually changing conditions, which causes the modification of probability density by the process. So the system will end with a final Hamiltonian, which is not equivalent to the initial Hamiltonian. Suppose $t_{initial}$ is the initial time at which the quantum system has some energy given by the Hamiltonian $H(t_{initial})$. The Hamiltonian will then keep on getting modified with changing conditions continuously. So after some time t_{final} evolves into a final Hamiltonian $H(t_{final})$ based on the Schrödinger equation. Here the system evolution critically depends on the time gap $t_{gap} = t_{final} - t_{initial}$.

In association with the process of adiabatic quantum computation, at first a ground state Hamiltonian is originated. The ground state of this Hamiltonian describes the solution to the problem in scope. Then a system with a simple Hamiltonian is prepared and initialized at ground state and allows this simple Hamiltonian to evolve, which results in a final required complicated Hamiltonian. And as per the adiabatic theorem, the system will remain in a ground state, and describes the solution to the problem in scope. The time complexity of the adiabatic quantum algorithms depends on the spectral gap in energy of the Hamiltonian. So to keep the system in ground state during evolution, the spectral gap gap_{min} should be very small. The run time of the quantum algorithm then can be considered to be bounded by $O(\frac{1}{gap_{min}^2})$.

14.5 Quantum K-Means Clustering

In very recent years, quantum machine learning has become a major field of research interest. The field includes the investigation of some of the computational hard problems like classification and clustering of big data. Clustering is a computational hard problem and so many research activities are going on around the world to find a better algorithm in terms of speed-up by using quantum properties. Seth Lloyd et al. [6] discussed a quantum version of the K-means clustering algorithm, which is exponentially faster than the classical counterpart.

The approach is similar to the classical one except that all the means will be expressed in quantum superposition. The algorithm starts by selecting k vectors as initial seeds for each of the clusters. This initial seed selection is very important in both classical and quantum algorithm as this projects the efficiency of the algorithm. These seeds may be chosen at random or in a way that maximizes the average distance between them followed by re-clustering. An efficient approach has been discussed in [6] for finding the seed quantum mechanically based on the adiabatic quantum process. In this approach, at first a quantum state for finding initial seed is constructed,

$$|\Psi\rangle = |\psi\rangle_1 \otimes, \cdots |\psi\rangle_k,$$

where $|\psi\rangle_k = (M)^{-\frac{1}{2}} \sum_{j=1}^{M} |j\rangle$ is the uniform superposition of vector k. The initial Hamiltonian is considered as

$$H_0 = 1 - |\Psi\rangle\langle\Psi|$$

and then use the following final Hamiltonian:

$$H_S = \sum_{j_1 \cdots j_k} F[|j_1\rangle\langle j_1| \otimes \cdots \otimes |j_k\rangle\langle j_k|],$$

where

$$F = f(\{|\vec{v}_{jz} - \vec{v}_{jz'}|^2\})$$

and

$$f = - \sum_{z,z'=1}^{k} |\vec{v}_{jz} - \vec{v}_{jz'}|^2.$$

The seed set that maximizes the average distance squared between seeds will be the ground state of the above final Hamiltonian. Also, the adiabatic algorithm to find the sets of g vectors that lie in the same cluster is represented as

$$H_c = \sum_{j_1 \cdots j_g} F[|j_1\rangle\langle j_1| \otimes \cdots \otimes |j_g\rangle\langle j_g|],$$

where

$$F = f(\{|\vec{v}_{jz} - \vec{v}_{jz'}|^2\})$$

and

$$f = \sum_{z,z'=1}^{k} |\vec{v}_{jz} - \vec{v}_{jz'}|^2 + \kappa\delta_{j_z, j_{z'}}, \; \kappa > 0.$$

This shows that the distance term now gives sets of vectors that are closely associated or clustered. The term $\kappa\delta_{j_z, j_{z'}}$ states that the vectors in z and z' positions are not the same.

Now move on to the next step in which we will first construct a quantum state that contains a uniform superposition of all the vectors. Each vector is assigned to its appropriate cluster. This quantum state will provide the information of the similarities of the states. The above required quantum adiabatic state is

$$|\chi\rangle = (M)^{-\frac{1}{2}} \sum_{c, j \in c} |c\rangle|j\rangle.$$

The process is very straightforward like in the case of the classical approach, except that all the means will be expressed in quantum superposition. The process is as follows:

Step 1: Choose k initial seeds, say i_c, for each k clusters such that maximizes the average distance between the vectors.

Step 2: Do re-clustering.

Step 3: Continue the re-clustering process until the cluster assignment state is unchanged.

Now we will discuss the implementation of these steps in detail. Begin with an initial quantum state

$$\frac{1}{\sqrt{Mk}} \sum_{c' j} \left|c'\right\rangle|j\rangle \left(\frac{1}{\sqrt{k}} \sum_{c} \left|c'\right\rangle|i_c\rangle\right)^{\otimes d}$$

where $\left(\frac{1}{\sqrt{k}} \sum_c \left|c'\right\rangle|i_c\rangle\right)^{\otimes d}$ are d seeds quantum states. As per earlier discussion, the distances $|\vec{v}_j - \vec{v}_{i_{c'}}|^2$ can be calculated in the $c^i j$ component of the superposition and applied to the phase $e^{-i\triangle t|\vec{v}_j - \vec{v}_{i_{c'}}|^2}$.

Now, consider an initial Hamiltonian $H_0 = 1 - |\phi\rangle\langle\phi|$, where $|\phi\rangle = \frac{1}{\sqrt{k}} \sum_{c'} \left|c'\right\rangle$. Next, adiabatically make this initial Hamiltonian to grow to the next-level Hamiltonian H_1 which will result in the following clustering state:

$$|\chi_1\rangle = (M)^{-\frac{1}{2}} \sum_{c, j \in c} |c\rangle|j\rangle.$$

At the next re-clustering step, we will consider d such above states $|\chi_1\rangle$. Calculate the average distance between $\vec{v}_j$ and the mean of cluster c, i.e.,

$$|\vec{v}_j - (M_c)^{-\frac{1}{2}} \sum_{k \in c'} \vec{v}_j|^2 = |\vec{v}_j - \vec{v}_c|^2,$$

and apply the phase $e^{-i\delta t|\vec{v}_j - \vec{v}_{c'}|^2}$ to each component $\left|c'\right\rangle|j\rangle$ of the superposition, where δ is the accuracy of the distance evaluation.

Now perform the adiabatic algorithm to the combined state

$$|\chi_2\rangle = \frac{1}{\sqrt{Mk}} \sum_{c', j} \left|c'\right\rangle|j\rangle|\chi_1\rangle^{\otimes d}$$

with initial Hamiltonian $1 - |\phi\rangle\langle\phi|$ and gradually deform to the final Hamiltonian H_f. Here, associate each cluster c' with the set of js by rotating $\left|c'\right\rangle$ that must be assigned to c'. The final state is then

$$\left(\frac{1}{\sqrt{Mk}} \sum_{c', j \in c'} \left|c'\right\rangle|j\rangle\right) |\chi_1\rangle^{\otimes d}.$$

The above algorithm can be used to allocate states to clusters by repeating d times iterations to create a quantum superposition of the cluster assignments at each step until the reassignment doesn't change the states.

14.6 Sequence Clustering in Classical Paradigm

Sequence clustering [15] is a clustering process where a group of sequences that are similarly based on some features are grouped. Sequence clustering has significant application usability in bio-informatics, in which the sequences can be either of protein origin or genomic.

14.6.1 Sequence-clustering methods

Sequence-clustering methods can be, in general, categorized into three classes: *proximity-based methods*, *feature-based methods*, and *model-based methods*. Let's discuss the methods one by one.

Proximity-based method:

In this type of approach, the key effort of the clustering process is in formulating similarity measures between the sequences. Any standard distance-based method can be applied with such measures. Here quantification of similarity between objects can be achieved by using a distance measure with a given set of objects in feature space. And naturally, the objects that are distant from others can be considered outliers. Proximity-based approaches assume that the nearness of an outlier object to its nearest neighbor objects significantly diverges from the nearness of the object to most of the other objects in the data set.

In general, there are two types of proximity-based outlier detection methods: distance-based and density-based methods.

Distance-based outlier detection method: A distance-based outlier detection method refers to the neighborhood of an object, which is defined by a given radius. An object is then considered an outlier if its neighborhood does not have enough other points.

Density-based outlier detection method: In this approach, we compare the density around a point vector with the density around its local neighbors. The relative density of a point vector compared to its neighbors is computed as an outlier score.

Featured-based method:

In this type of method, a set of features is extracted from each individual data sequence that captures historical information. The sequence-clustering problem is thus condensed to a more addressable vector of features clustering.

Model-based method:

In model-based methods, an analytical model is anticipated for each cluster; then a set of models are found which best fits the data.

14.6.2 Background on hidden Markov model

The hidden Markov model (HMM) is the backbone of our discussed sequence-clustering algorithms. So before beginning our discussion of classical sequence clustering we will briefly look into HMMs.

HMMs [16–20] are machines that advance randomly from one internal state to another by producing an output symbol in every time step. We can detect the output symbol only externally. The internal states of the machine remain hidden. Consequently, the time evolution of an HMM can be characterized through a set of transition matrices T_m, where m represents the output symbol generated during the respective time step. Technically, an HMM is defined by a triplet $\lambda = (T, E, \pi)$.

Here T is the transition matrix, which holds the probability values of state transitions:

$$T = \{T_{ij}, 1 \leq j \leq N, T_{ij} \geq 0\}, \text{and}$$

$$T_{ij} = P[S_j|S_i], 1 \leq i, j \leq N,$$

where $S = \{S_1, S_2, \ldots, S_N\}$, the finite set of hidden states, and T_{ij} is the probability of a state transition of S_i to S_j.

$E = \{e(O|S_j)\}$ is the emission matrix indicating the probability of the emission of observed symbol O when the system state is S_j.

The initial state probability distribution is $\pi = \{\pi_i\}$, and $\pi_i = P[S_i], 1 \leq i \leq N, \pi_i \geq 0$, $\sum_{i=1}^{N} \pi_i = 1$.

Once we have the HMM model, next we need to train it with given observable sequences. A well-known Baum–Welch [21] algorithm is used to train the HMM with the help of forward-backward procedures [22].

The Baum–Welch algorithm finds the maximum likelihood estimate of the parameters of an HMM, which is given a set of observed feature vectors or sequences. It finds a local maximum for λ, say λ^*, that maximizes the probability of the observation:

$$\lambda^* = argmax_{\lambda} P(O|\lambda).$$

The evaluation step of the Baum–Welch algorithm [23] is performed by forward-backward procedures. The forward-backward algorithm computes the posterior marginal of all the hidden state variables for a given observable sequence O. Forward-backward procedure is a fine example of dynamic programming in that it efficiently computes the values that are required to obtain the posterior marginal distribution.

14.6.2.1 Baum–Welch algorithm

The following steps describe the Baum–Welch algorithm [24]:

Step 1: Set the HMM model $\lambda = (T, E, \pi)$ with random initial conditions.

Step 2: Apply the forward procedure. Let us consider that

$$\alpha_i(t) = P(O_1, \cdots, O_t|Q_t = i, \lambda).$$

Here, $\alpha_i(t)$ is the probability of observing a subsequence of observables $O_1, \cdots, O_t$ at time t and state $Q_t = i$. Now recursively calculate $\alpha_i(t)$ as

1. $\alpha_i(1) = \pi_i E_i(O_1)$, and

2. $\alpha_j(t+1) = [\sum_{i=1}^{N} \alpha_i(t) T_{ij}][E_j(O_{t+1})]$.

Step 3: Apply the backward procedure. Let us consider $\beta_i(t) = P(O_{t+1}, \cdots, O_{t'}|Q_t = i, \lambda$. This is the probability of the ending sub sequence $O_{t+1}, \cdots, O_{t'}$. Now recursively calculate $\beta_i(t)$ as

1. $1.\beta_i(t^{'}) = 1$, and

2. $\beta_i(t) = \sum_{j=1}^{N} \beta_i(t+1) T_{ij} E_j(O_{t+1})$.

Step 4: Once we calculate the values $\alpha_i(t)$ and $\beta_i(t)$, we will update the λ. For this let us first calculate some temporary variables, say, $\gamma_i(t)$ and $\xi_{ij}(t)$:

$$\gamma_i(t) = P(Q_t = i|O, \lambda) = \frac{\alpha_i(t)\beta_i(t)}{\sum_{j=1}^{N} \alpha_j(t)\beta_j(t)}.$$

With the given HMM parameter λ and observed sequence O, this is the probability of being in state i at the time t.

Also,

$$\xi_{ij}(t) = P(Q_t = i, Q_{t+1} = j|O, \lambda) = \frac{\alpha_i(t)T_{ij}\beta_i(t)E_j(O_{t+1})}{\sum_{i=1}^{N}\sum_{j=1}^{N} \alpha_i(t)T_{ij}\beta_i(t)E_j(O_{t+1})}.$$

With the given HMM parameter λ and observed sequence O, this is the probability of being in state i and j at the time of t and $t + 1$ respectively.

Now, update the HMM parameter λ:

Step 1: Find the expected frequency spent in state i at time 1 as

$$\pi_i^{'} = \gamma_i(1).$$

Step 2: Find the updated expected number of transitions from state i to state j as

$$T_{ij}^{'} = \frac{\sum_{t=1}^{t^{'}-1} \xi_{ij}(t)}{\sum_{t=1}^{t^{'}-1} \gamma_i(t)}.$$

Step 3: Find the updated expected number of times the output observations have been equal to c_k while in state i over the expected total number of times in state i, i.e., $E^{'}$ as

$$E^{'}(c_k) = \frac{\sum_{t=1}^{t^{'}} I_{O_t=c_k}\gamma_i(t)}{\sum_{t=1}^{t^{'}} \gamma_i(t)}$$

where $I_{O_t=c_k} = \{^{1,\ if\ O_t=c_k}_{0,\ otherwise}$ is an indicator function.

Step 4: Iterate the steps until reaching a convergence.

14.6.3 Classical sequence-clustering algorithm

Let's begin our discussion with the significant classical sequence-clustering research work done so far. The problem in sequence clustering is to find an appropriate metric for measuring similarities or dissimilarities between sequences, and, as we can explore in [25,26], that HMMs are a widely used suitable tool for that purpose. The HMMs have not been explored extensively for sequence clustering, although HMM has been widely used in areas like speech recognition and natural language processing. The [27] demonstrates the use of HMM for clustering word sequences with a proximity-based approach. Most of the work for sequence clustering falls under the category of the proximity-based clustering approach. In most of these methods, HMMs were used to compute the similarities measurement within the sequences [22,28] and then standard distance matrix-based approaches were used to do the clustering. The algorithm for such a standard approach has been described as follows:

Step 1: At first, train one HMM λ_i for each sequence O_i.

Step 2: Develop a similarity measurement formulation between sequences. For that, the first one needs to compute a distance matrix $D = \{D(O_i, O_j)\}$. This distance matrix will represent the similarity measurement, and can be calculated by many methods. These methods are as follows:

a. calculation with forward probability $P(O_j|\lambda_i)$

b. calculation with Euclidean distances of the discrete observation probability

d. calculation with Bayes probability error, etc. [29] where, $O_1, \cdots, O_V$ are the given V observation sequences to be clustered

Step 3: Once the similarity measurements have been computed, perform the clustering by using any pairwise distance-matrix-based method.

In another discussion, the author in [30] shows a different and better approach to do sequence clustering with the featured-based approach, in which they extracted the set of features from each sequence with the help of HMM. Their proposed algorithm is given as follows:

Assume $S_1, S_2, \cdots, S_V$ are the given V observed sequences to be clustered and $\{D(S, S_i)\}$ is the similarity measurement between a sequence S and a reference sequence S_i. The computed similarity measurement is interpreted as the feature of the sequence S. The procedure continues as follows:

Step 1: Let $R = \{P_1, ..., P_R\}$ be a set of R representative objects; these objects may belong to the sequence set ($R \subseteq T$) or may be otherwise defined. In a simple case it could be $R = T$.

Step 2: For each sequence $P_R \in R$, train one HMM λ_R.

Step 3: Each sequence S_i of the data set is then represented by the set of similarities $D_R(S_i)$ to the elements of the representative set R, computed with the HMMs $\lambda_1 ... \lambda_R$ as

$$D_R(S_i) \begin{pmatrix} D(S_i,\ P_1) \\ D(S_i,\ P_2) \\ \vdots \\ D(S_i,\ P_R) \end{pmatrix} = \frac{1}{L_i} \begin{pmatrix} log\ P(S_i|\lambda_1) \\ log\ P(S_i|\lambda_2) \\ \vdots \\ log\ P(S_i|\lambda_R) \end{pmatrix}$$

where L_i is the length of the sequence S_i.

Step 4: Now carry out clustering in $\mathbb{R}^{|R|}$; here $|R|$ is the cardinality of R.

All the above approaches are very good in technical and implementation terminology but when the number of sequences to cluster grows significantly (big data or big sequences), the time complexity starts to suffer a lot. In the next section we propose a new approach to tackle this problem by designing a framework to work with the quantum paradigm. We demonstrated a quantum algorithm to address the sequence-clustering problem and show the significant exponential improvements in time complexity.

14.7 Sequence Clustering in Quantum Paradigm

The research focus on sequential clustering has increased tremendously but mostly in the classical paradigm. In this chapter we have proposed an interesting framework for sequence clustering in the quantum paradigm.

In our approach we have designed a framework based on the HMM and quantum K-means. We have used a new feature space, where each sequence is characterized by its similarity to all other sequences.

The central idea behind the proposed approach is to construct a new representative space using the similarity values between sequences obtained via the HMMs, and then perform the clustering in that space by using quantum K-means clustering.

The new representation space will be the set of point vectors V where each point vector will be the representation of corresponding set of sequences S. So mathematically, a point vector V_i will be the representation of a corresponding sequence S_i, i.e.,

$$f(S_i) \rightarrow V_i.$$

Here, each point vector has some features and so technically we can say that each point vector is actually a set of certain features called a *feature vector*. We will discuss how to develop the feature vector for each point vector, and to do so we will use classical HMM. The approach suggests selecting first a set of sequences, and with the help of these sequences to develop and train HMMs. Let us consider the training sequence data set T for HMM training, $T_l \in T$ and $T \subseteq S$. Here each $T_1, T_2, \cdots, T_l$ sequences are used to train l HMMs (λ_l models). At this moment we have l trained HMMs. Now all the sequences S_i which are to be clustered will be represented by the set of similarities $X_l(S_i)$, indeed a point vector V_i, to the elements of the representative set T, which are computed with the HMMs $\lambda_1, \lambda_2, \cdots, \lambda_l$. Symbolically we can define this as

$$V_i = X_l(S_i) = \frac{1}{L_i}\begin{pmatrix} log\ P(S_i|\lambda_1) \\ log\ P(S_i|\lambda_2) \\ \vdots \\ log\ P(S_i|\lambda_l) \end{pmatrix}$$

where L_i is the length of the sequence S_i.

A point vector with a set of some features now represents a corresponding sequence. Now we are done with the construction of the set of point vectors. We then transform this classical featured point vector (obviously considering big data) to quantum form.

Let's begin by representing classical large featured point vector data, where each vector is in a d-dimensional complex vector form in the classical sense. We can represent this data set over $log_2 d$ qubits onto quantum states in QRAM [31–36] with $O(log_2 d)$ steps. Once the data set is mapped to quantum states, we are ready to apply our quantum K-means clustering onto this data set.

The QRAM allows us to access the data in quantum parallel and perform memory access in coherent quantum superposition [7]. Suppose R is an address register and it contains a superposition of addresses $\sum_j \psi_j |j\rangle_R$, then the QRAM will return a superposition of data in a data register dr which is correlated with address register, i.e.,

$$\sum_j \psi_j |j\rangle_R \rightarrow \sum_j \psi_j |j\rangle_R |D_j\rangle_{dr}$$

where D_j is the jth memory cell content.

Now apply the quantum K-means clustering algorithm to cluster these point vectors. Let's summarize the algorithm:

Step 1: Let $T = \{T_1, T_2, \cdots, T_l\} \in S$ be a set of sequences for HMM training purpose.

Step 2: Train one HMM λ_l for each sequence T_l.

Step 3: Represent each sequence S_i of the data set by the set of similarities $X_l(S_i)$, i.e., a featured point vector, to the elements of the training set T, computed with the HMMs $\lambda_1, \lambda_2, \cdots, \lambda_l$:

$$V_i = X_l(S_i) = (x_1, x_2, \ldots, x_l)$$

where $x_l = (\frac{1}{L_i}(log\ (S_i|\lambda_l)))$

Step 4: Transform the classical featured point vectors into the quantum states.

Step 5: Perform the clustering on featured point vectors using quantum K-means clustering method.

The accuracy of our approach can be discussed very straightforwardly. There are three parts at a higher level to consider for discussing the correctness: implementation of classical HMM, transformation of classical point vectors data set to quantum form, and clustering with quantum K-means clustering algorithm. The classical HMM is a very well established and mature algorithm. In our approach the training sequences are first used to train HMM to generate point vectors which have some features and represent related sequences. This shows the correctness for implementation of classical HMM. In the discussion of the next part we see that we transform the classical point vectors into quantum form using QRAM [6], where d-dimensional point vectors have been mapped onto quantum states with $log_2 d$ qubits, so this mapping took $O(log_2 d)$ steps. For the last part that is clustering with quantum K-means algorithm we used quantum K-means clustering algorithm for clustering the point vectors. The clustering problem here has been considered as the quadratic programming problem and solved by adiabatic quantum processes [6]. The performance of the algorithm depends strongly on a decent choice of initial seeds, and should be as far apart as possible to each other's. So we assumed the same in our approach.

Definitely, optimal solution for the K-means clustering process is an $NP-complete$ problem and so we assumed that the approximate solutions of these hard problems are well within the grasp of the quantum adiabatic algorithm, instead of assuming to solve in polynomial time.

At present, the estimated data generation rate shows that it is crossing the order of 10^{18} bits and soon will pass the order of 10^{20} bits and beyond in a classical way. With quantum computing states, these volume of big data could be represented with 60 bits (with respect to 10^{18} bits), 67 bits (with respect to 10^{20} bits). This shows that the clustering over point vectors can be performed using fewer operations with a quantum computer, which exponentially reduces operations. So with a relatively small scale quantum computer we can achieve the target. Obviously a much more powerful quantum computer will be required for much more complex problems.

14.8 Complexity Analysis

In this section we are going to discuss the run time complexity of the sequence-clustering algorithms (classical as well as quantum). We have categorized the different sections to analyze the complexities of each section separately, and then combine the separate sections' complexity to obtain the total run time complexity of the algorithm. We have considered the following subsections to discuss:

a. Learning with classical hidden Markov model

b. Classical to quantum data transformation

c. Classical K-means clustering

d. Quantum K-means clustering

At first, we will discuss the complexity analysis of classical sequence clustering followed by quantum sequence clustering.

14.8.1 Complexity analysis in classical approach

In this case we will first consider the sections in scope, which are as follows:

a. Learning with classical hidden Markov model

b. Classical K-means clustering

Learning of HMM is done with the Baum–Welch algorithm, which we discussed in an earlier section. The algorithm uses forward and backward procedure algorithms. The forward procedure with brute force approach generates N^L states sequences. Then by considering the calculation of the joint probability of each state sequence, the total time complexity will be $O(LN^L)$, where L is the length of the sequences and N is the number of symbols in the state set. But in practical situation, the number of possible hidden states is very high which makes the time complexity $O(LN^L)$ infeasible. However, later research has helped in improving to $O(LN^2)$. For the backward procedure, the time complexity is $O(LN^2)$ too.

The K-means clustering problem is NP hard problem in finding the optimal solution. In classical sense the time complexity of K-means clustering algorithm is $O(ikVd)$, where V is the number of d-dimensional vectors and k is the number of desired clusters, and the algorithm converse after i iterations. In general the time complexity of the algorithm is $O(polynomial(Vdk))$.

The total time complexity is the sum of the time complexity of HMM training with Baum–Welch approach and the K-means clustering algorithm:

$$O(LN^2) + O(ikVd)$$

14.8.2 Complexity analysis in quantum approach

The analysis in the above section shows that the total time complexity of the algorithm heavily depends on the length of the sequences and the number of sequences. As these factors increase (referring the case of "big data" and "big sequences"), the time complexity will not be feasible for computation with a classical computer. We need a more sophisticated computer, a quantum computer, to handle this issue. Let's analyze the time complexity of the new proposed quantum sequence-clustering algorithm. In this case, the transformation of big sequences to point vectors process is done by classical HMM so the complexity analysis will be same for this part as discussed in previous section. Apart from that, we have the following two sections to discuss:

a. Classical to quantum data transformation

b. Quantum K-means clustering

As discussed in earlier sections in our new proposed approach, once the sequence data set is transformed into point-featured vectors set for clustering with quantum K-means algorithm. But before that the classical featured point vectors set needs to be transformed into a quantum featured point vectors set. Let's begin by representing a classical featured point vector (actually a big data set), which is in a d-dimensional complex vector in a classical way. These d-dimensional complex featured point vectors can be represented as $log_2 d$ qubits onto quantum states in QRAM as discussed in [21–27] with $O(log_2 d)$ steps. Once the data set is mapped to quantum states, the quantum K-means algorithm can further process this data for clustering. The QRAM lets us access the data in the quantum parallel and perform memory access in coherent quantum superposition [28]. Suppose Reg is an address register, and it contains a superposition of addresses $\sum_j \psi_j |j\rangle_{Reg}$, then the QRAM will return a superposition of data in a data register dreg which is correlated with address register, i.e.,

$$\sum_j \psi_j |j\rangle_{Reg} \rightarrow \sum_j \psi_j |j\rangle_{Reg} |D_j\rangle_{dreg}$$

where D_j is the jth memory cell content. In QRAM, it takes only $O(logVd)$ operations to access the data, and it uses $O(Vd)$ resources where V is the total number of featured point vectors.

Now let's discuss the complexity of the quantum K-means algorithm. The quantum K-means algorithm takes time in between $O(\varepsilon^{-1} logkVd)$, when clusters are well separated, and $O(\varepsilon^{-1} klogkVd)$ on a quantum computer [5], with accuracy ε to construct the resulting quantum state. Therefore, the analysis shows that the total time complexity of the proposed new big data quantum sequence-clustering algorithm will be

$$O(LN^2) + O(log_2 d) + O(\varepsilon^{-1} klogkVd)$$

and when the clusters are well separated,

$$O(LN^2) + O(log_2 d) + O(\varepsilon^{-1} logkVd),$$

where N is the number of symbols in the state set in HMM and L is the length of the longest sequence in the sequence set. Consider the fact that the minimum gap during the adiabatic stage for the algorithm is $O(1)$ when the clusters are well separated [6].

We can see that this is an exponential improvement in time complexity when compared to the classical big data sequence-clustering approach.

14.9 Conclusion

In this chapter, we have discussed how to implement a big data (or big sequence) sequence-clustering quantum mechanically in which run time is exponentially faster than the classical version of sequence-clustering algorithm. We have covered all the related topics to create a base to understand our proposed approach of quantum sequence clustering of big data (or big sequence). We started our discussion with an introduction of the new proposed approach. Then we explored topics like *clustering* where we emphasized the different types of clustering and their implementations with applicability importance. In this section, we also discussed about *big data* and clustering with big data. Next, the classical K-means algorithm was explained. Then in Section 14.4, *Introduction to Quantum Computing*, we learned some of the most important quantum computing concepts which helped us to understand the benefits of the quantum approach in comparison to the classical approach. We also developed a base to ease the discussion of our proposed approach. In Section 14.5, quantum K-means clustering, we focused on the recently published quantum K-means clustering algorithm. Then in next two sections we discussed the existing sequence-clustering approaches and our proposed big data quantum sequence-clustering algorithm respectively. Finally, in Section 14.8 we explained the *complexity analysis* of the sequence clustering in the classical as well as in the quantum paradigm, and showed that the quantum version exhibits exponential run time speed-up in comparison to the classical counterpart.

References

1. JA Hartigan and MA Wong. Algorithm as 136: A K-means clustering algorithm. *Journal of the Royal Statistical Society. Series C (Applied Statistics)*, 28(1):100–108, 1979.

2. H-P Kriegel, P Kröger, J Sander, and A Zimek. Density-based clustering. *Wiley Interdisciplinary Reviews: Data Mining and Knowledge Discovery*, 1(3):231–240, 2011.

3. M Ester, H-P Kriegel, J Sander, X Xu. A density-based algorithm for discovering clusters in large spatial databases with noise. In *KDD '96 Proceedings of the Second International Conference on Knowledge Discovery and Data Mining*, Portland, OR, vol. 96, pp. 226–231, 1996.

4. AD Mauro, M Greco, and M Grimaldi. A formal definition of big data based on its essential features. *Library Review*, 65(3):122–135, 2016.

5. K Shvachko, H Kuang, S Radia, and R Chansler. The Hadoop distributed file system. In *2010 IEEE 26th Symposium on Mass Storage Systems and Technologies (MSST)*, Washington, DC, pp. 1–10. IEEE, 2010.

6. S Lloyd, M Mohseni, and P Rebentrost. Quantum algorithms for supervised and unsupervised machine learning. *arXiv preprintarXiv:1307.0411*, 2013.

7. MA Nielsen and IL Chuang. *Quantum Computation and Quantum Information*. Cambridge, UK: Cambridge University Press, 2010.

8. A Einstein, B Podolsky, and N Rosen. Can quantum-mechanical description of physical reality be considered complete? *Physical Review*, 47(10):777, 1935.

9. JA Wheeler and WH Zurek. *Quantum Theory and Measurement*. Princeton, NJ: Princeton University Press, 2014.

10. R Eisberg, R Resnick, and J Brown. Quantum physics of atoms, molecules, solids, nuclei, and particles. *Physics Today*, 39:110, 1986.

11. PW Shor. Polynomial-time algorithms for prime factorization and discrete logarithms on a quantum computer. *SIAM Review*, 41(2):303–332, 1999.

12. LK Grover. A fast quantum mechanical algorithm for database search. In *Proceedings of the Twenty-Eighth Annual ACM Symposium on Theory of Computing*, Dallas, TX, pp. 212–219. ACM, 1996.

13. Adiabatic Quantum Computation. Adiabatic quantum computation—Wikipedia 2016. https://en.wikipedia.org/wiki/Adiabatic_quantum_computation [Online]. Accessed June 20, 2016.

14. KL Pudenz and DA Lidar. Quantum adiabatic machine learning. *Quantum Information Processing*, 12(5):2027–2070, 2013.

15. C Fraley and AE Raftery. How many clusters? Which clustering method? Answers via model-based cluster analysis. *The Computer Journal*, 41(8):578–588, 1998.

16. LE Baum and T Petrie. Statistical inference for probabilistic functions of finite state Markov chains. *The Annals of Mathematical Statistics*, 37(6):1554–1563, 1966.

17. LE Baum and JA Eagon. An inequality with applications to statistical estimation for probabilistic functions of Markov processes and to a model for ecology. *Bulletin of the American Mathematical Society*, 73(3):360–363, 1967.

18. LE Baum and G Sell. Growth transformations for functions on manifolds. *Pacific Journal of Mathematics*, 27(2):211–227, 1968.

19. LE Baum, T Petrie, G Soules, and N Weiss. A maximization technique occurring in the statistical analysis of probabilistic functions of Markov chains. *The Annals of Mathematical Statistics*, 41(1):164–171, 1970.

20. LE Baum. An equality and associated maximization technique in statistical estimation for probabilistic functions of Markov processes. *Inequalities*, 3:1–8, 1972.

21. LR Rabiner. A tutorial on hidden Markov models and selected applications in speech recognition. *Proceedings of the IEEE*, 77(2):257–286, 1989.

22. L Rabiner and B Juang. An introduction to hidden Markov models. *IEEE ASSP Magazine*, 3(1):4–16, 1986.

23. LR Welch. Hidden Markov models and the Baum-Welch algorithm. *IEEE Information Theory Society Newsletter*, 53(4):10–13, 2003.

24. Baum–Welch algorithm—Wikipedia 2016. https://en.wikipedia.org/wiki/Baum%E2%80%93Welch_algorithm [Online]. Accessed June 20, 2016.

25. P Smyth. Clustering sequences with hidden Markov models. *Advances in Neural Information Processing Systems*, 9:648–654, 1997.

26. A Panuccio, M Bicego, and V Murino. A hidden Markov model-based approach to sequential data clustering. In *Joint IAPR International Workshops on Statistical Techniques in Pattern Recognition (SPR) and Structural and Syntactic Pattern Recognition (SSPR)*, Windsor, ON, Canada pp. 734–743. Springer, 2002.

27. LR Rabiner. A tutorial on hidden Markov models and selected applications in speech recognition. *Proceedings of the IEEE*, 77(2):257–286, 1989.

28. C Bahlmann and H Burkhardt. Measuring HMM similarity with the Bayes probability of error and its application to online handwriting recognition. In *Proceedings of the Sixth International Conference on Document Analysis and Recognition, 2001*, Seattle, WA, pp. 406–411. IEEE, 2001.

29. IV Cadez, S Gaffney, and P Smyth. A general probabilistic framework for clustering individuals and objects. In *Proceedings of the Sixth ACM SIGKDD International Conference on Knowledge Discovery and Data Mining*, Boston, MA, pp. 140–149. ACM, 2000.

30. M Bicego, V Murino, and MAT Figueiredo. Similarity-based clustering of sequences using hidden Markov models. In *International Workshop on Machine Learning and Data Mining in Pattern Recognition*, Leipzig, Germany, pp. 86–95. Springer, 2003.

31. V Giovannetti, S Lloyd, and L Maccone. Quantum random access memory. *Physical Review Letters*, 100(16):160501, 2008.

32. S Lloyd, M Mohseni, and P Rebentrost. Quantum principal component analysis. *Nature Physics*, 10(9):631–633, 2014.

33. I Chiorescu, N Groll, S Bertaina, T Mori, and S Miyashita. Magnetic strong coupling in a spin-photon system and transition to classical regime. *Physical Review B*, 82(2):024413, 2010.

34. DI Schuster, AP Sears, E Ginossar, L DiCarlo, L Frunzio, JJL Morton, H Wu, GAD Briggs, BB Buckley, DD Awschalom et al. High-cooperativity coupling of electron-spin ensembles to superconducting cavities. *Physical Review Letters*, 105(14):140501, 2010.

35. Y Kubo, FR Ong, P Bertet, D Vion, V Jacques, D Zheng, A Dréau, J-F Roch, A Auffèves, F Jelezko et al. Strong coupling of a spin ensemble to a superconducting resonator. *Physical Review Letters*, 105(14):140502, 2010.

36. H Wu, RE George, JH Wesenberg, K Mølmer, DI Schuster, RJ Schoelkopf, KM Itoh, A Ardavan, JJL Morton, and GAD Briggs. Storage of multiple coherent microwave excitations in an electron spin ensemble. *Physical Review Letters*, 105(14):140503, 2010.

Index